Exploring Robotics with ROBOTIS Systems

Chi N. Thai

Exploring Robotics with ROBOTIS Systems

Second Edition

 Springer

Chi N. Thai
roboteer@comcast.net
www.cntrobotics.com

Additional material to this book can be downloaded from http://extras.springer.com.

ISBN 978-3-319-86712-0   ISBN 978-3-319-59831-4   (eBook)
DOI 10.1007/978-3-319-59831-4

Printed on acid-free paper

This Springer imprint is published by Springer Nature
The registered company is Springer International Publishing AG
The registered company address is: Gewerbestrasse 11, 6330 Cham, Switzerland

Preface

The second edition is a substantial revision of the original edition (about 190 additional pages), created in response to readers feedback, as well as to new hardware and software releases from ROBOTIS. All original 11 chapters had been revised or updated to reflect these changes, in particular:

- Chapter 3 includes additional descriptions of the newer CM-50, OpenCM-700, and X series Dynamixels.
- Chapter 4 offers new introductory sections for the ROBOPLUS V.2 tools such as R+SCRATCH which interfaces with MIT's SCRATCH 2 software and the OLLOBOT SDK that opens up Android Programming for Firmware 2.0 types of controllers such as CM-50/150/200 and OpenCM-7.00/9.04.
- Chapter 5 provides a closer look at the MANAGER and TASK V.2 tools regarding their use with the 485-EXP expansion board for the OpenCM-9.04/C system with the use of a new feature called "Dynamixel Channel". It also has new sections on usage of the IR Array Sensor for arbitrary single track maneuvers, new remote control capabilities via SCRATCH 2 event processing constructs, and also from the use of SmartPhone tilt sensors.
- The new Chap. 6 combines the former Chaps. 6 and 7 to provide a single chapter regarding Position Control Applications of ROBOTIS Dynamixels and to show complex interactions between parameters, such as Present Position, Goal Position, Goal Speed, Torque Limit, Present Load, Motion Page, and Joint Offset. New application projects of the PhantomX Reactor robotic arm are illustrated, such as an Avoider Arm and Remote Control of a Mobile Manipulator Platform using a CM-530 controller.
- Chapter 9 now covers Embedded C features with an OpenCM-9.04/C + 485-EXP controller platform with new applications to the Remote Control of the Mobile Manipulator Platform and its use of SmartPhone video cameras in conjunction with NIR distance sensors.
- Chapter 11 is entirely new and written to document the use of the R+SCRATCH tool and the PLAY 700 App (released by ROBOTIS-USA in December 2016), in conjunction with SMART commands from a TASK program or from Arduino-style

codes using the OpenCM IDE, to access various services from an iOS or Android device such as video camera, gesture sensor, touch areas, audio and video playback, text-to-speech, and speech recognition.

- Chapter 12 is also entirely new and describes the OLLOBOT SDK which was originally designed for the OLLOBOT Kit released in August 2016 to enable its control from an Android device using Android Studio. Fortunately for ROBOTIS users everywhere, this SDK turns out to be applicable to all Firmware 2.0 controllers such as CM-50/150/200 and Open-CM7.00/9.04. This chapter is for users interested in Java and Android Programming to create custom Android Apps that interact with ROBOTIS' robotic systems using the new Dynamixel Protocol 2.0.

Trademark Attributions

Content in this book made reference to the following trademarks:

ROBOTIS product names are trademarks or registered trademarks of ROBOTIS, Inc. Used with permission by ROBOTIS Co. Ltd.

Android®, YouTube®, Chrome® and Android Studio® are registered trademarks of Google Inc.

Atmel AVR®, WinAVR® and Atmel Studio® are registered trademarks of Atmel.

ARM® and Cortex® are registered trademarks of ARM Ltd.

PhantomX Reactor, ArbotiX-M and WidowX are trademarks of Interbotix Labs.

Scratch® is a registered trademark of the Scratch Team at the Massachusetts Institute of Technology.

Windows®, Kinect®, Visual Studio® are registered trademarks of Microsoft Inc.

iOS®, Mac OS® and iPhone® are registered trademarks of Apple Corporation.

Linux® is a registered trademark of Linus Torvalds.

Camtasia Studio® is a registered trademark of TechSmith.

Cyton® and Actin® are registered trademarks of Robai Corporation.

Pixl Board™ is a trademark of HumaRobotics.

Raspberry Pi® is a registered trademark of the Raspberry Pi Foundation.

Java® is a registered trademark of Oracle Corporation.

Arduino® is a registered trademark of Arduino AG.

USB2AX™ is a trademark of Xevelabs.

UP-Board™ is a trademark of the Up Board Organization.

ROS™ is a trademark of the Open Source Robotics Foundation.

Intel® Joule™ and RealSense™ are trademarks of Intel Corporation.

MRT3™ is a trademark of MRT International Limited.

LEGO NXT® is a registered trademark of the LEGO Group.

ABILIX® is a registered trademark of SHANGHAI XPARTNER ROBOTICS CO., LTD.

CUBELETS® is a registered trademark of Modular Robotics.

ZigBee® is a registered trademark of the ZigBee Alliance.
BlueTooth® is a registered trademark of Bluetooth SIG, Inc.
Eclipse® is a registered trademark of the Eclipse Foundation, Inc.
Unity® is a registered trademark of Unity Technologies.
Galaxy S4® is a registered trademark of SAMSUNG.
ZenPad S8® is a registered trademark of ASUS.
IFTTT® is a registered trademark of IFTTT Corporation.

Lawrenceville, GA, USA   Chi N. Thai

Contents

Chapter 1
Motivations and Instructional Approach

1.1 Motivations

My personal robotics journey began with RoboCup 2007 hosted by Ga Tech that summer. I was so impressed by the advancement in humanoid robotics on display there and thought that this could be an important component for engineering instruction at The University of Georgia, as at that time we were seeking to expand engineering beyond biological and agricultural engineering. Also at RoboCup 2007, I met Mr. Jinwook Kim from ROBOTIS and was introduced to the Bioloid Comprehensive and Expert kits which were about the only "in-depth" but "affordable" robotic systems at the time.

In 2008 when I started to draft out the instructional materials for my "Embedded Robotics" course, the only source of ROBOTIS technical information was from their "Quick Start", "Programming Guide" and "Expert" manuals. I was also much inspired by the works of Bekey (2005), Bräunl (2006), Matarić (2007), and Miller et al. (2008) which resulted in my adoption of a project-based approach to my course. Also during this time, I was going through some trainings by the UGA Center for Teaching and Learning which made me aware of "learner-centered" approaches (Fink 2003; Weimer 2002) and screen/classroom capture technologies such as TechSmith® Camtasia Studio® (Thai et al. 2008). I also had found the "Spiral" model from Dr. Joseph Bergin (Bergin 2012) very inspiring in designing the flow of presented topics and instructional materials in the "Embedded Robotics" course and also in this book. Unfortunately, not all classroom teaching approaches can be directly transferred into a "book", for example I modified the "contract" teaching style by allowing students to choose their own projects to suit their own interests and personal strengths (Thai 2014). However, in this book, I can only describe those projects and provide web links to show the students achievements.

I was also fortunate to have help from outstanding students during the development period of this course:

© Springer International Publishing Switzerland 2017
C.N. Thai, *Exploring Robotics with ROBOTIS Systems*,
DOI 10.1007/978-3-319-59831-4_1

- Mr. Alex Fouraker (pursuing his BSAE degree) was instrumental in exploring the limits of what the Bioloid kits can do: Alex's work on the GERWALK climbing stairs is still the most accessed item at the ROBOTIS Gallery web site with 255,192 views as of 12/10/2014 (Thai 2009).
- In Spring 2010, this course was taught for the first time and more instructional design details for this particular implementation can be found in this article (Thai 2010). During this course, Mr. Matthew Paulishen (also pursuing his BSAE degree) had extended many BIOLOID projects for PC-side Visual C++ and Machine Vision programming for bipedal robots (Thai 2011). Since then this course had been fine-tuned using student feedbacks and in accordance with the continuing hardware and software updates from ROBOTIS.

Between Spring 2012 and Fall 2013, I had the opportunity to repurpose my UGA classroom recordings with Drs. Yan-Fu Kuo and Ping-Lang Yen from National Taiwan University for a Distance Education project based on a Flipped Classroom model (Thai et al. 2013). This project has helped in refining the scaffolding of the presented materials and genesis of new course projects.

From their end, ROBOTIS had their Q&A web site since 2007 and added more web resources for users such as their e-Manual (http://support.robotis.com/) in early 2010 and RobotSource™ (http://www.robotsource.org/?na=en&pc_ver=1) in late 2011. ROBOTIS also maintained several tutorial channels on YouTube®:

- https://www.youtube.com/user/ROBOTISCHANNEL
- https://www.youtube.com/channel/UCJ9SxQelzEQobOEX_6a_8LA (ROBOTIS America).
- https://www.youtube.com/channel/UC0M8G5T34THKEgQbdtj1aug (ROBOTIS CS).

Furthermore, ROBOTIS continues to improve the instructional design of their paper-based manuals that are included in their BIOLOID STEM and PREMIUM robotic kits. At present, there are also several web-based Forums such as the ones from RoboSavvy (https://robosavvy.com/forum/viewforum.php?f=5) and Trossen Robotics (http://forums.trossenrobotics.com/) providing ROBOTIS user communities worldwide and they perform many useful functions. However, a reading of the postings from the above web sites would show that some beginners are still having a tough time learning and some folks are still having problems finding solutions to suit their needs.

Those were my motivations to write this book using a web-based multimedia approach for illustrating robotics concepts, code implementations and actual resulting robot behaviors. This approach also allows the "thinning" of the book (a video is worth many pages of text!), for easier updates and for users to skip subjects they do not need. This book is mainly designed for self-learners and presents in a more organized manner various information sources from ROBOTIS technical training manuals (paper-based and web-based), including my own materials used in teaching the "Embedded Robotics" course at the College of Engineering of The University of Georgia from 2008 to 2016.

In this 2nd edition, I have expanded/updated the original 11 chapters with appropriate materials that came forth since 2015 from ROBOTIS as well as from my teaching at UGA. I also reorganized the old Chaps. 6 and 7 into a single chapter dealing Position Control Applications. I also added two new chapters: Chaps. 11 and 12 on applications using mobile devices such as phones and tablets. Also with the 2nd edition, I have created a private website http://www.cntrobotics.com to serve as the companion website to this book (and its previous first edition). The video tutorials accompanying this 2nd edition can be accessed at this web link (https://www.youtube.com/playlist?list=PLtix7rPAJwqz85vK-9Dl1nNKnD0xFB8J9).

1.2 Instructional Approach

In keeping with the "Spiral" instructional model and a project-based approach, the rest of this book is structured as follows:

- Chapter 2 – Overall view of ROBOTIS robotics systems from 1999 through mid-2017.
- Chapter 3 – Hardware characteristics of the robot systems used in this book: controllers, actuators and sensors (i.e. PLAY700, DREAM, SMART III, IoT, BIOLOID, MINI and OLLOBOT).
- Chapter 4 – Description of the evolution of ROBOTIS software tools since 1999 and illustration of their basic use for firmware updating and evaluation of wired and wireless control performances.
- Chapter 5 – Foundational robotics concepts using wheeled robots and the Manager and Task tools: basic use of actuators and sensors, autonomous behaviors achieved via reactive control and behavior control approaches, introductory remote communication and control concepts. Projects implemented: Sequence Commander, Line Tracer and Smart Avoider.
- Chapter 6 – Multi-link robots and actuator position control concepts using the Task and Motion V.1 tools: Proportional-Integral (PI) implementations, Motion Pages, and required modifications to remote control programming application, advanced position control features such as 'Load Limit", "Present Load" and "Join Offset", and use of Callback functions. Projects implemented: BugBot, Bipedal Bots, Rover Bot with Arm and NIR sensors, GERWALK and load-sensing Gripper.
- Chapter 7 – Communications programming using "Remocon" packets with ZigBee hardware: 1-to-1 and broadcast modes. Projects implemented: Embedding special signals into standard packets for bot-to-bot and PC-to-bots application programming, Leader and Follower grippers, Multiple-users control scheme.
- Chapter 8 – Integrating advanced sensors such as gyroscope, inertial measuring unit, foot pressure sensor, and color video camera. Projects implemented: Balance of Humanoid Robot, Color Object Search by Carbot and GERWALK.

- Chapter 9 – Embedded C options and implementations on BIOLOID and OpenCM9.04-A/B/C systems using ROBOTIS SDK and OpenCM IDE examples: Remote Control Rover Bot with Arm and using Phone Camera via SMART commands.
- Chapter 10 – ROBOTIS (DARWIN)-MINI system: PC wireless communications options, new motion concepts in Motion V.2, sensor integration with TASK and MOTION, choreography application to two MINIs.
- Chapter 11 – Using Multimedia with Firmware 2.0 systems: via MIT's SCRATCH two framework, ROBOTIS' SMART framework.
- Chapter 12 – ROBOTIS' OLLOBOT framework and its applications to CM-50/150/OpenCM9.04 systems. Remote Control, Voice Control, Accessing on-board Sensors.

In closing for this chapter, I would like to quote Dr. Ben Shneiderman from his book "Leonardo's Laptop" p. 113 of Shneiderman (2002):

> … we might rethink education in terms of collect-relate-create-donate:
> COLLECT – Gather information and acquired re-sources.
> RELATE – Work in collaborative teams.
> CREATE – Develop ambitious projects.
> DONATE – Produce results that are meaningful to someone outside the classroom…

References

Bekey GA (2005) Autonomous robots. The MIT Press, Cambridge

Bergin J (2012) Pedagogical patterns: advice for educators. Create Space Independent Publishing Platform, North Charleston, p 115

Bräunl T (2006) Embedded robotics. Springer, Heidelberg

Fink LD (2003) Creating significant learning experiences. Jossey-Bass, San Francisco

Matarić MJ (2007) The robotics primer. The MIT Press, Cambridge

Miller DP et al (2008) Robots for education. In: Siciliano B, Khatib O (eds) Springer handbook of robotics, 1st edn. Springer, Heidelberg, pp 1283–1301

Shneiderman B (2002) Leonardo's laptop. The MIT Press, Cambridge, pp 113–114

Thai CN et al (2008) Robotics-based curriculum development for an immigration course into computer systems engineering. (https://docs.wixstatic.com/ugd/714442_903b93c33d8b45bf94e33f7788520d50.pdf). Accessed 1 Jan 2017

Thai CN (2009) Bioloid GERWALK robot going up stairs steps. (https://youtu.be/aKBx1nc2jfU). Accessed 1 Jan 2017

Thai CN (2010) Teaching robotics to students with mixed interests. (https://docs.wixstatic.com/ugd/714442_b06aff6abf704b2f814ffc9d4a3c69ae.pdf). Accessed 1 Jan 2017

Thai CN (2011) Biped robot going up stairs steps. (https://youtu.be/-9eN432Uuwc). Accessed 1 Jan 2017

Thai CN et al (2013) Cooperative teaching in a distance education environment (https://docs.wixstatic.com/ugd/714442_b06aff6abf704b2f814ffc9d4a3c69ae.pdf). Accessed 1 Jan 2017

Thai CN (2014) Syllabus CSEE/ENGR-4310: Embedded Robotics (Fall 2014). (https://docs.wixstatic.com/ugd/714442_f95e6cb3e9314bc09b9eff12f2c14577.pdf). Accessed 1 Jan 2017

Weimer M (2002) Learner-centered teaching. Jossey-Bass, San Francisco

Chapter 2
ROBOTIS' Robot Systems

2.1 General Systems Description

ROBOTIS Co., Ltd. (Seoul, South Korea) was founded by Bill Byoung-Soo Kim in 1999 along with two other engineers. The current CEO (Byoung-Soo Kim) and Vice-President (In-Yong Ha) were from the original team. The ROBOTIS name derived from the question "What is a robot?" and their vision statement can be read at (http://en.robotis.com/index/company_01.php). In 2009, ROBOTIS opened their USA office in Irvine, California and in 2016, the USA office moved to Lake Forest, CA. Currently ROBOTIS has more than 200 partners in 40 countries worldwide.

"Dynamixel" was the brand name uniquely connected to ROBOTIS. "Dynamixel" encapsulated several modularization and standardization concepts applied to both sensors and actuators equipped with embedded computing and communications capabilities (see Fig. 2.1). Actually, the Main Hardware Controller was also considered as a Dynamixel with a reserved ID = 200, and a co-controller such as a Smart Phone would have a reserved Dynamixel ID = 100 (see Chap. 12 for more details).

In 1999, ROBOTIS launched their first product called Didi and Titi (see Fig. 2.2). This web link shows a TV commercial for Didi and Titi (http://en.channel.pandora.tv/channel/video.ptv?c1=05&ch_userid=do7minate&prgid=49759295&ref=na).

Since then, ROBOTIS has released 27 more products:

1. Toma (2002 – released in Korea only).
2. Dynamixel – AX-12 (2003).
3. Cycloid (2004 – released in Korea only).
4. BIOLOID – Beginner and Comprehensive (2005).
5. URIA (2006 – released in Korea only).
6. Dynamixel – RX-64 (2006).
7. OLLO (2008).
8. Dynamixel – EX-106 (2008).
9. BIOLOID PREMIUM (2009).
10. Dynamixel – MX series (2011).

© Springer International Publishing Switzerland 2017
C.N. Thai, *Exploring Robotics with ROBOTIS Systems*,
DOI 10.1007/978-3-319-59831-4_2

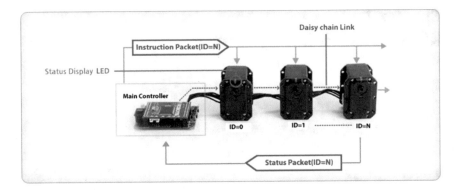

Fig. 2.1 "Dynamixel" concept (Courtesy of ROBOTIS)

Fig. 2.2 Didi and Titi
(Courtesy of ROBOTIS)

11. DARWIN-OP (2011), renamed ROBOTIS-OP in 11/2014.
12. BIOLOID STEM (2012).
13. IDEAS (2013).
14. THOR-MANG (2013).
15. Dynamixel-Pro H-series (2013).
16. Dynamixel – XL-320 (2014).
17. ROBOTIS MINI and OpenCM-9.04 (2014).
18. Dynamixel-Pro L-series (2014).
19. DREAM and PLAY (2014).
20. SMART I (2015) and SMART II (2016).
21. IoT (2016).
22. OLLOBOT (2016).
23. Dynamixel X-series (2016).
24. TurtleBot3 and OpenCR (2016).
25. PLAY 700 system (2016).
26. SMART III (2017).
27. MiCom Educational Kit, based on the OpenCM-9.04 (2017).

This list shows ROBOTIS' commitment to continuing development and improvement and to serve a very broad clientele in terms of age as well as technical level.

Fig. 2.3 DREAM robots

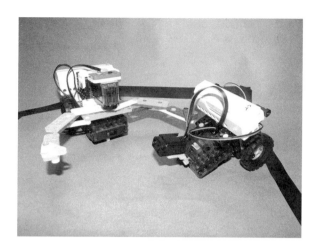

These products had been adopted by hobbyists of all ages as well as teachers and researchers worldwide (please visit http://www.ROBOTIS.com for more details).

Currently, the systems designed for young children are: PLAY, DREAM, SMART and IoT. They are colorful and use a quick-connect system adapted from the standard rivet concept to ease hands-on creative activities for young children (see Fig. 2.3). They can be constructed to be simple motorized robots (PLAY 200–600 series) as well as programmable robots (PLAY 700, DREAM, SMART and IoT). Furthermore PLAY 700, DREAM, SMART and IoT system were designed to be used with mobile devices such as cell phones and tablets (https://play.google.com/store/apps/developer?id=ROBOTIS&hl=en). Their embedded controllers are based on the STM32F103C8 and STM32L151C8 from STMicroelectronics.

The PLAY 700 system was introduced in late 2016 and it used the CM-50 controller which could be used with <u>either one</u> of two software packages: the MIT SCRATCH V.2 (via the ROBOTIS' R + SCRATCH tool acting as an HTTP extension), or with the ROBOTIS' R + TASK tool for the creation of programs that could execute on the CM-50 directly and interact with a PC or a mobile phone/tablet in either Android® or iOS® operating system (see Fig. 2.4 and playlist at https://www.youtube.com/playlist?list=PLtix7rPAJwqzqvrI6nr4qI0FILYrLqbgb).

The DREAM system used a controller named CM-150, while the SMART systems used the CM-200 (see Fig. 2.5). Although visually similar to each other and both capable to run TASK programs, the CM-200 was slightly larger and had more capabilities than the CM-150 such as eight GPIO ports (CM-200) instead of two (CM-150) and only the CM-200 could additionally support the MOTION programming features provided by the R + Motion V.2 tool (see Chap. 4 for more details on the R + Motion tool). The DREAM system can also use the R + m.PLAY700 Mobile App to access the SMART facilities on the smart phone (see video at https://www.youtube.com/watch?v=-O-vZhAH-bQ and Chap. 11 for more details). The SMART III system has its own Mobile App package and the SMART III system would be

Fig. 2.4 DOG robot from the PLAY 700 series

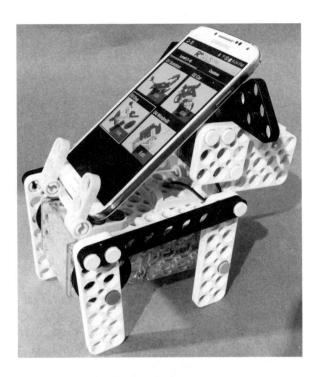

Fig. 2.5 CM-150 (*left*) and CM-200 (*right*)

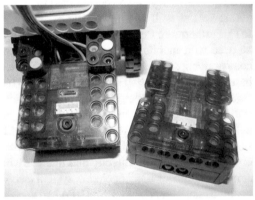

suitable for a First-Year Odyssey type of course for freshman engineering students, alas currently this system is only available for Korean speaking users.

The author had used the PLAY 700, OLLO, DREAM and BIOLOID STEM systems to teach introductory robotics to middle and junior high school students via week-long and week-end short courses. The instructional approaches and materials used are described at the following web links:

- Summer Academy at UGA (https://www.cntrobotics.com/sauga)
- Duke TIP Academic Adventures (https://www.cntrobotics.com/duke-tip-aa)
- Duke TIP Scholars Weekend (https://www.cntrobotics.com/duke-tip-sw)

The IoT system used the OpenCM7.0 as controller (i.e. STM32L151C8). Similarly to the CM-50, it could be interfaced either via the SCRATCH V.2 tool from MIT for "IoT Level 1" (see Sect. 4.6 for more details) or via the usual ROBOTIS R + TASK V.2 tool for "IoT Level 2" and its own mobile app in both Android and iOS flavors (see Chap. 4 also for more details on the R + TASK tool). As of late 2016, the IoT systems are only available for Korean speaking users (see Fig. 2.6).

The BIOLOID systems (BEGINNER, COMPREHENSIVE, STEM and PREMIUM) are designed to be various entry points (depending on one's budget) into the robotics field for those interested in taking a more comprehensive journey into this knowledge area. They use standard screws and nuts for a sturdier fastening method, with some parts using the DREAM/SMART's rivet system also. The older kits (BEGINNER and COMPREHENSIVE) were designed off the Atmel AVR chip while more recent systems (STEM and PREMIUM) rely on the ARM architecture using such embedded controllers as STM32F103C8 and STM32F103RE from STMicroelectronics (see Fig. 2.7).

In 2012, ROBOTIS made a strategic shift into the open hardware-software movement with their OpenCM systems whereas users can collaborate on hardware and software development worldwide (see Fig. 2.8).

ROBOTIS is also well known for its humanoid robots in the competitive and research arenas such as the ROBOTIS GP (http://en.robotis.com/index/product. php?cate_code=121510), ROBOTIS OP (http://en.robotis.com/index/product. php?cate_code=111010) and ROBOTIS OP2 (http://en.robotis.com/index/product. php?cate_code=111310), and THOR-MANG3 (http://en.robotis.com/index/prod-uct.php?cate_code=111410). In Spring 2014, ROBOTIS released a more affordable humanoid robot called ROBOTIS MINI (http://en.robotis.com/index/product. php?cate_code=121310) based on the OpenCM9.04-C controller and the XL-320 servo motor. Outwardly, the ROBOTIS MINI (see Fig. 2.9) is roughly a half-scale version of the ROBOTIS OP, but with reduced capabilities and a different kinematic linkage solution for its legs. It is designed to be operated and programmable from mobile devices (Android® and iOS®) as well as personal computers (MS Windows®). The MINI system can also use the R + m.PLAY700 mobile app to

Fig. 2.6 ROBOTIS IoT system (level 1)

Fig. 2.7 A bioloid STEM
robot

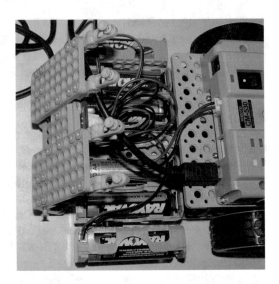

Fig. 2.8 OpenCM-9.04-B
controller (*left*), XL-320
servo motor (*center*), and
USB2Dynamixel
communication converter
(*right*)

access the SMART facilities on the smart phone (see video at https://www.youtube.
com/watch?v=-O-vZhAH-bQ and Chap. 12 for more details).

The OLLOBOT was introduced to developers at the SHAPE AT&T Tech Expo
in July 2016 (see Fig. 2.10). It can be controlled via a cell phone (Android or iOS)
with an Android App (https://github.com/ROBOTIS-GIT/OLLOBOT) or with an
IFTTT IF® App (https://play.google.com/store/apps/details?id=com.ifttt.ifttt).

The TurtleBot 3 was introduced at the ROSCon 2016 conference for developers
and university graduate students who wanted to use ROS™ (http://www.ros.org/)
and currently (Spring 2017) it came in two models: "Burger" and "Waffle" (see
Fig. 2.11). It used the OpenCR controller and was designed to be modular with an
open hardware and software paradigm (http://turtlebot3.readthedocs.io/en/latest/).

Fig. 2.9 ROBOTIS MINI
action figure

Fig. 2.10 The OLLOBOT
system with a Samsung S4®
phone

As this book is geared towards university undergraduate students or self-learners, the following systems will be considered in further details: BIOLOID BEGINNER, COMPREHENSIVE, STEM and PREMIUM, OpenCM, PLAY700, ROBOTIS MINI and OLLOBOT.

Fig. 2.11 Current TurtleBot 3 models: "Burger" (*left*) and "Waffle" (*right*) (Courtesy of ROBOTIS)

Fig. 2.12 CM-5 controller
(discontinued)

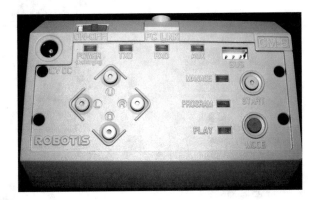

2.2   Robotics Kits Considered in Book

The BIOLOID BEGINNER and COMPREHENSIVE systems use the CM-5 (see
Fig. 2.12) as its main controller which is an Atmel ATmega™ @ 16 MHz and with
128 KB of flash memory. It connects to ROBOTIS' sensors and actuators via the
Dynamixel TTL bus (three-pin) and has ZigBee wireless communications capabilities via the ZIG-100 modules.

Actually there was another CM-5 based system called BIOLOID EXPERT that
was available between 2005 and 2009. It used the same basic hardware as the
COMPREHENSIVE, but had a wireless color video camera (Fig. 2.13) and a
Visual C++ V.6 library with functions to control sensors, actuators, video camera
and also to perform machine vision processing routines and ZigBee wireless
communications.

When the BIOLOID PREMIUM system first came out in 2009, it was shipped
with the CM-510 controller (see Fig. 2.14-*left*) which was a 16-MHz ATmega with
256 KB flash memory, but since 2013 it is shipped with the CM-530 (Fig. 2.14-*right*)

Fig. 2.13 Wireless color
video camera from
BIOLOID EXPERT kit
(discontinued)

Fig. 2.14 CM-510 (*left*-discontinued) and CM-530 (*right*) controllers

which is based on the 72-MHz STM32F103RE with 512 KB flash memory.
Otherwise, visually and functionally, the CM-510 and CM-530 are identical to each
other. Both are also capable of handling embedded C applications via the WinAVR
(CM-510) or WinARM (CM-530) tool chains (see further discussions in Chaps. 4
and 9).

It should be mentioned that there are also two bare bone ATmega-based controllers, CM-2+ (discontinued) and CM-700 (see Fig. 2.15), which were designed for
custom needs when the user has to mix two types of Dynamixel modules together
in the same controller (i.e. three-pin TTL and four-pin RS-485, see Chap. 3 for more
details). Embedded C is also available for the CM-700.

The latest BIOLOID system (2012) is the STEM kit which combines hardware
construction approaches from the previous BIOLOID kits (i.e. screws and nuts) and
from the DREAM kits (i.e. plate and rivets). The STEM also has new hardware to
create more secure pin joints (Fig. 2.16) and an IR Sensor Array (to be described
later in Chap. 3). The STEM kit uses the CM-530 controller. It comes as two separate kits, Standard and Expansion, and the Standard kit is required for the proper use
of the Expansion kit.

Fig. 2.15 CM-2+ (*left* – discontinued) and CM-700 (*right*) controllers (Courtesy of ROBOTIS)

Fig. 2.16 BIOLOID
STEM pin-joint hardware

All ROBOTIS BIOLOID systems use the RoboPlus software suite consisting of four tools:

1. MANAGER – for general hardware troubleshooting and obtaining firmware update for controllers and dynamixels.
2. TASK – general programming environment for the user.
3. MOTION – motion programming environment for multi-links robots.
4. DYNAMIXEL WIZARD – for troubleshooting and updating firmware on Dynamixel actuators and sensors.

The MANAGER, TASK and MOTION tools exist in two versions depending on which controllers are being used (see Chaps. 3 and 4 for more details).

Starting in 2014, the OpenCM platform is the vehicle for ROBOTIS to accomplish its open-hardware and software goals in the near future for operating systems such as MS Windows®, Mac OSX® and Linux®. The CM-900 was the beta platform first available in 2012 but is no longer sold by ROBOTIS. It had two editions, ES and V.1.0, which are based on the STM32F103C8 microcontroller (Fig. 2.17). They carried 64 KB of flash memory and support many hardware interface standards such as USB (1), CAN (1), USART (3), I2C (2) and SPI (2). The ES version supported both AX/MX-TTL and RS-485 Dynamixel ports, the V.1 edition additionally supported the new XL-TTL Dynamixel port. Both supported software

Fig. 2.17 CM-900 ES (*left*) and V. 1 (*right*)

Fig. 2.18 OpenCM-9.04-A/B/C controllers (*left* to *right*)

development using an Arduino-based IDE (called "OpenCM IDE") and wireless communications programming via ZigBee and BlueTooth. Real-time debugging (SWD and JTAG) was also available using additional tools such as ST-LINK/V2 and Keil µVision.

Currently, only the OpenCM-9.04 series is commercially available and it comes in three versions A, B and C (Fig. 2.18). The 9.04 series has the same hardware features previously listed for the 9.00 series, however they have a smaller physical format and 128 KB of flash memory.

The B version is "ready-to-go" if the user plans to use a mixture of AX/MX-TTL and XL-TTL Dynamixel modules (but it is no longer available commercially). The A version is essentially a user-customizable B version whereas the user can install only the needed headers. Both A and B versions are completely open hardware and software controllers whereas users can adapt their own firmware and boot loader as they wish. They use the OpenCM IDE as the programming interface.

Fig. 2.19 Expansion
board OpenCM-485-EXP
for the OpenCM-9.04
series (beta version –
green, release
version – *blue*)

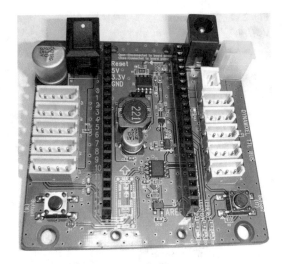

Currently the C version comes with the ROBOTIS MINI kit and is also available by itself (http://www.robotis.us/opencm9-04-c-with-onboard-xl-type-connectors/). It has only the XL-TTL connectors installed. Most importantly, it has a proprietary firmware so that it can operate with the R + Task and R + Motion V.2 software suite. Alternately, the user can use the OpenCM IDE with version C but this mode would effectively erase the proprietary firmware, thus if the user wants to use the R + Task and R + Motion packages again, a firmware recovery process must be performed (see Chap. 4).

If the user requires more TTL and RS-485 Dynamixel ports, the expansion board OpenCM-485-EXP can be used (Fig. 2.19).

The CM-50 (PLAY700 system) and OpenCM-9.04/C + 485-EXP combo would also be used extensively to illustrate the SMART structure that ROBOTIS created to let its Firmware 2.0 controllers to collaborate with mobile devices running on Android and iOS systems (see Fig. 2.4 and Chaps. 9 and 11).

2.3   MiCom Training Kit

In 2017, ROBOTIS plans to release the "MiCom Training Kit". It is based on the OpenCM9.04-C and oriented towards the DIY type of user. Figure 2.20 shows its main components:

- One OpenCM-9.04-C with 2-mm header mounted.
- Three NIR sensors (5-pin type).
- Two XL-320 actuators.
- One solderless breadboard and some 100 mm jumper wires.
- Some electrical and electronics components such as resistors (10 KΩ and 100 KΩ), 1 red LED, matching NIR LED emitter/receiver, 1 toggle switch, 1 variable resistor, a microphone and a 7-segment LED display.

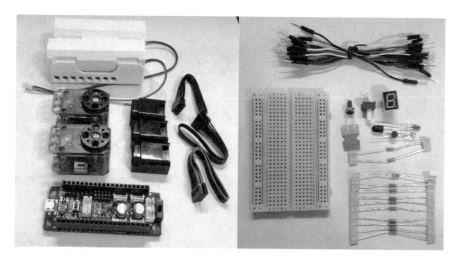

Fig. 2.20 Micom training kit (due in 2017)

- Additional OLLO frame and wheel parts to make a wheeled robot using NIR sensors so that it can follow a black track (for example).

With this kit, the DIY user can start programming using the TASK tool for a quick learning approach, then later can switch to the Arduino based ROBOTIS IDE for a low-level programming style using the C language and the breadboard for experimental circuits such as:

- Direct pin access and digital output (PWM and duty cycle).
- Analog input and A/D conversion.
- Serial communications.
- Dynamixel control and instruction packet management.
- Memory access, etc.…

A review of this system is described in "Video 2.1" and the mentioned Arduino project "MiCom_1.ino" is included in the Springer Extras materials.

2.4 OLLOBOT

Relatively speaking, the OLLOBOT can be considered as the "odd ball" of the ROBOTIS family, but it still uses the familiar ARM controller STM32F103RET6 (72 MHz, 512 KB Flash RAM) – see Fig. 2.21. Please note the four sets of five-pin connector at the top of the circuit board which could be used for GPIO sensor interfaces in the future perhaps?

Figure 2.22 represents the "current" (Fall 2016) Control Table for the OLLOBOT controller listing its various built-in functions that are programmable. The reader can see that only two servo motors are accessible, thus the sensing and high-level logic would have to be done at the cell phone level. So working with the OLLOBOT

Fig. 2.21 Controller board for OLLOBOT

Register	Address	Size [byte]	Name	Description	Access	Default Value	Min Value	Max Value
0	0x0	2	Model Number	Model number	R	440	--	--
7	0x7	1	ID	DYNAMIXEL ID	R	200	--	--
79	0x4F	1	Green LED	Green LED on/off	RW	0	0	1
80	0x50	1	Blue LED	Blue LED on/off	RW	0	0	1
97	0x61	1	Input Voltage	Battery input voltage (unit: 0.1V)	R	--	0	255
112	0x70	1	Controller X-Axis Value	X-axis coordinate of joystick	RW	0	-100	100
113	0x71	1	Controller Y-Axis Value	Y-axis coordinate of joystick	RW	0	-100	100
128	0x80	1	Port 1 Servo Mode	Change left motor mode (wheel or joint)	RW	0	0	1
129	0x81	1	Port 2 Servo Mode	Change right motor mode (wheel or joint)	RW	0	0	1
136	0x88	2	Port 1 Motor Speed	Speed of left wheel	RW	0	-1024	1024
137	0x89	2	Port 2 Motor Speed	Speed of right wheel	RW	0	-1024	1024
156	0x9C	2	Port 1 Servo Position	Left wheel servo position	RW	--	-10240	10240
157	0x9D	2	Port 2 Servo Position	Right wheel servo position	RW	--	-10240	10240

Fig. 2.22 Control table for OLLOBOT controller (Courtesy of ROBOTIS)

will involve developing apps for Android devices (see Chap. 12 for more details). The various ROBOTIS SDKs for the OLLOBOT can be obtained here – http://www.robotis.us/ollobotsdk/.

2.5 System(s) Selection Criteria

When I first prepared the instructional materials for my "Embedded Robotics" course in early 2009, the only option was the CM-5 based systems: BIOLOID COMPREHENSIVE and BIOLOID EXPERT. But nowadays, as shown in the previous sections, we have many more choices for entry points into this "Robotics" journey. Furthermore, ROBOTIS has been putting in diligent efforts for maintaining backwards compatibilities for their software updates and their newly developed actuators and sensors so that the CM-5 based systems are still useful. For example, the following devices are compatible with the CM-5, although they may not function at their full capacities due to the 16 MHz clock speed on the Atmel AVR® chip:

- IR Array Sensor (2012) (http://support.ROBOTIS.com/en/product/auxdevice/sensor/ir_sensor_array.htm)
- Servo motor MX-28T (2011) (http://support.robotis.com/en/product/actuator/dynamixel/mx_series/mx-28at_ar.htm)
- Color video camera HaViMo2 (2010) (https://www.havisys.com/?page_id=8)

So let me go out on a limb and share with you these recommendations:

1. For the DIY user, the MiCom Training Kit can be an economical approach whereas the user can start with TASK and progress to ROBOTIS IDE. This kit could turn out to be a popular entry kit for a first look at the ROBOTIS robotics ecology.

2. The CM-530 STEM or PREMIUM kits would offer a fast MCU @ 72 MHz and the 5-pin GPIO connectors, and Embedded C is readily available (via WinARM and Eclipse). If you have little background in computer programming, you can get started on the RoboPlus Suite V.1 or V.2 and transition to Embedded C with WinARM or purchase additional CM-9.04-A's or B's for an Arduino™-style IDE, as all your hardware would still be compatible. ROBOTIS' e-Shop carries a Programming Guide showing how to use the RoboPlus V.1 software suite (http://www.robotis-shop-en.com/?act=shop_en.goods_view&GS=1486&GC=GD080400).

3. If you are already familiar with robotics and Arduino®, there is good reason to use the OpenCM9.04-C from the beginning. But you will have to purchase all the other components separately as you need them (http://www.robotis-shop-en.com/?act=shop_en.goods_list&GC=GD0803). The assembly instructions and parts lists for practically all ROBOTIS robots (except for the BIOLOID COMPREHENSIVE and PREMIUM series) are now available in the software tool called R + Design (http://support.robotis.com/en/software/roboplus2/r+design/r+design.htm) which essentially is a 3-D CAD tool (see Fig. 2.23). The OpenCM9.04-C is designed to work with the second generation of the RoboPlus software suite which has four tools: R+Design, R+Task, R+Motion and R+Manager. This option would afford the users the most flexibility in switching between ROBOTIS-style and Arduino-style user interfaces. If the user is further interested in interfacing the robot with a smart phone to access its facilities such as camera, audio-visual and messaging services, the OpenCM9.04-C can be programmed to work with the phone app R+m.PLAY700 which is available for Android and iOS systems.

4. When the SMART III system is available internationally, it would be a lower cost system for a freshman 1-credit course in introductory robotics, but it does not have room for expansion into more rigorous topics.

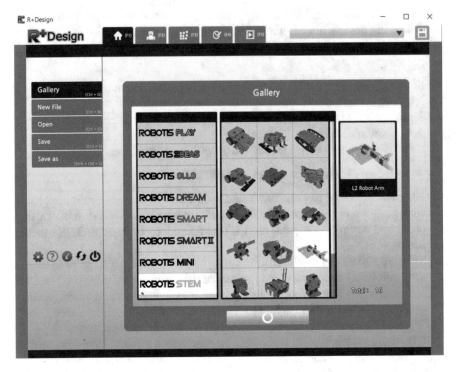

Fig. 2.23 R + Design tool

2.6 Review Questions for Chap. 2

1. What is the product brand name uniquely attributed to ROBOTIS Co.?
2. What was the name of the first commercial product from ROBOTIS?
3. Which are the robotic system(s) from ROBOTIS that use a rivet-like system for building robots?
4. Which ROBOTIS systems can be interfaced with SCRATCH 2?
5. Which ROBOTIS systems can be interfaced with R + MOTION?
6. Which ROBOTIS systems can be interfaced with a smart phone?
7. Can the phone app R + m.PLAY700 work with the DREAM system?
8. What is the name of the tool that displays the mechanical assemblies for most of ROBOTIS robots?
9. What are the micro-controllers currently supported by ROBOTIS?
10. List the robotic system(s) that are based on the STM32 family from STMicrolectronics Co.
11. List the robotic system(s) that are based on the ATmega AVR family from Atmel Co.
12. What is the name of the controller(s) associated with the open hardware-software initiative from ROBOTIS?

13. Which **CM-XXX** controller(s) can implement C-based computer programs from the user?
14. Which **CM-XXX** controller(s) run at 72 MHz?
15. List the four software tools that come with the RoboPlus Software Environment.
16. List three possible communication/interface protocols that are available to the user for use with ROBOTIS systems.
17. Which controller(s) support user-based firmware and boot loader?
18. Which controller(s) support an Arduino-based IDE?
19. Which controller(s) support an expansion board?
20. Name the four types of linkages that are found in a typical four-bar linkage system.
21. A _____ linkage is called a _____ linkage when it can turn a full 360°.
22. **Identify** (i.e. draw) the correct links and **name** them properly for the four-bar linkage system as shown for the Cricket robot's front leg picture below.

Chapter 3
Hardware Characteristics

In this chapter, the goal is to go over the main hardware characteristics of ROBOTIS' controllers, actuators and sensors so that the user can evaluate their hardware options in designing a robotic system to suit their needs.

Current ROBOTIS systems can be broadly divided into two groups, "Firmware 1.0" and "Firmware 2.0". The microcontrollers used in "Firmware 1.0" systems are based on the Atmel AVR® chip, except for the CM-530 which is based on an ARM Cortex M3® chip from STMicroelectronics (see Fig. 3.1). "Firmware 2.0" systems are all based on ARM Cortex M3 chips from STMicroelectronics (see Fig. 3.2).

In general, all "Firmware 1.0" controllers would operate with the original RoboPlus software suite V.1: MANAGER, TASK, MOTION, TERMINAL & DYNAMIXEL WIZARD (http://support.robotis.com/en/software/roboplus_main. htm), while all "Firmware 2.0" controllers would operate with the V.2 of the RoboPlus suite: R+MANAGER, R+TASK and R+MOTION (http://support.robotis. com/en/software/roboplus2.htm). However both R+TASK V.2 and R+MOTION V.2 tools work with both Firmware series (see Fig. 3.3), essentially users have to be more discerning when needing to use the MANAGER V.1 and DYNAMIXEL WIZARD tools.

Via the TASK tool, all "Firmware 2.0" controllers (CM-50, CM-150, CM-200, OpenCM9.04 and OpenCM7.00) have extensive interfaces with mobile devices (see more details in Chap. 11).

This ROBOTIS e-Manual web link has a comprehensive table of all compatible hardware devices: controllers, sensors, actuators, communications and batteries power (http://support.robotis.com/en/product/controller_main.htm).

© Springer International Publishing Switzerland 2017
C.N. Thai, *Exploring Robotics with ROBOTIS Systems*,
DOI 10.1007/978-3-319-59831-4_3

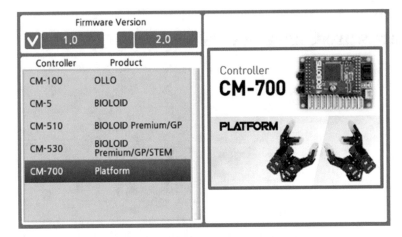

Fig. 3.1 "Firmware 1.0" systems

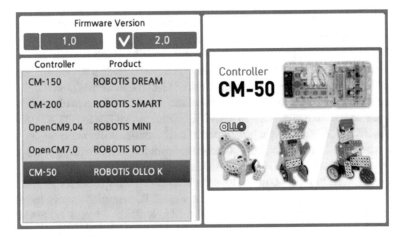

Fig. 3.2 "Firmware 2.0" systems

3.1   The Atmel AVR® Family

3.1.1   CM-5 (Discontinued)

The CM-5 controller (c. 2005) is based on the ATmega128 clocked at 16 MHz and has 128 KB of flash memory (see Fig. 3.4). It comes with the BIOLOID BEGINNER and COMPREHENSIVE (BBC) kits.

It has three TTL-Dynamixel (3-pins) connections, thus ones can use:

1. AX and MX types of actuators, but only the AX-12A actuators come with the BBC kits.

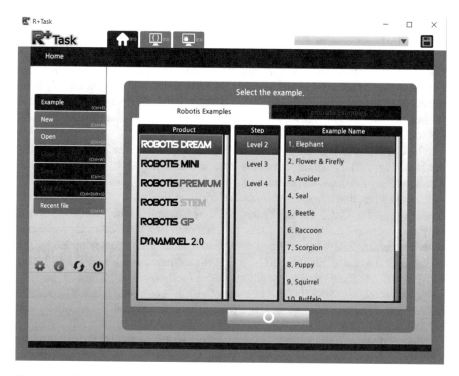

Fig. 3.3 R+TASK V.2 software tool

Fig. 3.4 Internal view of the CM-5

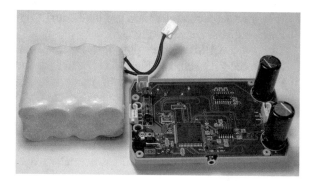

2. Also available with the BBC kits is the AX-S1, an integrated sensor, with 3 NIR sensors (right, center, left) which can be used in both active and passive modes. It has a buzzer which can also serve as a sound clap counter. It also has an NIR receiver (IR-10) which can be used with the RC-100 remote controller in NIR mode.

3. Other Dynamixel-based sensors (3-pin) do exist for other types of measurements such as foot pressure, 3-axes acceleration or color video camera which will be discussed in more details in Sect. 3.3.

It uses a mini-jack connector/cable (BSC-10) to communicate with the PC via a 9-pin RS-232 connector. If your PC has only USB ports, you need to use the USB2Dynamixel module to serve as the interface between the USB port and the CM-5 communication port. More advanced users can also use the USB2Dynamixel to control 3-pin (TTL) and 4-pin (RS-485) types of actuators and sensors directly from the PC (C/C++ programming knowledge is required and several web sites have related information and tutorials such as "softwaresouls.com", "forum.trossenrobotics.com" or "robosavvy.com/forum").

The CM-5 is also capable of wireless communications via the ZIG-100 daughter card (see Fig. 3.5). However it can only accommodate only ONE mode of communication at any one time, i.e. either wired or wireless but not both. With the "wired" option, ones can download microcontroller codes as well as send "remote control" commands to the CM-5 via the Virtual RC-100 Controller. The "wireless" option only allows remote control commands via a physical RC-100 Controller (equipped with a matching ZIG-100) or via another CM-5 (also with a matching ZIG-100). Please note that since 2012, the RC-100 got discontinued and was replaced by the RC-100A/B to accommodate additional BlueTooth® communications features (http://www.robotis.us/rc-100b/). Only the RC-100B is currently available commercially. More detailed ZigBee® and BlueTooth® communications programming will be discussed in Sect. 5.6 and Chap. 7.

If your PC has a 9-pin RS-232 connector (assumed to be COM1), you can adopt the following scheme to use the "wired" connection for programming and debugging, and then switch to "wireless" for operating your robot during run-time:

1. Using the black 9-PIN to mini-jack serial cable hooked up to COM1, keep on using the TASK V.1 tool via COM1 to create your TSK program and download it to the CM-5 controller and use the Virtual Controller RC-100 as normal to check out your code (see Fig. 3.6).
2. Use a spare USB port to plug in a combination module (USB2Dynamixel-Zig2Serial-ZIG-100, see Fig. 3.7) – in my particular case, it turned out to be COM34. Next, UNPLUG the mini-jack serial cable FROM the CM-5 (if you leave this mini-jack connector plugged in, the CM-5 disables its ZigBee circuit) (review Fig. 3.4).
3. Make sure that you switch to COM34 in TASK tool (V.1), then click on "View Output of Program" on the menu bar to open up the Output Window (see Fig. 3.8) and then use the Virtual RC-100 as normal, except now it is really routed wirelessly via COM34 – also make sure that you actually have the CM-5 in play mode and had started the downloaded program as normal.

Fig. 3.5 CM-5 with (*left*) and without (*right*) daughter card ZIG-100

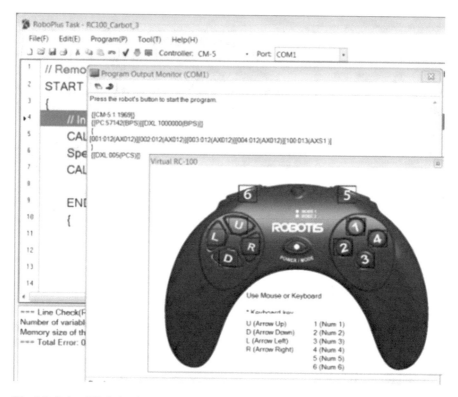

Fig. 3.6 Using CM-5 via wired connection (COM1)

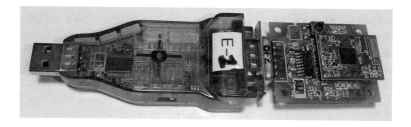

Fig. 3.7 USB2Dynamixel-Zig2Serial-ZIG-100 combo

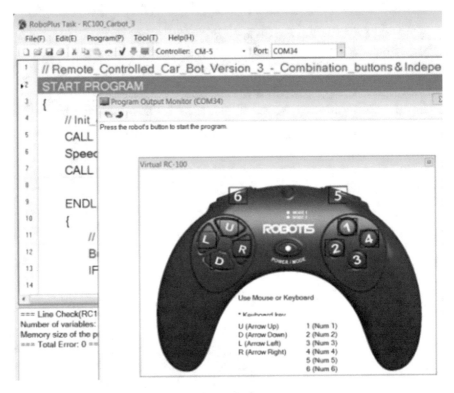

Fig. 3.8 Using CM-5 via wireless connection (COM34)

3.1.2   CM-510 (Discontinued)

The CM-510 controller (c. 2009) is based on the ATmega256, still clocked at 16 MHz but has 256 KB of flash memory (Fig. 3.9). It comes with the BIOLOID PREMIUM (BP) kit from 2009 to 2012. After 2012, the CM-510 is replaced by the CM-530 for the BP kit.

It has an extended suite of Dynamixel, I/O and communications ports:

1. Five 3-pin TTL Dynamixel ports (work with AX and MX actuators and other Dynamixel based sensors).
2. Six 5-pin GPIO ports (work with sensors such as NIR (intensity-based, distance measurement (NIR reflected-angle-based), gyroscope, touch, color, magnetic, with temperature, ultrasonic and motion recognition sensors).
3. One 4-pin communication port for the ZIG-110A which is the equivalent of the ZIG-100 (Fig. 3.10). All previously discussed PC communication options for the CM-5 still apply for the CM-510 (Sect. 3.1.1). The maximum throughput for the ZIG-100/110A is 250 kbps.

Fig. 3.9 Internal views of the CM-510

Fig. 3.10 ZIG-100 and
ZIG-110A ZigBee
communication modules

4. Starting in 2012, BlueTooth communications devices were added: BT-100/110A
   and BT-210 (Fig. 3.11). The maximum throughput for the BT-100/110A/210 is
   250 kbps. They can be used with the same 4-pin communication port on the
   CM-510 but with different features unique to the BlueTooth protocol. The BT-
   100/A will only work in the RC-100A/B remote controller. When using BT modules, the user must take note that the PC would associate 2 COM ports to each
   BT module used, one OUT-GOING and one IN-COMING (the user can check
   this information via Windows® Device Manager). Thus when using ROBOTIS'
   application software on the PC side, the user needs to make sure to choose the
   OUT-GOING COM port.
5. The CM-510 has a sound generator and detector built in its circuit board, and
   still uses the serial mini-jack cable to communicate with the PC for the "wired"
   option.

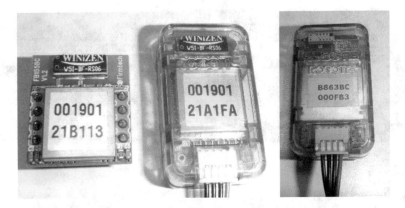

Fig. 3.11 BlueTooth communication devices BT-100/110A and BT-210

3.1.3   CM-700

The CM-700 controller (c. 2009) is also based on the ATmega256 (16 MHz, 256 KB) and is still commercially available. It is composed of 2 boards: the CM-700 controller board per se, and an expansion board that allows it to communicate at 2 protocols, TTL (3-pin connector) and RS-485 (4-pin connector) (see Fig. 3.12).

Functionally, the CM-700 is equivalent to the CM-510 (see Sect. 3.1.2), except that the CM-700 can additionally control DX/RX/EX Dynamixels (RS-485 protocol). The CM-700 is also similar in functionality to the combo OpenCM-9.04/C + 485EXP board (see Sect. 3.2.5), except that the CM-700 runs on an Atmel controller while the OpenCM-9.04/C runs on an ARM Cortex M3 controller.

3.2   The STM ARM® Cortex M3® Family

3.2.1   CM-150

The CM-150 controller (c. 2014) is based on the STM32F103C8, clocked at 72 MHz and has 256 KB of flash memory (Fig. 3.13). It comes with the DREAM system but is also available singly. It uses the USB port directly to communicate with the host PC via a USB2 cable (A to micro B) for programming. It also has a 4-pin communication port so that it can use other communication devices such ZIG-110A (ZigBee), BT-110A/210/410 (BlueTooth), and the LN-101 (Fig. 3.14) for wired communications.

The CM-150 runs at 3.7 V using a single LBS050 Li-ion battery. It has three internal NIR distance sensors and two dedicated ports (Port 1 and Port 2) for continuous-turn actuators such as the GM-12A, and also two GPIO ports (Port 3 and 4) for various sensors (such as NIR, Touch) and servo motors (such as SM-10A).

Fig. 3.12 CM-700
Controller and its
expansion board (*left*) and
CM-510 (*right*)

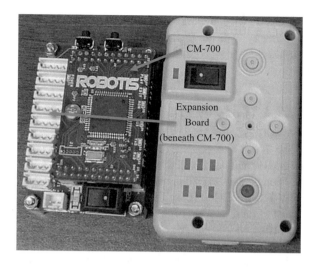

Fig. 3.13 CM-150
controller

Fig. 3.14 LN-101 USB
downloader

Software wise, it works with the R+MANAGER and R+TASK tools (V.2), and thus indirectly with the R+m.PLAY700 mobile app installed on either Android or iOS systems.

3.2.2   CM-200

The CM-200 controller (c. 2015) is also based on the STM32F103C8, clocked at 72 MHz and has 256 KB of flash memory (Fig. 3.15). It only has a 4-pin communication port so that it can only use communication devices such ZIG-110A (ZigBee), BT-110A/210/410 (BlueTooth), or the LN-101 for wired communications with the host PC and/or other robots as needed.

The CM-200 can run on either one or two LBS050 Li-ion batteries depending on the actuator and sensor load. Compared to the CM-150, it also has three internal NIR distance sensors and two dedicated ports (Port 1 and Port 2) for continuous-turn actuators such as the GM-12A, however it has eight GPIO ports (Ports 3 to 10) for various sensors (such as NIR, Touch) and servo motors (such as SM-10A). Software wise, it works with all three tools R+TASK, R+MANAGER and R+MOTION (V.2) and also with the mobile app R+m.PLAY700. It comes with the SMART I, II and III systems and currently it is not available commercially as a separate item.

Fig. 3.15 CM-200
controller

Fig. 3.16 Internal views of the CM-530

3.2.3 CM-530

The CM-530 controller (c. 2012) is based on the STM32F103RE, clocked at 72 MHz and has 256 KB of flash memory (Fig. 3.16). It uses the USB port directly (i.e. no need for the USB2Dynamixel module) and it needs a USB2 cable (A to Mini B). Otherwise it is functionally and visually identical to the CM-510.

It has the same extended suite of Dynamixel, I/O and communications ports as the CM-510:

1. Five 3-pin TTL Dynamixel ports.
2. Six 5-pin GPIO ports.
3. One 4-pin communications port, and one USB-mini port.

All actuators, sensors, ZigBee and BlueTooth modules previously discussed for the CM-510 (Sect. 3.1.2) work with the CM-530 with a boost in MCU clock rate (72 MHz instead of 16 MHz).

It should also be noted that the procedures described in Sect. 3.1.1 for simultaneous use of "wired" and "wireless" communications on the CM-5 will not work with the CM-530, vis-à-vis its USB connection and its wireless connections (whether ZigBee or BlueTooth). In other words, for a CM-530, the user could only have one option, either wired via USB, or wireless via BlueTooth/ZigBee (also noting that only BlueTooth could download TASK programs from the PC to the robot).

3.2.4 CM-900 (Discontinued)

The CM-900 series was ROBOTIS' entry into the open hardware and software movement and the CM-900 controller was available to the beta testing community between October 2012 and July 2013. It had two versions, ES and V.1 (see Fig. 2.17) and both were clocked at 72 MHz. V.1 was only commercially available during 2013.

The main technical specifications for the CM-900 V.1 are listed below:

	CM-900 V.1
CPU	STM32F103C8 (ARM Cortex-M3)
Op voltage	5–35 V (USB 5 V, DXL 12 V, 7.4 V)
Flash	64 KB
3-Pin DXL TTL	2
4-Pin DXL RS485	2
Mini 3-Pin DXL TTL	1

From the DXL connector and operating voltage options shown above, ones can see that the CM-900 was quite possibly envisioned to control the complete family of Dynamixel actuators from AX to Dynamixel Pro. It used an Arduino-based IDE called OpenCM IDE, currently at version 1.0.4 (see Fig. 3.17).

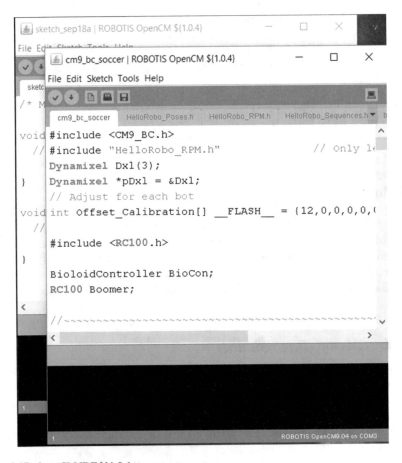

Fig. 3.17 OpenCM IDE V.1.0.4

Included with this book is a zipped file containing user manuals and other technical information for the CM-900 series (OpenCM-900.zip). These files originated from the CM-9 Developer's World Circle™ at http://www.robotsource.org/bs/bd. php?bt=forum_CM9DeveloperWorld Unfortunately this web site was closed in July 2017.

With feedbacks from the beta test community, the CM-900 was redesigned into the OpenCM-9.04 in July 2013.

3.2.5   OpenCM-9.04 Series

The OpenCM-9.04 series illustrates a segmented-market view by ROBOTIS which originally offered 3 versions of the OpenCM-9.04 (A, B and C) with the same main technical features listed below:

	CM-9.04-A/B/C
CPU	STM32F103CB (ARM Cortex-M3)
Op voltage	5–16 V (USB 5 V, DXL 12 V, 7.4 V)
Flash	128 KB

The OpenCM-9.04-A (Fig. 3.18) allows the user complete freedom in mounting only the components that are needed for his or her project from the power switch to the types and numbers of interface headers needed for AX/MX-TTL or XL-TTL (3-pin) and GPIO (5-pin), and even JTAG/SWD connection.

The OpenCM-9.04-B (Fig. 3.19) comes pre-connected for power switch, battery connections and connectors for AX/MX-TTL (2), XL-TTL (2), GPIO (4), JTAG/ SWD (4-pin) and wired/wireless communications (4-pin).

The A and B versions have open firmware and are designed to work only with the ROBOTIS IDE software. Included with this book is a zipped file containing user manuals and other technical information for the OpenCM-9.04-A/B series (OpenCM-904.zip) which originated from the ROBOTIS e-Manual web site http:// support.robotis.com/en/product/controller/opencm9.04.htm where interested users can get more updated information. Advanced users may also want to invest in the in-circuit debugger and programmer ST-LINK/V2 (http://www.st.com/web/en/catalog/tools/PF251168) and Keil μVision IDE (http://www.keil.com/arm/mdk.asp) to modify firmware and bootloader programs. The B version is no longer available commercially.

The OpenCM-9.04-C (Fig. 3.20) comes pre-connected for power switch, battery connections and connectors for XL-TTL (4), GPIO (4) and JTAG/SWD (4-pin) and wired/wireless communications (4-pin).

The C version comes with a proprietary firmware to make it function with the new generation of ROBOTIS software (V.2): R+Task, R+Motion and R+Manager. Alternately, the user can use the OpenCM IDE with version C but this mode would

Fig. 3.18 OpenCM-9.04-A
controller

Fig. 3.19 OpenCM-9.04-B
controller

Fig. 3.20 OpenCM-9.04-C
controller

Fig. 3.21 OpenCM-9.04-C
and 485-EXP set up

effectively erase the proprietary firmware, thus if the user wants to use the R+
packages again, a firmware recovery process must be performed.

If the user requires more AX/MX-TTL and/or RS-485 Dynamixel ports (with additional power needs), an expansion board can be used (OpenCM-485-EXP) and the
user also needs to solder extra headers (from an accessory set) onto the OpenCM-9.04
(see Fig. 3.21), please visit the web links below for more information:

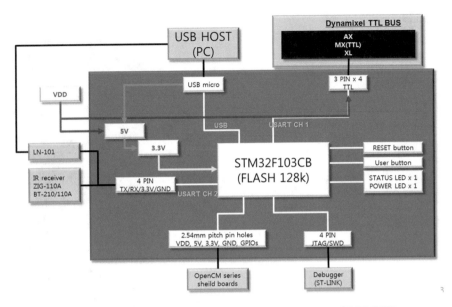

Fig. 3.22 Functional block diagram for OpenCM-9.04 series (Courtesy of ROBOTIS)

- http://support.robotis.com/en/product/controller/opencm_485_exp.htm
- http://www.robotis-shop-en.com/?act=shop_en.goods_view&GS=1623 &GC=GD080201.
- http://www.robotis-shop-en.com/?act=shop_en.goods_view&GS=1490&GC =GD080201.

As compared to the CM-5XX series, the OpenCM-9.04 series have more options regarding connections from the PC: plain USB2 cable (A to micro), or LN-101, ZigBee/BlueTooth via Serial 2 (USART CH 2) (see Fig. 3.22). Serial 1 (USART CH 1) was still reserved to the Dynamixels on the local OpenCM-904 Dynamixel TTL bus. Serial 3 was "new" and connected to the OpenCM 485 Expansion Board for access to AX/MX/DX/RX/PRO types of Dynamixel actuators.

Fortunately, a compatibility chart for all ROBOTIS products can be found at http://support.robotis.com/en/product/controller_main.htm, however a word of caution when using that list, for example:

- A plain USB connection would work best for an OpenCM-9.04-A/B controller if you use the ROBOTIS IDE (Arduino-based), but it won't work if you try to use it with the RoboPlus Manager (V.1) tool to update its firmware. However, the user can update the OpenCM-9.04-C firmware via this USB cable and RoboPlus Manager (V.1 & 2) (see more details in Sect. 4.2.2). A more versatile module is the LN-101 that would allow the Dynamixel Wizard tool to update firmware on a CM-9.04-C and the XL-320s connected to it, at the same time (see Sect. 4.1.2.1). The LN-101 can also be used with both MANAGER's versions and is highly recommended by the author (http://www.robotis-shop-en.com/?act=shop_ en.goods_view&GS=1277&keyword=LN-101).

Fig. 3.23 BT-410 series

- The ZigBee (ZIG-110A) and BlueTooth modules (BT-110A and BT-210) are compatible with the 9.04-C, but so far I had found that ZigBee performed better than BlueTooth in terms of connection latency and packet performance (more on this in Chap. 4).

Ironically, part of these issues is due to the fact that ROBOTIS is committed to continuing development of their hardware and software systems, so the software team is always playing catch-up to the hardware team! In Chap. 5, "particular" usages will be described in more details.

Furthermore, since Summer 2015, ROBOTIS released a new series of BlueTooth modules (BT-410) based on BlueTooth 4.0 Low Energy (BLE) standard. This series will be able to handle 1 to 1 (1:1) and 1 to N (1:N) communications (not yet implemented as of Spring 2017) and targeted towards mobile devices (see Fig. 3.23). This web link http://support.robotis.com/en/product/auxdevice/communication/bt-410main.htm compares the capabilities of ZigBee and BlueTooth communications devices offered by ROBOTIS. The maximum throughput rate for the BT-410s is 128 kbps and currently (Fall 2016) the BT-410s are not compatible with the MS Windows® environments.

3.2.6   OpenCM-7.00 (IoT)

The OpenCM-7.0 controller (not to be confused with the CM-700 – Sect. 3.1.3) comes with the ROBOTIS IoT system which is designed for educational home automation applications (see Fig. 3.24).

It uses the STM32L151C8 chip and its top side is shown in Fig. 3.25. It has a Power switch (middle and left side) and a Software Reset switch (top right corner), one 4-pin communication port (white connector), 1 USB2 port (micro B – left edge), a set of RGB LEDs (top left) and a microphone (bottom right).

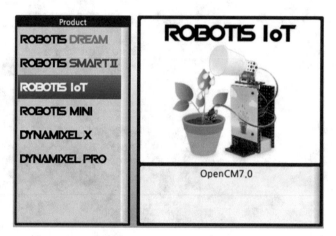

Fig. 3.24 ROBOTIS IoT system

Fig. 3.25 ROBOTIS OpenCM-7.0 controller (beta version)

It also has a total of 6 I/O ports:

- Ports 1 and 2 (2-pin black connectors, on right of mid-line) are reserved for continuous-turn type of motors such as GM-12A.
- Ports 3 through 6 are 4 GPIO ports (5-pin black connectors, 2 above and 2 below of component labeled "BTN 0327"). They can interface with servo motors (SM-10A) and a variety of sensors such as temperature, humidity, illuminance, passive IR, color, ultrasonic, magnetic and "user device". In particular, Port 3 can also operate an IR Sensor Array with 5 sensors (not yet available commercially).

3.2.7   CM-50 (PLAY 700)

The CM-50 controller (not to be confused with the CM-5 – Sect. 3.1.1) came with the ROBOTIS PLAY 700 system which was designed for the younger beginners (see Fig. 3.26). This system came with a BT-410 which was compatible only with

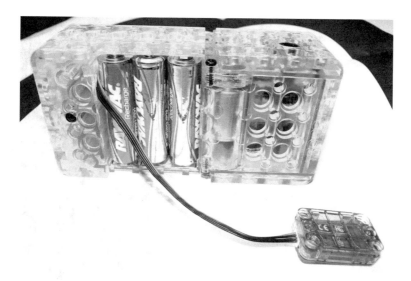

Fig. 3.26 ROBOTIS CM-50 controller with BT-210

mobile Android and iOS devices. To program the CM-50 on MS Windows environments, ones would need to use either the BT-210 or the LN-101 module, but ones would have to leave the LN-101 connecting cable with the finished robot due to its present connector design which was located deep in the battery well (see Fig. 3.26). The BT-210 was also compatible with Android phones and tablets.

It had built-in sound and NIR sensing capabilities and could support up to two continuous-turn motors. It could be programmed either with the SCRATCH V.2 tool (via the R+SCRATCH helper app and on MS Windows only) or with the TASK V.2 tool (mobile or Windows versions) which would let it access the camera, multimedia and gesture recognition features of an attached mobile device such as a phone or a tablet. Unfortunately, both software packages cannot be used at the same time due to incompatibility of the communication protocols used between the two software tools R+SCRATCH and TASK.

3.3 The Dynamixel Actuators Family

ROBOTIS has an extensive repertoire of actuators and maintains up-to-date technical information at this web link (http://support.robotis.com/en/product/actuator_main.htm). Currently most of ROBOTIS actuators are using Communication Protocol 1 (http://support.robotis.com/en/product/actuator/dynamixel/dxl_communication.htm), except for the Dynamixel X and Dynamixel PRO series which are using Communication Protocol 2 (http://support.robotis.com/en/product/actuator/dynamixel_pro/communication.htm).

Fig. 3.27 MIKATA™
robot (With permission
from creator Yoshihiro
Shibata)

The goal of this section was not to provide duplication of the above information but to provide the reader with some complementary hardware information using the AX-12 and MX-28 as examples.

The MX-28 was chosen as a representative of the types that can perform 0–360 degrees in position control mode and with PID control features, while the AX-12 was chosen as a representative of the types that have a more limited range of motion from 0 to 300 degrees when in position control mode and with only proportional control.

The latest actuators come from the X series (http://support.robotis.com/en/product/actuator/dynamixel_x_main.htm) such as the XL-320 used on the ROBOTIS MINI, and the XM/XH family to be used on the next generation of robots such as the MIKATA™, designed by Yoshihiro Shibata (see Fig. 3.27 and also the video at https://www.youtube.com/watch?v=focanpe3eok). The XM/XH actuators are being used in the development of the upcoming ROBOTIS OP3 humanoid robot (https://www.youtube.com/watch?v=9YtiereseaI).

3.3.1   The AX Series

The first ROBOTIS Dynamixel actuator was the AX-12 (c. 2003) which had a stall torque of 1.5 Nm and ran around 59 RPM. Throughout the years, it got updated to AX-12+ and then to AX-12A (current version). It has two "cousins":

Fig. 3.28 AX-12 PCB with Murata SV01A103 (*blue hexagonal box*) and main gear shaft

1. AX-18F/AX-18A which has a slightly higher stall torque of 1.8 Nm and runs faster at 97 RPM.
2. AX-12 W which is designed more for continuous rotation as wheels. It can rotate in wheel mode at 470 RPM and in joint mode at 54 RPM. Its stall torque is not provided by ROBOTIS.

They all use DC motors and have proprietary firmware implemented on an Atmel AVR microcontroller (16 MHz and 8 KB memory). Their firmware can be updated via the Dynamixel Wizard software tool, using a USB2Dynamixel module and requiring separate power (via the SMPS2Dynamixel, for example). See this web link for more details (http://support.robotis.com/en/software/roboplus/dynamixel_wizard.htm). In operations, the actuators get updated every 8 ms by the main controller via the Dynamixel bus. The TTL communication speed for the AX family can be set between 7.8 Kbps and 1 Mbps.

The good folks at Irish Robotics have provided circuit diagrams for the AX-12 motor and processor, and even for the CM-5 at the following links:

http://robosavvy.com/Builders/Chris.H/AX12Motor.pdf
http://robosavvy.com/Builders/Chris.H/AX12Processor.pdf
http://robosavvy.com/Builders/Chris.H/CM5.pdf

One AX-12 feature that users found most intriguing was how could the AX-12 rotate completely in 360 degrees when in wheel mode but was restricted to a 300 degrees range when in position control mode. This was due to the position encoder used on the AX-12 which is a Murata SV01A103, effectively a potentiometer with an effective range of only about 333 degrees (see Fig. 3.28)

During assembly at the factory, the main gear shaft (black plastic) is inserted into the semi-circular white rotor of the Murata part and then through a gear train it is connected to the DC motor (silver cylinder on the right in Fig. 3.28). Thus when the motor rotates, the white rotor also rotates but at a slower speed due to the gear ratio. The white rotor "wipes" against a resistive element inside the SV01A103 and referring back to the AX12Motor.pdf document, ones can see that when at a position 60 degrees left of the "12 o'clock" position, this rotor (Pin 2) would provide a 5 V read-

Fig. 3.29 "I" mark on horn of AX-12 actuator

ing to the AX-12 microcontroller which would then digitize it to a numerical value of 1023. Similarly, when this rotor reaches 60 degrees left of the "6 o'clock" position, Pin 2 would provide a 0 V reading which gets digitized to a numerical value of 0. And thus there is a ±30 degrees zone around the "9 o'clock" position where Pin 2 is not connected to the resistive element, and therefore no voltage reading is possible, and consequently the AX-12 microcontroller has no valid digitized value also (i.e. it does not know where it is at). This is why on the horn of the AX-12, there is an "I" mark (see Fig. 3.29) so that the user can use it to put the servo into the vicinity of the "512" position (3 o'clock position) in the assembly instructions for most robots, to be used as an initial pose before they perform any programmed action.

Ones can also recognize that with wears due to usage, the rotor may provide "unreliable" voltage readings to the AX-12 controller at certain positions and as a consequence of this, ones can observe that an older servo tends to "tremble" when it is commanded to stay at those "unreliable" positions.

As plastic gears can break off sometimes, users can order replacement sets at http://www.robotis-shop-en.com/?act=shop_en.goods_view&GS=1397&GC=GD080307. DIY videos on YouTube described the detailed steps plus tools needed to perform these maintenance tasks (http://www.youtube.com/watch?v=W1sOavdmIus).

Thus on higher-end actuators such as the MX-28, metallic gears and shafts are selected and most importantly contact-less magnetic position encoders are used.

3.3.2 The MX Series

The MX series started with the MX-28 which was designed for the ROBOTIS-OP (c. 2011). It has a stall torque of 2.5 Nm at 12 V and 55 RPM. This series use the STM32F103C8 @ 72 MHz as controllers and they has all metallic gears (see Fig. 3.30a). Once the gear train is removed, ones can see the magnetic encoder chip AS5045 inside the shaft cavity (Fig. 3.30b).

On the right in Fig. 3.30a, the big silver gear (where the servo's horn would be mounted to) has on the other end of its shaft a 2-pole magnet that would hover over the encoder chip AS5045 shown in Fig. 3.30b (see Fig. 3.17 on page 24 of the data

Fig. 3.30 (**a**, **b**) Internal views of MX-28T

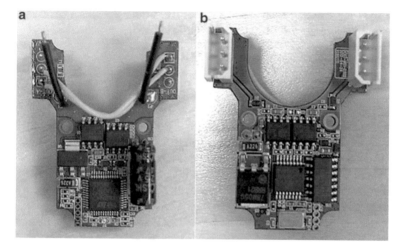

Fig. 3.31 (**a**, **b**) Two views of the PCB of an MX-28T

sheet available at http://www.ams.com/eng/Products/Position-Sensors/Magnetic-Rotary-Position-Sensors/AS5045). When the motor turns, this magnet also turns via the gear train, and thus changes the magnetic field characteristics as sensed by the AS5045 which uses its own firmware to output a corresponding PWM signal over a full turn of 360 degrees.

Figure 3.31a, b shows two views of the PCB of a typical MX-28T. In Fig. 3.30a ones can see the main STM controller, and on Fig. 3.31b the magnetic encoder AS5045 is shown. This magnetic encoder allows the MX actuators to have position control for the complete 0–360 degrees range at 0.088 degree resolution (i.e. 12 bits).

Information for the complete MX family can be found at this web link: (http://support.robotis.com/en/product/actuator/dynamixel/dxl_mx_main.htm). The TTL communication speed for the MX family can be set between 7.8 Kbps and 4.5 Mbps. The MX-28 has three "cousins":

1. MX-12 W, designed to be used as a wheel running at 470 RPM.
2. MX-64, with stall torque at 6.0 Nm and 63 RPM.
3. MX-106, with stall torque at 8.4 Nm and 45 RPM.

The effects of PID control features (MX-28) or the lack thereof (AX-12) on the position control behavior of these example actuators will be explored in more details in Chap. 6.

For a more general understanding of servo motor design and control, I would like to refer the readers to Clark and Owings (2003) or Kanniah et al. (2014). The "Humanoid Robot Lab" in Portugal had done detailed studies on the Bioloid Comprehensive kit and AX-12 (http://humanoids.dem.ist.utl.pt/humanoid/over-view.html). And there are many more Internet sites sharing others' knowledge and experiences with ROBOTIS products. ROBOTIS also maintains a ROBOTIS channel on YouTube showing many how-to videos for various maintenance tasks (http://www.youtube.com/user/ROBOTISCHANNEL/videos).

3.3.3   The X Series

The first member of the X family was the XL-320 actuator which debuted with the launch of the ROBOTIS MINI in 2014. It had a similar mechanical design to the AX-12A using a 300 degrees mechanical position encoder (see Fig. 3.32).

However the XL-320 was capable of PID control similarly to the MX-18, whereas each respective P-I-D control parameter could be changed on the fly from within the user computer program (see Fig. 3.33 for a TASK tool example), along with the setting of three operating parameters Goal Position, Goal Velocity and Goal Torque.

Fig. 3.32 PCB of an XL-320

Position Encoder

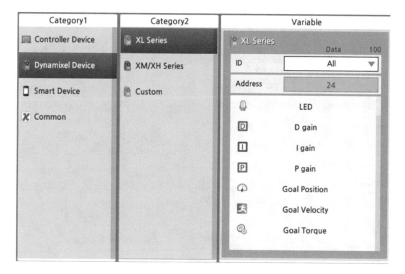

Fig. 3.33   PID capabilities of an XL-320

Fig. 3.34   XM-430 actuator

The latest members of the X-series were the XM and XH actuators, coming out in early 2016 (see Fig. 3.34). They used the RS-485 protocol (4-pin connector) and were compatible with the USB2Dynamixel module and 485-EXP board.

The XM/XH actuators offered six modes of operation from a high-level control such as Torque Control to a low-level control such as PWM control (http://support. robotis.com/en/product/actuator/dynamixel_x/xm_series/xm430-w210.htm):

- Torque Control
- Position Control
- Velocity Control

Fig. 3.35 Control parameters for XM-430 actuator available inside TASK tool

- Extended Position Control (multi-turn)
- Current-based Position Control
- PWM Control

They used magnetic encoders similar to the ones for MX-28s with a 12 bit resolution and a 360 degrees position control range. They also offered Profile Control for Smooth Motion Planning. A snapshot of the control parameters that could be set inside a TASK program is shown in Fig. 3.35.

3.4 ROBOTIS Sensors Family

Currently available sensors from ROBOTIS and third-party partners can be sorted into two groups:

1. Dynamixel-compliant sensors (3-pin TTL) can be used with controllers such as CM-5/510/530/700 and also on the 485-EXP expansion board (to be controlled by the OpenCM9.04-C via the ROBOTIS IDE software tool). They can be daisy-chained on the Dynamixel bus and are individually assigned with a unique ID (0–253, however ID 200 is reserved for the main microcontroller), and ID 100 is reserved for the remote device such as a smartphone:

    (a) AX-S1, Integrated Sensor from ROBOTIS, NIR intensity-based sensor (short and long range active modes or passive mode), sound claps detector and NIR remote control sensor (http://support.robotis.com/en/product/aux-device/sensor/dxl_ax_s1.htm).

(b) IRSA, IR Sensor Array from ROBOTIS, 7 NIR intensity-based sensors with buzzer features (http://support.robotis.com/en/product/auxdevice/sensor/ir_sensor_array.htm).

(c) AX-S20, first Inertia Measuring Unit from ROBOTIS, 3-D acceleration components and resulting XYZ rotational angles, plus a magnetic heading sensor. In beta tests in 2010, but it never got to full production.

(d) Foot Pressure Sensor (from HUV Robotics). Not commercially available currently.

(e) HaViMo2™ and HaViMo3™, color video cameras with onboard processing capabilities (https://www.havisys.com), commercially available in Europe, Mexico and Southeast Asia.

2. GPIO-based sensors (5-pin). These sensors have a open architecture and are attached to specific ports available on various CM and OpenCM controllers, for example:

(a) CM-510 and CM 530 have six GPIO ports, while OpenCM7.00 and OpenCM9.04 have only four GPIO ports.

(b) From ROBOTIS, we have a growing supply of these types of sensors (IR, Distance Measurement DMS, Gyroscope, Touch, Color, Magnetic) and in the near future (Temperature, Ultrasonic and Gesture Recognition).

3.4.1 AX-S1 and IRSA

Both sensors work on the principle of sending out 38-KHz pulses of NIR light (invisible to the human eyes) and then to process the NIR intensities reflected off objects in its path. Thus an object closer to the sensor would reflect more NIR energy (represented as a bigger numerical value to the microcontroller) than an object further away. In Fig. 3.36, the "clear" LED (right LED) sends out the NIR pulses, and the "black" LED (left LED) receives the reflected NIR intensities. Please note the "recessed" mounting of these LEDs to reduce interferences from each other.

This type of sensor does have a drawback when your robot has to deal with objects with different brightnesses as darker objects will be "perceived" in being "further away" than lighter objects although their "true" physical distances may be

Fig. 3.36 Forward looking set of NIR LEDs for AX-S1

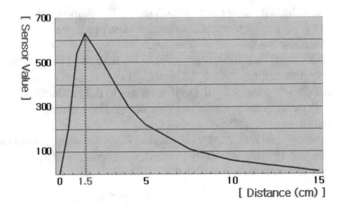

Fig. 3.37 Non-linear response of a typical NIR-LED sensor vs. distance

Fig. 3.38 Application of IRSA on a challenging track

the same. Furthermore their response is non-linear, including a reversal in response at extreme close range (see Fig. 3.37).

However this technique is quite effective in detecting white from black areas on a complicated track using an IR Sensor Array (IRSA) as shown in Fig. 3.38 (more details in Chap. 5). Currently the IRSA only comes with the BIOLOID STEM STANDARD kit and it is not available separately.

The AX-S1 can switch between a short range detection (up to 1–3 cm) and a long range detection (up to 10–12 cm) using a software parameter. It also has a passive mode to detect light sources. The AX-S1 can use its microphone to sample at 3.8 KHz sound events such as hand claps, but they have to be at least 80 ms apart.

These Dynamixel compliant sensors are also well documented at these ROBOTIS web links:

- http://support.robotis.com/en/product/auxdevice/sensor/dxl_ax_s1.htm.
- http://support.robotis.com/en/product/auxdevice/sensor/ir_sensor_array.htm.

More detailed applications programming of these sensors will be shown in Chap. 5.

3.4.2 AX-S20 (Discontinued)

Although this product never saw full production, the reader may be interested in its working. The AX-S20 used an ATmega8 for controller (i.e. Dynamixel compliant) and the sensor AMS0805WAH from Amosense (http://www.motionsense.net/kor/index/AMS0805WAH%20DataSheet%201.3%20%28AMOSENSE%29.pdf) for measuring 3-Axis Magnetic and Acceleration fields. The AMS0805WAH has an embedded calibration algorithm which eliminates the need for initial calibrations and supports precise calculation of motion data under all environments. It has 1 degree resolution for its Roll, Pitch and Azimuth angles. In practice, it supports about a 20 Hz sampling rate on a CM-510. Figure 3.39 shows an AX-S20 installed in the head piece of a PREMIUM Humanoid A robot and the resulting coordinates system (*as a side note, the AX-S20, with its magnetometer, was mounted in the robot's head so as to keep it away from the electromagnetic interferences of the operating actuators as much as possible*).

Figure 3.40 is a screen capture of a RoboPlus Manager V.1 session on a Humanoid A robot with an AX-S20 (ID = 105), and ones can see the list of various inertia parameters that ones can use in its programming.

So far we only got the opportunity to test the AX-S20 via the RoboPlus software suite and these application programming projects are discussed later in Chap. 8.

A newer IMU (MPU-9150) with gyroscope, accelerometer and compass components (each having three axes) is available at Spark-Fun Electronics (https://www.sparkfun.com/products/11486) and has been adapted to work with the OpenCM-9.04 framework.

Fig. 3.39 AX-S20 installed in the head of a PREMIUM Humanoid A robot and the resulting coordinates system

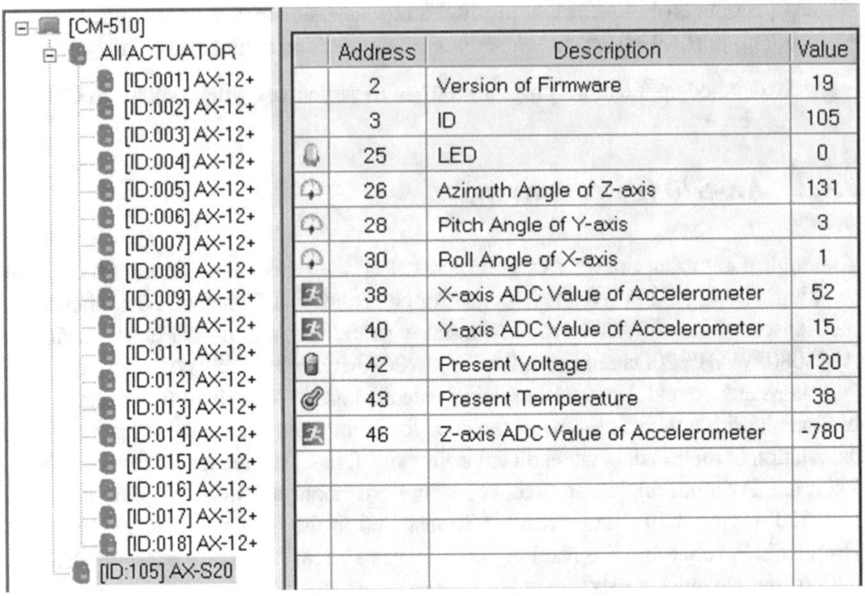

	Address	Description	Value
	2	Version of Firmware	19
	3	ID	105
	25	LED	0
	26	Azimuth Angle of Z-axis	131
	28	Pitch Angle of Y-axis	3
	30	Roll Angle of X-axis	1
	38	X-axis ADC Value of Accelerometer	52
	40	Y-axis ADC Value of Accelerometer	15
	42	Present Voltage	120
	43	Present Temperature	38
	46	Z-axis ADC Value of Accelerometer	-780

Tree view:
- [CM-510]
  - All ACTUATOR
    - [ID:001] AX-12+
    - [ID:002] AX-12+
    - [ID:003] AX-12+
    - [ID:004] AX-12+
    - [ID:005] AX-12+
    - [ID:006] AX-12+
    - [ID:007] AX-12+
    - [ID:008] AX-12+
    - [ID:009] AX-12+
    - [ID:010] AX-12+
    - [ID:011] AX-12+
    - [ID:012] AX-12+
    - [ID:013] AX-12+
    - [ID:014] AX-12+
    - [ID:015] AX-12+
    - [ID:016] AX-12+
    - [ID:017] AX-12+
    - [ID:018] AX-12+
  - [ID:105] AX-S20

Fig. 3.40 Screen capture of RoboPlus Manager used on a PREMIUM Humanoid A robot with AX-S20 (ID = 105)

3.4.3 Foot Pressure Sensor (FPS: From HUV Robotics)

This HUV Robotics product is also Dynamixel compliant, but it is not currently available commercially. Figure 3.41 shows how a set of FPS was mounted at the four corners of a foot of a PREMIUM Biped.

Each pressure sensor relies on a material that would change its electrical resistance (i.e. its voltage values) depending on the pressure imposed on it. These FPS require a flat hard surface for a satisfactory performance. Figure 3.42 shows the addressing scheme to access these values from inside a typical TASK program.

So far we only got the opportunity to test the HUV FPS via the RoboPlus software suite and these application programming projects are described later in Chap. 8.

ROBOTIS provides a similar product to the HUV FPS but only for the ROBOTIS-OP (DARwIn-OP) called FSR (http://www.robotis-shop-en.com/?act=shop_en.goods_view&GS=1442&keyword=FSR).

3.4.4 Color Video Cameras HaViMo 2.0 and 3.0

These color video cameras (Figs. 3.43 and 3.44) are available via e-shops located in the United Kingdom and Southeast Asia (https://www.havisys.com/?page_id=8 and https://www.havisys.com/?page_id=222).

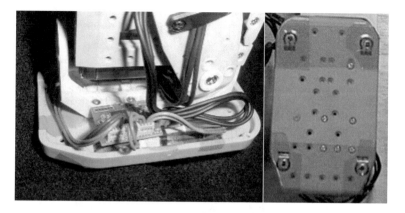

Fig. 3.41 HUV FPS mounted on foot of a PREMIUM Biped robot

```
                              LOOP FOR ( I = 0 ~ 2 )          121 or 122
                              {
                                  IFFSR = IFFSR +  ID[CurID]: ADDR[26(w)]
                                  OFFSR = OFFSR +  ID[CurID]: ADDR[28(w)]
                                  IRFSR = IRFSR +  ID[CurID]: ADDR[30(w)]
                                  ORFSR = ORFSR +  ID[CurID]: ADDR[32(w)]
                              }
```

Fig. 3.42 Addressing scheme to access FPS data from a TASK program

Fig. 3.43 HaViMo 2.0 video cameras mounted on PREMIUM BiPed

Fig. 3.44 HaViMo 3.0 video camera

The HaViMo 2.0 is Dynamixel compliant and it has a video frame resolution at 160×120 pixels (19 fps) with a color depth of 12 bits YCrCb. It has many powerful embedded image processing functions for tracking multi-colored blobs.

So far we have tested this camera successfully on CM-510 and CM-530 using RoboPlus TASK as well as Embedded C on them. This camera performed better on the newer controllers such as OpenCM-9.00 and OpenCM-9.04-B due to the faster MCU clock rate.

The HaViMo 3.0 is also Dynamixel compliant and it has a video frame resolution at 160×120 pixels (20 fps) with a color depth of 12 bits YCrCb. It has additional image processing functions over the HaViMo 2.0 such as on-line Gradient Vector Gridding and Image Averaging for lower resource processing needs such as Line Following.

Application programming projects for these cameras will be described later in Chap. 8.

3.4.5 GPIO (5-Pin) DMS Sensor

This sensor uses a triangulation technique and a Position Sensitive Detector to determine the angular displacement of the reflected light beam and thus can compute the distance to the object. It can achieve a longer detection range (10–80 cm) than the AX-S1 and is mostly independent of objects' brighnesses (see Fig. 3.45).

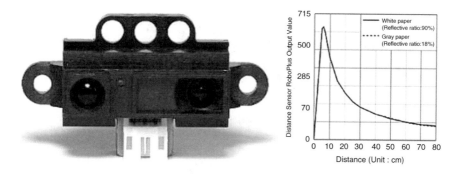

Fig. 3.45 DMS and output vs. distance graph (Courtesy of ROBOTIS)

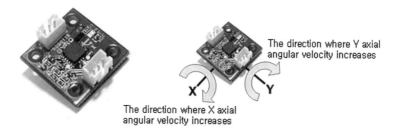

Fig. 3.46 MEMS based Gyro Sensor (GS-12) (Courtesy of ROBOTIS)

3.4.6   GPIO (5-Pin) Gyroscope Sensor GS-12

This sensor (see Fig. 3.46) is a MEMS based 2-Axis gyroscope with digital output values between 45 and 455, corresponding respectively to −300 deg/s and +300 deg/s (i.e. we are dealing with very fast motion rates for your robot).

It can be used within a CallBack routine to help stabilize Humanoid robots doing fast walking moves. It is usually mounted near the Center of Gravity of the robot, and because it is MEMS based, it is immune to electromagnetic interferences from the actuators. Some gyro applications programming are presented in Chap. 8.

3.4.7   Other GPIO (5-Pin) Sensors and Output Devices

When the OLLO system was coming out in 2008, a few GPIO sensors, actuators and output devices also became available: IR, Touch, Servo Motor and LED display (see Fig. 3.47).

The latest crop of GPIO sensors came with the ROBOTIS-MINI in 2014: Color, Magnetic, Temperature, Ultrasonic, Object Detection (see Fig. 3.48).

Fig. 3.47   GPIO modules: IR, Touch, Servo Motor & LED display (Courtesy of ROBOTIS)

Fig. 3.48   Color, magnetic, temperature, ultrasonic, object detection sensors (Courtesy of ROBOTIS)

3.4.8   Recent Adaptations of Smartphone Features

Circa 2014, ROBOTIS started to integrate technologies from mobile devices such as Smartphones and Tablets into "Firmware 2.0" controllers such as the CM-50, CM-150, CM-200, OpenCM-7.00 and OpenCM-9.04 to obtain a whole host of new features as shown in the screen-captures below (Fig. 3.49). Application programming projects using these features on Android and iOS devices will be presented in Chap. 11.

3.5   Other Compatible and Upcoming Hardware

Other companies had developed various technologies using ROBOTIS hardware and below is only a partial list:

1. "Open-Hardware" controllers such as the ArbotiX-M™ (see Fig. 3.50) and ArbotiX-Pro Robocontrollers™ – ArbotiX-M was based on the Atmel AVR chip while the ArbotiX-Pro was based on the STM32F103 chip (https://interbotix.com/arbotix-robocontroller). Trossen Robotics carries two series of robotic arm that use the ArbotiX controllers – PhantomX™ (using AX-12A actuators) and WidowX™ (using a mixture of MX-64, MX-28 and AX-12 actuators) (http://www.trossenrobotics.com/robotic-arms.aspx). Vanadium Labs has software support materials at http://vanadiumlabs.github.io/arbotix/. Applications using the PhantomX robotic arm will be described in Chaps. 6 and 9.
2. The USB2AX dongle is the "little brother" to the USB2Dynamixel module (http://www.xevelabs.com/doku.php?id=product:usb2ax:usb2ax). However it

Fig. 3.49 Integration of Smart mobile technologies into R+Task tool (partial list)

can only support the TTL (3-pin) protocol. The ROBAI® company used the
USB2AX® in their Cyton® robotic arms (http://www.robai.com/) and developed their own sophisticated control software called ACTIN® (http://www.
robai.com/robots/actin-robot-control-software/). For an application of their dual-
arm Epsilon system with the KINECT 1®, please visit this YouTube video
(https://www.youtube.com/watch?v=OFMmZg-YiZM).

Fig. 3.50 ArbotiX-M controller

Fig. 3.51 Pixl board – XL-320 shield for the Raspberry Pi

3. The Pixl® board (Fig. 3.51) was designed to interface and control up to 6 XL-320 actuators with the Raspberry Pi® single board computer (http://www.genera-tionrobots.com/en/402420-carte-pixl.html).
4. ROBOTIS also has their own RPi shield for TTL and RS-485 protocols, to be used in conjunction with an OpenCM-9.04 as subcontroller for the Dynamixels used (see Fig. 3.52). This RPi shield had been in beta test, so it may be released to the general public in 2017.

Fig. 3.52 ROBOTIS RPi shield (*green*) and OpenCM9.04 (*blue*)

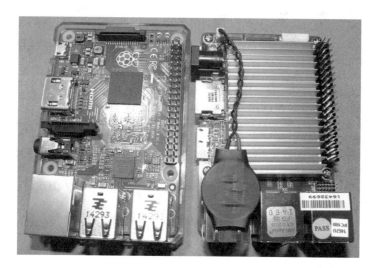

Fig. 3.53 Raspberry Pi 3 (*left*) and UP (*right*) single board computers

5. Among the compatible single board computers (see Fig. 3.53), we have the Raspberry Pi 3 for use with Linux® and the UP® board for running either Linux® or full Windows 10®.
6. The ROBOTIS OpenCR board was designed to be a hardware controller capable of supporting the ROS 2 Embedded framework on the TurtleBot 3 (https://github.com/ROBOTIS-GIT/OpenCR; https://github.com/ROBOTIS-GIT/turtlebot3).

It uses a STM32F746NGH6 for controller (1 MB Flash memory, 320 KB RAM) and has integrated accelerometer and gyroscope sensors. Two versions (Basic and Premium) would be available by late Spring 2017 (http://turtlebot3.readthed-ocs.io/en/latest/about/index.html#). The Basic version would use a Raspberry Pi 3 Model B for its main controller (https://www.raspberrypi.org/products/rasp-berry-pi-3-model-b/), while the Premium version would use Intel®'s Joule™ for its main controller (http://ark.intel.com/products/96414/Intel-Joule-570x-Developer-Kit) and Intel®'s Realsense™ R200 for its RGBD sensor (https://software.intel.com/en-us/RealSense/R200Camera).

3.6   Review Questions for Chap. 3

1. Which Firmware group does the CM-530 belong to?
2. Which Firmware group does the CM-200 belong to?
3. Which Firmware group does the OpenCM-9.04 belong to?
4. Which controller(s) use a mini-jack connector (BSC-10) to connect with the PC?
5. Which controller(s) use a micro-B USB connector to connect with the PC?
6. Which controller(s) use a 4-pin connector to connect with the PC?
7. Which controller(s) allow "wired" and "wireless" communications to operate at the same time?
8. List controller(s) derived from the OpenCM9.04.
9. Which ROBOTIS controller came with the Turtle Bot 3?
10. Which ROBOTIS controller came with the PLAY 700 system.
11. Which CM controller(s) allow simultaneous wired and wireless communications from the PC?
12. How many "wired" interface modules are supported by ROBOTIS? And what are their names?
13. How many "wireless" interface modules are supported by ROBOTIS? And what are their names?
14. Which wireless protocol would open two COM ports on the PC side for each wireless module used?
15. Which module allows direct interfacing to the Dynamixels?
16. Which member of the CM-5XX family of controllers has 128 KB of RAM?
17. Which member of the CM-5XX family of controllers clocks at 72 MHz?
18. Which controller(s) use the Atmel AVR controller?
19. Which controller(s) use an ARM Cortex M3 architecture?
20. Which Dynamixel hardware interface uses a 4-pin connector? And which one is using 3-pin connector?
21. How many pins does a typical GPIO port have?
22. How many pins does the communication port have? This port is used to connect modules such as LN-101, ZIG-110A or BT-210.
23. On which controller(s) can the user access wired and wireless connections at the same time?

24. Hardware wise and software wise, how can the user connect wirelessly from a PC to a CM-5 based robot?
25. Hardware wise and software wise, how can the user connect wirelessly from a PC to a CM-510 based robot?
26. Hardware wise and software wise, how can the user connect wirelessly from a PC to a CM-530 based robot?
27. Hardware wise and software wise, how can the user connect wirelessly from a PC to an OpenCM-9.04 based robot?
28. How many ways can the user access the Virtual RC-100 Controller? From which tools?
29. Which sensor modules can be used to measure distances?
30. Which controller(s) have a built-in microphone? And which one(s) do not?
31. Which version(s) of the OpenCM-9.04 controller can be used with the RoboPlus Software Suite?
32. Which controller(s) have real-time debugging capabilities (JTAG/SWD)?
33. What are the difference(s) between the three versions A-B-C of the OpenCM-9.04?
34. How many serial communication ports can the OpenCM9.04 support?
35. Which sensor(s) interface to the GPIO ports of the ROBOTIS controllers?
36. Which sensor(s) are used to acquire positional data of a typical actuator when it is operating, such as the AX-12, MX-28 or XL-320?
37. List three sensors that are Dynamixel-compliant.
38. List three actuators that are Dynamixel-compliant.
39. List three sensors that are designed for use on the GPIO ports.
40. Which sensor module has two programmable ranges for its distance measurement task?
41. From a hardware circuitry point of view, how can a servo motor switch its rotation direction?
42. How would a user design a system that can control DYNAMIXEL-PRO actuators?
43. Which Dynamixel actuators can be controlled with a PID approach?
44. What does the specification "W" stand for in the naming of ROBOTIS actuators?
45. What is the hardware refresh cycle time for Dynamixels?
46. What is the type of the physical device used to transduce the angular position of the AX-12 actuator?
47. What is the type of the physical device used to transduce the angular position of the MX-28 actuator?
48. Why does the AX-S1 have a reverse-response to the distance when objects are too close to its NIR-LED sensing element?
49. What is the minimum time interval between sound claps that the AX-S1 needs in order to distinguish the sound claps as distinct audio events?
50. What are the current options for a user who wants to measure inertia-related parameters such as accelerations and rotational rates?

51. What are the current options for a user who wants to measure contact pressures/ forces between the robot and the supporting surface?
52. What embedded vision hardware is current available for the CM-5XX and OpenCM systems?

References

Clark D, Owings D (2003) Building robot drive trains. McGraw-Hill, New York
Kanniah J et al (2014) Practical robot design: game playing robots. CRC Press, Boca Raton

Chapter 4
Main Software Tools

In this chapter, the goal is to go over the main software tools provided by ROBOTIS using a simple demonstration system having two servo motors but swapping out different controllers (CM-5, CM-50, CM-510, CM-530 and CM-9.04-B/C) as needed. ROBOTIS also provided other software tools as mobile apps on Android and iOS devices (please see Chaps. 11 and 12 for more details about their applications).

ROBOTIS provides their software tools for free and most tools have proprietary source codes and/or firmware, except for the ROBOTIS OpenCM IDE for the CM-9.04-A/B/C systems (which is based on Arduino).

As previously mentioned in Chap. 3, current ROBOTIS systems are divided into two groups:

- The "Firmware 1.0" group includes controllers such as CM-100, CM-5, CM-510, CM-530 and CM-700.
- While the "Firmware 2.0" group includes controllers such as CM-50, CM-150, CM-200, OpenCM9.04 and OpenCM7.00, and also Dynamixel X and PRO series.

Consequently, there are also two sets of software tools designed for each specific firmware group:

- The "Firmware 1.0" group should use the original RoboPlus suite (from now on called RoboPlus V.1 in this book – see Fig. 4.1).
- The "Firmware 2.0" group should use the Version 2 of this software suite (from now on called R+ V.2 in this book – see Fig. 4.2).

Historically speaking, the very first ROBOTIS software suite was the BIOLOID suite (c. 2005) with "Behavior Control Programmer" and "Motion Editor". It got replaced with RoboPlus V.1.0.8 in 2009 along with the introduction of BIOLOID PREMIUM. Currently at version 1.1.3.0 (7/8/2014), the RoboPlus V.1 suite runs on MS Windows only or "mostly" on a good emulator of it on Mac OS devices. It is available for download at (http://en.robotis.com/BlueAD/board.php?bbs_

© Springer International Publishing Switzerland 2017
C.N. Thai, *Exploring Robotics with ROBOTIS Systems*,
DOI 10.1007/978-3-319-59831-4_4

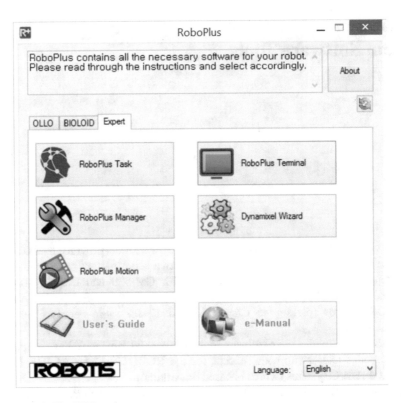

Fig. 4.1   RoboPlus V.1 Portal

id=downloads&mode=view&bbs_no=1152561&page=1&key=&keyword=&sort= &scate=). When installed, it is a portal to five software tools (MANAGER, TASK, MOTION, TERMINAL and DYNAMIXEL WIZARD) as shown in Fig. 4.1. Furthermore, once installed, each of these five tools will get automatic updates on their own whenever ROBOTIS pushes them out. Currently in Spring 2017, the V.1 tools are at V.1.0.34.1 for MANAGER, V.1.1.2.7 for TASK, V.1.0.29.0 for MOTION, and V.1.0.19.21 for DYNAMIXEL WIZARD.

The R+ V.2 tools are currently at V.2.1.12 for MANAGER, V.2.1.4 for TASK, V.2.4.1 for MOTION, V.1.2.3 for DESIGN™ and V.1.0.6 for R + SCRATCH™. They are available for download at http://en.robotis.com/BlueAD/board.php?bbs_ id=downloads&page=1&key=&keyword=&bbs_opt1=&scate. After installation, they are also set to be automatically updated when available.

There are also mobile versions of these V.2 software tools (Android and iOS) which are available for download at (https://play.google.com/store/apps/ developer?id=ROBOTIS&hl=en, see Fig. 4.3) or (https://itunes.apple.com/us/ developer/robotis-co-ltd/id948481761). These applications, especially R + m.Task and R + m.Motion, leverage extensively on existing computing and multimedia technologies available for current Smartphones and Tablets. These mobile apps such as SMART, MINI, IoT and PLAY700 were designed so that users could operate

Fig. 4.2 RoboPlus (R+) V.2 Portal (i.e. R+ LAUNCHER V.2)

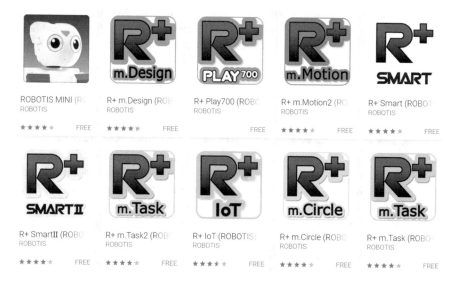

Fig. 4.3 Mobile versions of ROBOTIS software tools (Courtesy of ROBOTIS)

specific example robots with a minimum of computer programming knowledge, while others such as TASK and MOTION were more general purpose programming tools. These mobile tools also extended multimedia capabilities to the TASK tool and they would be further explored in Chaps. 11 and 12 using Android and iOS devices.

In the Embedded C area, there are two options for the users. Option 1 is more for professional programmers as it requires quite a bit of hardware and software knowledge from the user and it uses more complex tool chains based on compilers provided by the manufacturers of the microcontrollers themselves. These software tools, installation procedures and some tutorials are provided at the following link (http://support.robotis.com/en/software/embeded_c_main.htm) and they are designed for the CM-510/CM-700/CM-530 only and use several ROBOTIS SDKs. More video tutorials are available on YouTube's channel ROBOTIS CS (https:// www.youtube.com/playlist?list=PLEf1s0tzVSnSgVzf4AREpat_P_HRLSiDn).

Option 2 has a lower technical bar for entry as it is based on the Arduino interface and it is called OpenCM IDE which is available for download at (http://support. robotis.com/en/software/robotis_opencm/robotis_opencm.htm). Other video tutorials are available on YouTube's channel ROBOTIS CS https://www.youtube.com/pla ylist?list=PLEf1s0tzVSnRgF2Cu9r91bU3bHuZU8Fxu. Technical manuals and tutorial materials for OpenCM were provided previously in Chap. 2 as ZIP files. Both Embedded C options will be discussed further in Chap. 9.

In Summer 2016, ROBOTIS released the OLLOBOT SDK which could be used as the foundation of Android App Development (i.e. Java Programming) for the OLLOBOT (see Sects. 2.4 and 4.7). This SDK turned out to be also applicable to Firmware 2.0 systems in general (see Chap. 12 for more details).

4.1 DYNAMIXEL WIZARD Tool

You should not have to use this tool at all if you recently bought your new controllers and Dynamixel actuators or sensors, as they would have the latest firmware installed from the factory. In general, the Dynamixel's firmware is updated via the DYNAMIXEL WIZARD and the Controller's firmware is updated via the MANAGER tool, but see Sect. 4.1.2.1 for specific procedures for the OpenCM-9.04-C.

First you will need to acquire the USB2Dynamixel module (http://www.robotis-shop-en.com/?act=shop_en.goods_view&GS=1289&GC=GD080300) or the LN-101 module(http://www.robotis-shop-en.com/?act=shop_en.goods_view&GS=1277&GC= GD080300) depending on the type of controllers used (see chart at http://support. robotis.com/en/product/controller_main.htm).

4.1.1 TTL (3-Pin) and RS-485 (4-Pin) Dynamixels

Your best option is to go with the USB2Dynamixel module and you also need to arrange for independent power of the Dynamixels via the SMPS2Dynamixel (http://www.robotis-shop-en.com/?act=shop_en.goods_view&GS=1267&GC=GD080303) or just use the controller module as the power source (see Fig. 4.4 for an example setup with a CM-530 and 2 AX-12As). If you are using a CM-5, CM-510 or CM-700, you can use the same setup and just swap out the controller.

Figure 4.4 shows that the LiPo battery is powering the CM-530 controller and indirectly the AX-12As via the 3-pin Dynamixel bus, while the PC (i.e. Dynamixel Wizard) communicates to the AX-12As via the chain "USB >> USB2Dynamixel >> AX-12As". Please note that the side switch on the USB2Dynamixel module needs to be slid into the TTL (or RS-485) position (as appropriate) for this setup to work properly.

If you are only "updating" the firmware of your Dynamixels, you can string them up in a daisy-chain fashion as shown in Fig. 4.4, as the Wizard tool can update several Dynamixels (of the same type, to be on the safe side) during the same cycle. However, if something "real bad" had happened to a particular Dynamixel, and as part of the troubleshooting process you are trying to see if "recovering" the firmware on it will help, then you will need to hook up only ONE Dynamixel at a time. Procedures for updating, recovering and testing firmware are accessible at this link (http://support.robotis.com/en/software/roboplus/dynamixel_wizard.htm). For a more recent video of the process, please review enclosed video file "Video 4.1". If you are "recovering" successfully, remember to set the Dynamixel ID back to the one you were using before, as the "recovery" process will reset the ID to "1".

Fig. 4.4 Typical setup for Dynamixel firmware update using a CM-530 and 2 AX-12As

4.1.2   XL-TTL (3-Pin) Dynamixels

Currently, the only XL-TTL Dynamixel in existence is the XL-320 operating with the OpenCM-9.04-A/B/C controllers. At the time of writing of the first edition of this book, ROBOTIS was still working on their RoboPlus tools to work with the OpenCM systems, thus the procedures described herein were obtained from past experimentation (pre-2016) by the author. Due to the difference in firmware used on the A/B version and the C version, the respective firmware update processes were also different. However since mid-2016, with the introduction of MANAGER V.2, the firmware update process for the XL-320 had gotten simpler.

4.1.2.1   OpenCM-9.04-C

The 9.04-C controller's firmware and the firmware of XL Dynamixels attached to it can be updated at the same time via the 4-pin communication port. The user has several options for communications hardware, such as "wired" via the LN-101 or "wireless" via ZIG-110 and BT-110/210 (please note that currently the BT-410 is not compatible with the MS Windows environments). The LN-101 route is the most reliable for this rather delicate operation, but the reader should try other options just for learning experiences. Currently, the firmware update task can be performed either via the DYNAMIXEL WIZARD tool (i.e. RoboPlus V.1) or via the R+ MANAGER V.2 tool.

Figure 4.5 describes a setup using an LN-101 with a 9.04-C and 2 XL-320 servos.

If the DYNAMIXEL WIZARD tool (currently at V.1.0.19.21) is used and once it is started, the user can click on the icon for "XL-320 Firmware Management" (see Fig. 4.6) and choose either "Update" or "Recovery" depending on the need and then follow the on-screen instructions.

If the R+ MANAGER V.2 tool is used and once started, the user would choose the product "ROBOTIS MINI" on the main menu and follow the on-screen instructions. Please see video clip "Video 4.2" for more details on both procedures.

Please note that these tools will additionally update the firmware of the 9.04-C controller to its most current version.

In Chap. 3, Fig. 3.21 showed that the OpenCM9.04-C controller could be expanded with the 485-EXP board to allow it to additionally control AX/MX-TTL and RS-485 Dynamixels, the current version 1.0.19.21 of the Dynamixel Wizard tool could access "Protocol 1.0" Dynamixels attached to the 485-EXP board once the Dynamixel Channel got set to Channel 1 via MANAGER V.2. At present (Spring 2017), only the MANAGER V.2 tool can access the Dynamixels physically attached to the 485-EXP board (and only XM/XH Dynamixels – see further details in Sect. 5.2.2). Furthermore, only the ROBOTIS IDE can concurrently control a mixture of Dynamixels that are attached to both OpenCM9.04-C and 485-EXP boards (see Sect. 9.2). In short, the users still need to wait for all the software tools to catch up with the new hardware.

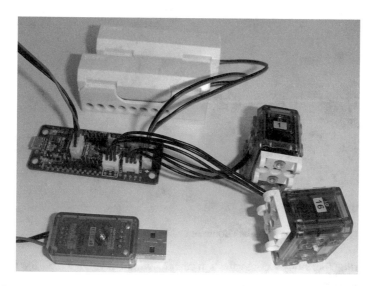

Fig. 4.5 Typical setup for Dynamixel firmware update using an OpenCM-9.04-C

Fig. 4.6 Pull-down menu
for the "XL-320 Firmware
Management" icon

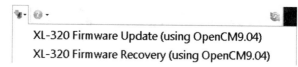

XL-320 Firmware Update (using OpenCM9.04)

XL-320 Firmware Recovery (using OpenCM9.04)

4.1.2.2   OpenCM-9.04-B

If the user only has the OpenCM-9.04-B controller, the Dynamixel firmware update
process is much more complex because the B version does not have a "nice" firmware to interact with the previous "XL-320 Firmware Management" tool (Sect.
4.1.2.1). In a way, the user will have to use similar steps shown in Sect. 4.1.1 with
the USB2Dynamixel module. Figure 4.7 shows the required hardware setup:

- The USB2Dynamixel module serves as the interface between the PC's USB port
  and the Dynamixel Port on the OpenCM-9.04-B (USB2Dynamixel's side switch
  set on TTL as per Sect. 4.1.1).
- The OpenCM-9.04-B controller acts a common Dynamixel bus because it has
  both AX/MX-TTL and XL-TTL connectors. It also serves as the power source
  for the XL-320 actuators.

This setup **also requires** that the "DxlTosser" sketch be preloaded and running
on the 9.04-B (see Chap. 9 for more details). Then ones can get the Dynamixel
Wizard tool started and apply similar procedures described in Sect. 4.1.1 for the
AX/MX-TTL Dynamixels to the XL-320 actuators, i.e. "updating" several
Dynamixels of the same type or "recovering" only one Dynamixel (see video "Video
4.3" for more details).

Fig. 4.7 Typical setup for Dynamixel firmware update using an OpenCM-9.04-B

Updating the 9.04-B controller's firmware would involve JTAG/SWD proce-dures that are outside the scope of this book.

4.2 MANAGER Tool

4.2.1 CM-5, CM-510, CM-530

The MANAGER V.1 tool, currently at V.1.0.34.1, is fully functional for the CM-5, CM-510 and CM-530 controllers and the user **must choose** wired connections via the USB2Dynamixel > > BSC-10 (CM-5 and CM-510) or USB-mini cable (CM-530).

The video file "Video 4.4" highlights how Manager can perform a quick check on a Bioloid Premium Humanoid A equipped with some extra sensors such as AX-S20, Gyro and Foot Pressure Sensors (see Fig. 4.8).

More on-line resources for the MANAGER tool can be found at http://support. robotis.com/en/software/roboplus/roboplus_manager_main.htm.

4.2.2 OpenCM-9.04-A/B/C

First, the OpenCM-9.04-A/B controllers were designed to work with the ROBOTIS Arduino-like IDE so they do not work with the MANAGER V.1 or V.2 tools at all. The OpenCM-9.04-C will work with both tools but with some differences as shown below.

The MANAGER V.1 tool could **only update** the 9.04-C's firmware (which could be done via the DYNAMIXEL WIZARD tool also – see Sect. 4.1.2.1). Through

Fig. 4.8 "Bal'Act" Humanoid A robot

experimentations, the author has determined that the LN-101 module should be used for the most reliable results with this procedure.

Figure 4.9 shows a hardware setup needed to update the firmware of an OpenCM-9.04-C controller using an LN-101.

The next step is to start the MANAGER V.1 tool. At the main menu bar, choose the correct COM port corresponding to the LN-101 and click on the "Controller Firmware Management" icon (left of the "?" icon) and follow on-screen instructions to get the 9.04-C controller updated to its latest firmware version (see video file "Video 4.5" for more details).

If the user had bought a ROBOTIS MINI kit which came with a BT-210, the above procedure also worked well, as long as the OUT-GOING COM port was chosen for the connection between the PC and the BT-210.

If the user had bought a "loose" OpenCM-9.04-C (http://www.robotis.us/opencm9-04-c-with-onboard-xl-type-connectors/) which came with a USB A-to-micro-B cable, the above procedure also worked quite satisfactorily when using the appropriate COM port.

Starting in 2016, you can also use the hardware combination OpenCM-9.04-C + OpenCM-485EXP to update firmware on the Dynamixels from the X and PRO series via R + MANAGER V.2 (see Fig. 4.10).

The author recommends to use the newer MANAGER V.2 tool as it is fully functional with the OpenCM-9.04-C + 485EXP combination (as shown in second half of "Video 4.5"). However from personal experiences, the author had found that the MANAGER V.1 tool was more reliable than MANAGER V.2 to recover the firmware for the OpenCM-9.04-C if the user had previously switched to using the ROBOTIS IDE on his or hers OpenCM-9.04-C module, but now wanted to go back to the ROBOTIS proprietary firmware so that TASK and MOTION could be accessed again.

Fig. 4.9 Typical setup for controller firmware update for an OpenCM-9.04-C

Fig. 4.10 R + MANAGER V.2 can update firmware on Dynamixel X and PRO

4.3   TASK Tool

The TASK V.1 tool, currently at V.1.1.2.7, is a cross-over between a standard text-based IDE and a pure icon-based IDE to help beginners with a more controlled syntax. It has all the standard control structures (sequential, repetition and conditional) and provides variables and functions definition. It even supports one CallBack function, but it does not support variable arrays.

Since 2013, the TASK V.1 tool is well documented in the Software Programming Guide included in the Bioloid Premium kit. If you got an earlier edition of this kit that did not have this user's guide, I would recommend first-time users to get it at http://www.robotis-shop-en.com/?act=shop_en.goods_view&GS=1486&GC=GD080400. The ROBOTIS e-Manual web site also has much technical information for the Task tool at http://support.robotis.com/en/software/roboplus/roboplus_task_main.htm. Thus instead of repeating all this information here, I would like to share a "comparative study" across different controllers using a demonstration setup with two AX-12As or XL-320s using "wired' and "wireless" (ZigBee and BlueTooth) communications using the Virtual RC-100.

4.3.1 TASK V.1 and CM-5/510/530 Controllers

The demonstration setups for the CM-5XX family is shown in Fig. 4.11.
The test TASK V.1 code (TestVRC-100_CMs.tsk) is shown in Fig. 4.12.
It is quite simple:

- Just an overall Endless_Loop with an initial Wait_While loop checking on whether the Virtual RC-100's buttons had been pushed or not (line 5).
- When a button had been pushed, check it to see if it was:

  - "Up" then set the Goal_Position for servos 1 and 16 to "1023" (lines 8–12).
  - "Down" then set the Goal_Position for servos 1 and 16 to "0" (lines 13–17).
  - "Nothing", i.e. upon release of any button of the RC-100's buttons, tell servos 1 and 16 to go to Goal_Position "512" (lines 18–22). The author chose to treat this condition explicitly because it was known that the RC-100 would send the "0" signal (statement 18) only **once**, thus any "worthy" communication hardware/software implementation would have to be able to pick up this one-time event. This feature was used to evaluate the performance between wired and wireless communications as shown in sections below.

Fig. 4.11 Demonstration setups for TASK V.1 "comparative study" (left to right – CM-5, CM-510 and CM-530)

```
1    START PROGRAM
2    {
3        ENDLESS LOOP
4        {
5            WAIT WHILE ( ➔ Remocon Data Received  == FALSE )
6            Button = ➔ Remocon RXD
7
8            IF ( Button == ➔U )
9            {
10               ● ID[1]: ⊙ Goal Position = 1023
11               ● ID[16]: ⊙ Goal Position = 1023
12           }
13           ELSE IF ( Button == ➔D )
14           {
15               ● ID[1]: ⊙ Goal Position = 0
16               ● ID[16]: ⊙ Goal Position = 0
17           }
18           ELSE IF ( Button == ➔-- )
19           {
20               ● ID[1]: ⊙ Goal Position = 512
21               ● ID[16]: ⊙ Goal Position = 512
22           }
23       }
24   }
```

Fig. 4.12 Demonstration test code (TestVRC-100_CMs.tsk)

The video file "Video 4.6" showed how this test code was performing on a CM-5 using a "wired" connection (i.e. USB > > USB2Dynamixel > > BSC-10 > > CM-5) or a "wireless" connection (i.e. USB > > USB2Dynamixel > > Zig2Serial > > ZIG-100-1 > > ZIG-100-2 on CM-5 > > CM-5. Both options worked well without any discernable performance issue.

This test code was also tried on a similar system with a CM-510 using "wired" and "wireless" connections. For a CM-510, there were three "wireless" options: ZIG-110, BT-110 or BT-210. The video file "Video 4.7" showed that all three wireless options were functional but there were some subtle performance differences:

- ZigBee connections were made within 1 s of executing of the program, while BT connections could take up to 3–4 s to be done, thus users need to take these BT delays into account for their own applications.
- ZigBee data buffers seemed to be updated much more efficiently than BT data buffers, as BT systems seemed to be "stuck" when a particular VRC-100 button was held in for 2–3 s. It is not known whether this is an issue on the PC side or on the BT modules side, or that this is an expected behavior of BlueTooth protocols.
- BT-210's performance was less than BT-110's in the above test conditions.

When the same tests were performed on a CM-530 system ("wired" plus the previous three "wireless" options), all wired and wireless connections were functional but the same issues discussed previously were also found on the CM-530 (see video file "Video 4.8").

4.3.2   TASK V.1 and OpenCM-9.04-C

To complete the comparative study, the same TASK code was tested on an OpenCM-9.04-C system with the same four communication protocols as in Sect. 4.3.1, but with XL-320 servos instead:

• Wired, through LN-101.
• Wireless, through ZIG-110, BT-110 and BT-210.

The video file "Video 4.9" showed that the previous wireless issues existed for the OpenCM-9.04-C also.

(Note: since the V1.1.2.4 update for TASK (9/5/2014), the previous BlueTooth issues are resolved satisfactorily).

Thus, at present, ZigBee is the best performing wireless option for controlling ROBOTIS robotic systems, however it is also recognized that BlueTooth is more widely implemented on PCs and mobile devices thus requiring no extra hardware for the users to buy. It was also shown that a TSK code is portable across existing RoboPlus-compatible platforms as long as the "correct" controller was chosen before compiling and downloading the codes.

4.3.3   R+ TASK V.2 New Features

To wrap up Sect. 4.3, the author had included in the "EXTRA MATERIALS.ZIP" file the TSKX versions (i.e. for the R + TASK V.2 tool), corresponding to the previous TSK test codes used in Sects. 4.3.1 and 4.3.2 for the reader to try out and to verify the same conclusions mentioned at the end of Sect. 4.3.2. Please note that TASK V.2 can read in the old TSK files and then can convert them into the new TSKX format permanently, i.e. TASK V.2 works with Firmware 1.0 and 2.0 controllers.

TASK V.2's new user interface has some useful features such as:

• The often-used tools "INSTRUCTIONS", "VARIABLES" and "FUNCTIONS" are readily accessible as tabs on the right hand side (Fig. 4.13).
• The new "FIND" tool is a handy feature (Fig. 4.14), and so is the new "ERROR" tool (Fig. 4.15).
• The "DEBUG" screen is completely redesigned showing the Virtual RC-100 Controller at all time (Fig. 4.16).

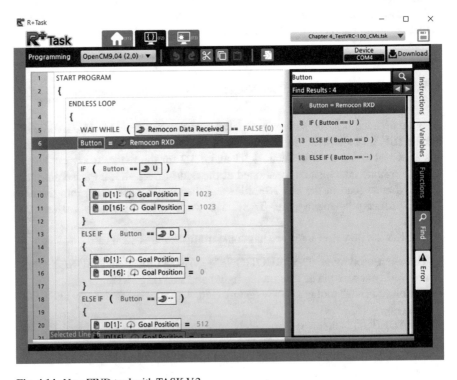

Fig. 4.13 New User Interface with TASK V.2

Fig. 4.14 New FIND tool with TASK V.2

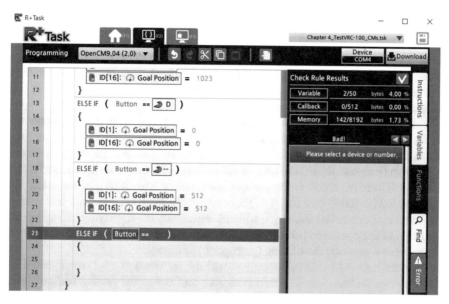

Fig. 4.15 New ERROR tool with TASK V.2

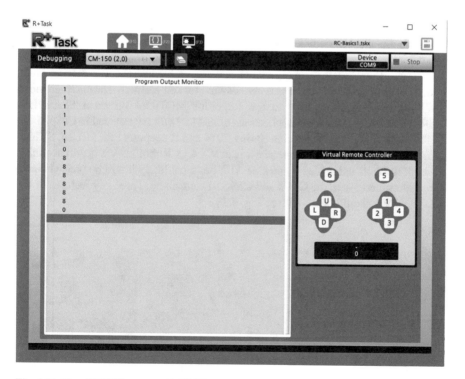

Fig. 4.16 New DEBUG screen with TASK V.2

4.4 MOTION Tools (V.1 and V.2)

The TASK tool is well suited for reading and writing data/commands from/to a small number of actuators and sensors as on a wheeled robot, as the inherently "sequential" execution of those commands (line by line as shown in Fig. 4.11) still has an acceptable performance. However on a humanoid robot with 18 actuators that have to work together and "simultaneously" in order to perform a maneuver such as walking, a different way of generating and executing sets of actuator goal positions in "parallel" on those 18 actuators in a timely manner has to be devised.

For this purpose, ROBOTIS created the Motion tools, from the original BIOLOID Motion Editor (c. 2005) to the RoboPlus Motion V.1 (c. 2009), to the latest R+ Motion V.2 (c. 2014) (see Fig. 4.17). Conceptually, these Motion tools were based on cartoons (or frames) animation which is essentially the re-playing of a sequence of poses of the considered character (robot) within a time line, with slight variations in each pose to create the illusion (perception) of motion.

The BIOLOID Motion Editor only provided front views of a 3-D robot, and they were shown "relative" to the main controller and not to the ground surface. With RoboPlus MOTION V.1, 3-D graphics of the robot are available (still relative to the controller/body) but were used more for animation check of the robot poses, rather than for creating and editing them. Since 2013, the Software Programming Guide (SPG) included in the Bioloid Premium kit has very good tutorial materials for learning how to use the MOTION tool (V.1). I would recommend beginners to obtain it from http://www.robotis-shop-en.com/?act=shop_en.goods_view&GS=1486&GC=GD080400 and use it well. The ROBOTIS e-Manual also has much information for the MOTION V.1 tool (http://support.robotis.com/en/software/roboplus/roboplus_motion_main.htm), but the materials in the SPG is better suited for self-learners. The enclosed video file "Video 4.10" shows how to use RoboPlus MOTION V.1 on a PREMIUM GERWALK robot. Additional applications of this tool will be provided in Chap. 6.

R+ Motion V.2 was released in Spring 2014 and it supports PREMIUM, STEM, SMART & MINI systems (currently it is at V.2.4.1). RoboPlus V.1 is still available to support robots using CM-5 and CM-510, but most likely it will be phased out in a few years along with the CM-5 and CM-510 controllers. Three key features of R+ MOTION V.2 should be noted (see Fig. 4.18):

Fig. 4.17 Evolution of the MOTION tools from 2005 to 2014

Fig. 4.18   Key features of R+ MOTION V.2

1. The display of a global time line for each "motion-unit".
2. The use of a "physics" engine so that the robot 3-D frame can be manipulated relative to the "ground" surface, and not with respect to the robot's body as in V.1. However this "physics" engine won't be able to accommodate highly acrobatic moves like body flipping or rolling!
3. The robot motions can be edited and executed on an existing 3-D model without the need of the real robot actually being connected to the PC or mobile device during that time.
4. On fast graphics display hardware, the graphics engine will be able to synchronize the simulated robot moves to the real robot moves in the physical world.

The web-based user manual for the R+ MOTION tool V.2 is available at http://support.robotis.com/en/software/roboplus2/r+motion2/rplus_motion2.htm. Enclosed is a video illustrating some basic uses of R+ MOTION V.2 on an OpenCM-9.04-C with two XL-320 servos (Video 4.11). More applications of this tool will be shown in Chap. 10.

4.5   R+ DESIGN Tool

Spring 2014 also saw the first release of the R+ DESIGN tool (V.1.1) for MS Windows PCs (see Fig. 4.19), but it was originally designed for mobile systems and for the OLLO system as part of the suite of R + m.Task and R + m.Design (c. 2012).

The current R+ DESIGN V.1.2.3 supports the following systems: OLLO, PLAY, IDEAS, DREAM, SMART I/II, STEM and ROBOTIS-MINI (unfortunately the PREMIUM kit is not included at present). Currently, R+ DESIGN only supports 3-D design and assembly of ROBOTIS parts, but the R+ DESIGN and MOTION tools together show ROBOTIS' vision of integration from design to simulation/control/testing (on the computer) for individual users as well as for communities of robotics designers and instructors. At present, those activities are beginning via

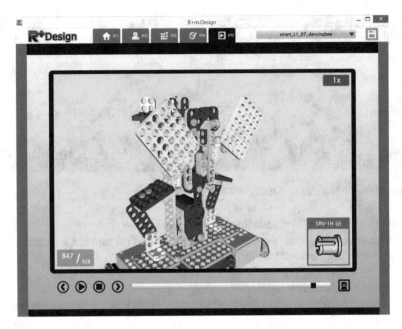

Fig. 4.19 R+ DESIGN tool with a SMART robot

websites such as STEAM Education Association http://www.steamcup.org/new/?mid=main_eng#).

Currently, only the Korean version of the user manual for R+ DESIGN is available at http://support.robotis.com/ko/software/r+design_main.htm. However, if the English-reader happens to be using Google Chrome® for web browser, he or she can just "right-click" on those web pages and choose to have them translated into English (as best as Google Chrome can).

Personally, I have used R+ DESIGN on OLLO/DREAM kits with my younger students (5–6 years old) and they definitely preferred R+ DESIGN over the instruction pages of the paper-based manuals, because they can "see" the 3-D assemblies from different view angles on the computer screen. However, there was one issue that required the students to refer back to the paper-based manuals and it was about cable routing (at present, only the paper manuals have illustrations of "proper" cable routing during assembly).

4.6 R+ SCRATCH Tool

The R+ SCRATCH tool was made available in 2016 and was effectively a "helper app" serving as an HTTP extension to the SCRATCH 2.0 software tool from MIT (https://scratch.mit.edu/scratch2download/). Currently, R+ SCRATCH (v1.0.6)

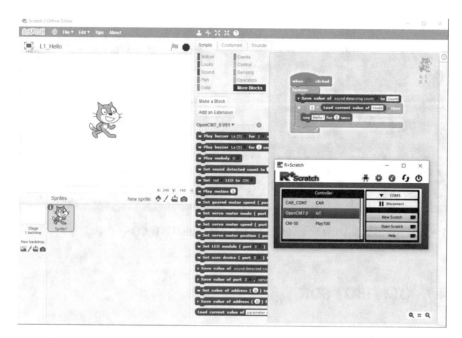

Fig. 4.20   R+ SCRATCH tool and MIT's SCRATCH 2.0 Offline Editor

supports three controllers: CAR_CONT, OpenCM-7.0 and CM-50, it is the smaller blue-gray window at the middle right of Fig. 4.20 which showed that it was connecting to an OpenCM-7.0 controller via COM5 (as an example).

The background image in Fig. 4.20 was the SCRATCH 2.0 Offline Editor with a small programming script, in the top-right area, which communicated to the OpenCM-7.0 board via the R+ SCRATCH app through HTTP port 8099. Once the user clapped his or her hands, the OpenCM-7.0 controller would pick up this event and send the "sound detecting count" parameter's value to the SCRATCH 2.0 script (to be saved in variable "count") which then would display the text balloon "Hello!" for 2 s in the "Stage" area near the "Cat" sprite, if and only if "count" was greater than "1". The bottom middle column in Fig. 4.20 showed all the current code blocks that were part of the "OpenCM7_0 V01" extension that the programmer could use in crafting a SCRATCH 2.0 script. The user could see that the buzzer and sound detector were supported, along with continuous motors and "port" devices such as servo motors and other GPIO sensors.

Figure 4.21a, b showed the HTTP extension blocks that are available for the controllers CM-50 (*left*) and CAR_CONT (*right*) – much more limited as compared to the OpenCM7.00.

The on-line manual for R+ SCRATCH is available at http://support.robotis.com/en/software/roboplus2/r+scratch/r+_scratch.htm.

More applications of the R+ SCRATCH tool and SCRATCH 2 will be described in Chaps. 5 and 11.

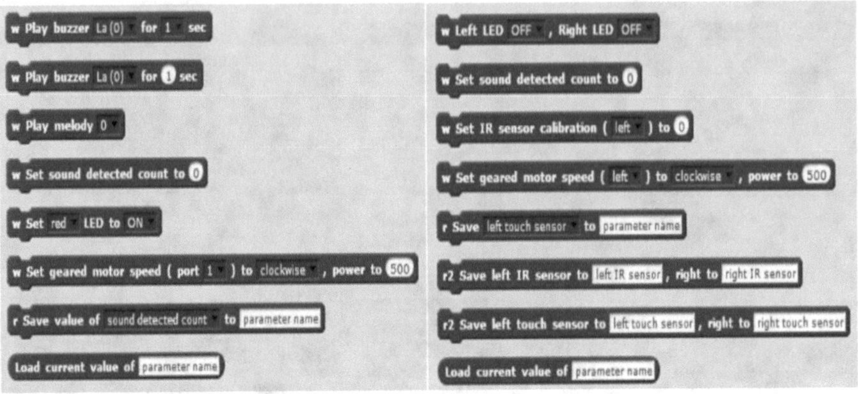

Fig. 4.21 (**a**, **b**) SCRATCH Extensions for CM-50 *(left)* and CAR_CONT *(right)*

4.7   OLLOBOT SDK

In the Summer of 2016, ROBOTIS released the OLLOBOT SDK for advanced users who had some background in Android Programming (https://github.com/ROBOTIS-GIT/OLLOBOT). This tool turned out to be more broadly applicable to Firmware 2.0 Controllers such as the CM-50/150/200 and OpenCM-7.00/9.04. Chapter 12 described selected applications of this SDK to the OLLOBOT itself and selected Firmware 2.0 controllers such as the CM-50 and CM-150, and also to an "OLLOBOT MINI" constructed from an OpenCM-9.04 with a couple of XL-320s and some OLLO sensors.

4.8   If I Were to Restart …

Recapping Chaps. 3 and 4 and if I were to restart my robotics journey at the present time, I would start with the BIOLOID PREMIUM system using the CM-530 controller. Next I would replace the CM-530 with the OpenCM-9.04-C controller combined with the OpenCM-485 shield so that I could keep using my 12V LiPo battery and the AX/MX Dynamixels. But now I could access multi-media features provided by Android tablets and Smart Phones via the R+ V.2 software suite. Later, I could switch to OpenCM IDE for closer access to the ARM controller. Lastly, I could go further with the OLLOBOT SDK to get more hands on experience with Android Programming and get into the intricacies of Dynamixel and Remocon packets.

4.9   Review Questions for Chap. 4

1. Which option(s) does a user have to run ROBOTIS software on Mac OS based computers?
2. What are the software tools that are available with the ROBOPLUS software suite on Windows platforms? Or on Android/iOS mobile platforms?
3. What tasks can the MANAGER tool provide?
4. What tasks can the TASK tool provide?
5. What tasks can the DYNAMIXEL WIZARD tool perform for the user in conjunction with a USB2Dynamixel or LN-101 module?
6. Which controllers can have their firmware updated using the MANAGER tool?
7. How can firmware update/recovery be performed for a Dynamixel actuator?
8. Which way(s) can be used to set the IDs of the ZigBee modules? Via MANAGER tool? Or via TASK tool?
9. From which tool can the user perform ZigBee management tasks with the ZIG2SERIAL module?
10. Which kind of signal (if any) does the "physical" RC-100 Remote Controller send out when the user releases one of its buttons?
11. Which kind of signal (if any) does the "virtual" RC-100 Remote Controller send out when the user releases one of its buttons?
12. What are the differences between versions 1 and 2 for ROBOTIS' MOTION tools?
13. What graphics engine is the R + MOTION tool V. 2 based on?
14. List key features of the R + MOTION V. 2 tool.
15. What can the R + DESIGN tool be used for?
16. Which ROBOTIS robotics system does the R + DESIGN tool support?
17. What can the R+ SCRATCH tool be used for?
18. Which HTTP port is the MIT's SCRATCH 2.0 software tool using to communicate with the R+ SCRATCH app?
19. What are the controller types currently supported by R + SCRATCH.
20. What are the Embedded C options for the user?
21. What is the name of the SDK that works with Android Studio?

Chapter 5
Foundational Concepts

For this chapter, the assumptions are that the reader is a beginner to robotics but has some exposure to computer programming using icon-based or text-based development environments. This chapter's main topics are listed below:

- "Sense-Think-Act" paradigm and illustration of its diverse interpretations in the design of hardware and software systems using selected robotics systems.
- "Sequence Commander" CarBot to explain "Sequential-Repetition-Selection" controls and "Functional Decomposition" concepts used in basic robotic programming.
- "Smart Avoider" CarBot to illustrate "reactive" and "behavior-based" control architectures.
- "Line Tracer" CarBot to demonstrate "timed" and "conditional" maneuvers, the concept of "self-localization" and the feedback between robot design and robot performance tuning.
- Introductory Remote Control concepts using Virtual and Physical remote controllers, and also using a Smart Phone's Tilt Sensor.

The above topics will be developed using the ROBOTIS MANAGER and TASK tools, and also using MIT's SCRATCH 2 software (https://scratch.mit.edu/scratch-2download/). Two good references for SCRATCH 2.0 programming are Ford (2014) and Warner (2015). For more advanced topics in wheeled robots, the reader is referred to Campion and Chung (2008), Dudek and Jenkin (2010), and Cook (2011).

5.1 "Sense-Think-Act" Paradigm

Winfield (2012) defined a robot as:

1. *An artificial device that can **sense** its environment and **purposefully act** on or in that environment;*
2. *An **embodied** artificial intelligence; or*
3. *A machine that can **autonomously** carry out useful work.*

© Springer International Publishing Switzerland 2017
C.N. Thai, *Exploring Robotics with ROBOTIS Systems*,
DOI 10.1007/978-3-319-59831-4_5

Nourbakhsh (2013) pointed out that robotics research and innovation had been inspired from human intelligence which depends on two aspects (Fig. 5.1):

1. A meaningful 2-way connection with the world; this connection has inputs termed "perception" and outputs back to the world termed "action" (from our own body or via tools that we create).
2. An internal decision making process termed "cognition" that maps our sensory inputs into deliberate actions.

When translated to the realm of world-robot interactions, we got the familiar "Sense-Think-Act" paradigm (Fig. 5.2).

Ones should note the "endless loop" shown in Figs. 5.1 and 5.2 as this is the fundamental characteristic found in all robotics computer programs. This usually is the first "conceptual" obstacle to overcome for beginners, as we all learned how to do basic computer programming by designing and creating programs that would run

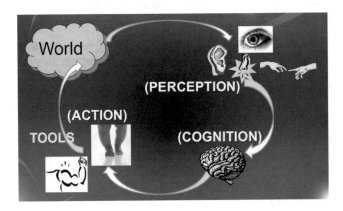

Fig. 5.1 World-human interactions

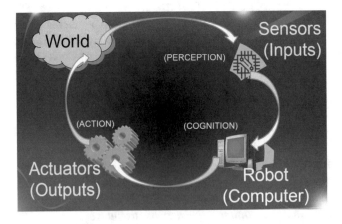

Fig. 5.2 World-robot interactions

only once. Some recent robotics systems designed for "very young" roboteers even "hide" this implicit "endless loop", such as the MRT3® system from My Robot Time (http://www.myrobottime.com/#!proudt/c8hd).

Figure 5.3 was a screen capture of the MRT3 programming interface where a fairly constrained Input >> Output approach (i.e. Reactive Control) was used. The interested reader can also watch this video clip describing the software programming aspects of the MRT3 system (Video 5.1).

For years, Lego NXT® has been the leader in providing a "drag & drop" graphical programming interface for young robot enthusiasts, but a recent robotics product called ABILIX® introduced a new graphical approach using standard flowchart icons (http://www.abilix.com/support.php) along with a parallel C language development interface (see Fig. 5.4 and Video 5.2).

Another innovative robotics product is called CUBELETS® from Modular Robotics (http://www.modrobotics.com/cubelets/), and in a way it manages to get rid of the software interface entirely by "transferring" enough of the "Think" capacity into the "Sense" and "Act" components of the standard "Sense-Think-Act" paradigm, so as to enable a second "inner" loop between their "Sense" and "Action" Cubelets (see Fig. 5.5 and Video 5.3).

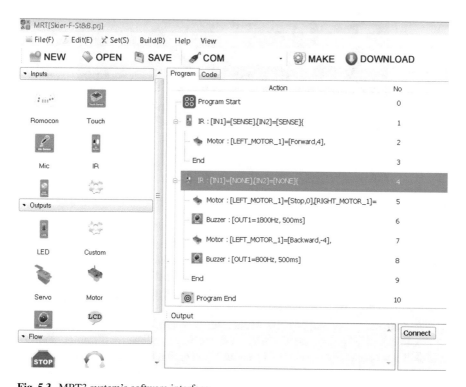

Fig. 5.3 MRT3 system's software interface

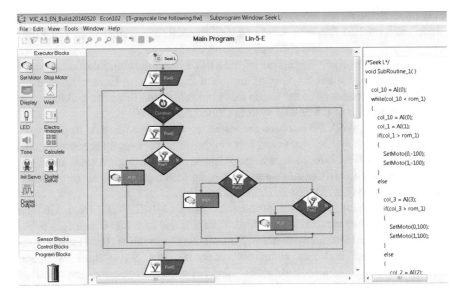

Fig. 5.4 ABILIX system's software interface

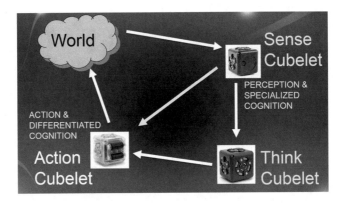

Fig. 5.5 CUBELETS robotics system

The ROBOTIS' TASK tools (V.1 and V.2) could be characterized as a "context-sensitive" text editor and it is closer in programming style to professional tools such as Visual Studio® (see Figs. 5.6 and 5.7). Thus it would take comparatively longer to be proficient at it, but then ones can shift to an Arduino-type interface rather quickly (see Chap. 9).

From an end-user point of view, I think that it is good that we have such a range of possible "entry" points into the "robotics" journey to adjust for different maturity and skill levels of every beginner who should realize that this journey can be very far and wide, requiring continual adjustment of the tools that we would use throughout this journey.

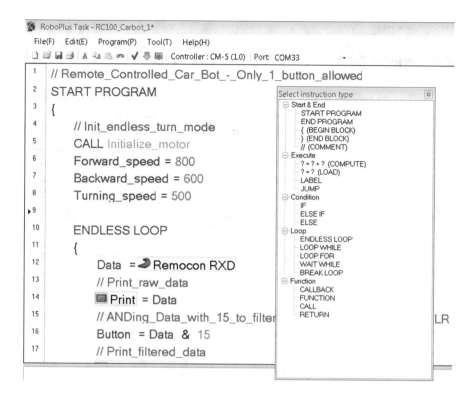

Fig. 5.6 ROBOTIS' TASK V.1 tool interface

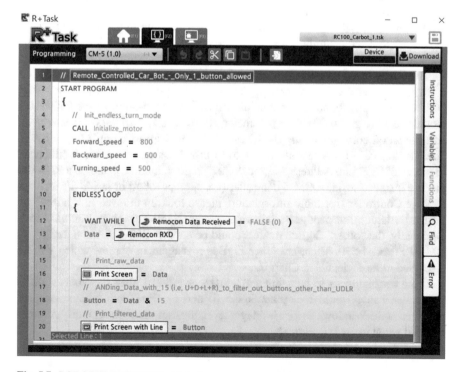

Fig. 5.7 ROBOTIS' TASK V.2 tool interface

5.2   Primer for MANAGER and TASK Tools

In practice, MANAGER and TASK work together very well. MANAGER is an efficient tool to do a thorough but quick check on all hardware components with some special operations such as controller firmware update and managing ZigBee communication settings. TASK is the main algorithm development tool supporting all the standard logical structures (sequential, conditional and repetition) and functions, but it does not support arrays and advanced mathematical functions (for those features, the user will need to use Embedded C tools or the OpenCM IDE described in Chap. 9).

As previously mentioned in Chap. 4, MANAGER V.1 was designed to handle the older CM controllers such as CM-5, CM-510, CM-530 and CM-700, while MANAGER V.2 would only handle the newer controllers such as CM-50, CM-150, CM-200, OpenCM 7.0 and OpenCM 9.04.

5.2.1   MANAGER V.1 Capabilities

For this section, a CM-510 CarBot (Fig. 5.8) was chosen to illustrate the use of the MANAGER V.1 tool. This CarBot used four AX-12As in wheel mode (ID 1 and 3 on left side; ID 2 and 4 on right side) and two IR sensors, one looking forward (Port 1) and one looking down (Port 2). It also had a ZigBee module (ZIG-110A) set to a 1-to-1 communication mode.

Once MANAGER V.1 was executed and connected to the appropriate COM port, the user would see a similar interface to the one shown in Fig. 5.9 (depending on one's actual robot of course). All Dynamixel-compliant (3-pin) actuators and sensors would be listed on the left panel below the Controller item.

In the "Controller" subpanel on the right, the parameter "My Remote ID" (Address 36) corresponded to the actual ZigBee ID (read-only) of the ZIG-110A module used (e.g. 11090). The parameter "Remote ID" (Address 34) was user-editable and it corresponded to the ID of the "other" ZigBee module (e.g. 11091). To make this ZIG-110A module switch to its broadcast mode, the user would need to input "65535" into Address 34. The "other" ZigBee module could be a ZIG-100 module installed inside a CM-5 controller (i.e. another robot) or inside the RC-100 Remote Controller (for more information, please look up the web link http://support.robotis.com/en/product/auxdevice/communication/zigbee_manual.htm). Optionally, the "other" ZigBee module could be another ZIG-110A connected to a CM-510 or CM-530. There are other ZigBee management options which will be discussed in more details in Chap. 7.

If the user happens to be using a BT-110A or BT-210, then the MANAGER's Controller Panel would look like Fig. 5.10 and as BlueTooth connections are managed at the Windows Device Manager level, the user should **not** modify Address 34 at all when using BT modules.

Fig. 5.8 CM-510 CarBot used

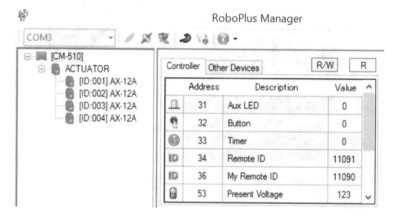

Fig. 5.9 MANAGER V.1 controller panel (with ZIG-110A in 1-to-1 mode)

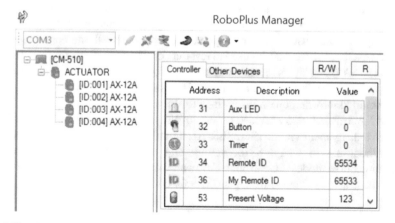

Fig. 5.10 "Controller" subpanel (with BT-110A or BT-210).

Fig. 5.11 "Other Devices" subpanel (with IR sensors on Ports 1 and 2)

Controller	Other Devices		R/W	R

	Address	Control	Value
Port1	80	IR	12
Port2	81	IR	321
Port3	N/A	NONE	0
▶ Port4	N/A	NONE	0
Port5	N/A	NONE	0
Port6	N/A	NONE	0

[CM-510]
ACTUATOR
[ID:001] AX-12A
[ID:002] AX-12A
[ID:003] AX-12A
[ID:004] AX-12A

Address	Description	Value
3	ID	1
6	CW Angle Limit (Joint / Wheel	0
8	CCW Angle Limit (Joint / Whe...	0
11	the Highest Limit Temperature	70
12	the Lowest Limit Voltage	60
13	the Highest Limit Voltage	140
17	Alarm LED	36
18	Alarm Shutdown	36
24	Torque ON/OFF	0
25	LED	0
26	CW Compliance Margin	1
27	CCW Compliance Margin	1
28	CW Compliance Slope	32
29	CCW Compliance Slope	32
30	Goal Position	883

[CM-510]
ACTUATOR
[ID:001] AX-12A
[ID:002] AX-12A
[ID:003] AX-12A
[ID:004] AX-12A

Address	Description	Value
24	Torque ON/OFF	0
25	LED	0
26	CW Compliance Margin	1
27	CCW Compliance Margin	1
28	CW Compliance Slope	32
29	CCW Compliance Slope	32
30	Goal Position	883
32	Moving Speed	0
34	Goal Torque	1023
36	Present Position	883
38	Present Speed	0
40	Present Load	0
42	Present Voltage	121
43	Present Temperature	38
46	Moving	0

Fig. 5.12 "AX-12" built-in ROM procedures

In the "Other Devices" subpanel, the user could check on the status of the GPIO (5-pin) actuators and sensors used on the six ports available on the CM-510 and CM-530 (Fig. 5.11). In this particular case, IR sensors were used on Port 1 and Port 2. If MANAGER V.1 was used for the FIRST time with any GPIO hardware configuration, the user would have to manually assign the correct sensor to the specific port used for that sensor in order to update the controller flash memory contents.

MANAGER V.1 can also be used to test out actuators functions for troubleshooting purposes (see Fig. 5.12 and watch Video 5.4). The web link (http://support. robotis.com/en/product/actuator/dynamixel/ax_series/dxl_ax_actuator.htm) has more detailed information on the built-in ROM procedures.

MANAGER V.1 also provided access to Windows Device Manager (Fig. 5.13), ZigBee management via the ZIG2SERIAL module (Fig. 5.14 and more details will be forthcoming in Chap. 8 for setting and using one-to-one and broadcast communication modes), Controller Firmware Management (Fig. 5.15 and previously illustrated in Chap. 4 for OpenCM-9.04 controllers).

Fig. 5.13 MANAGER V.1 access to windows device manager

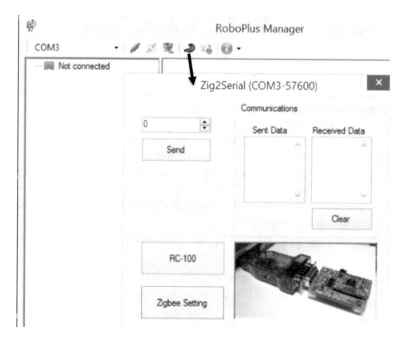

Fig. 5.14 MANAGER V.1 access to ZigBee management

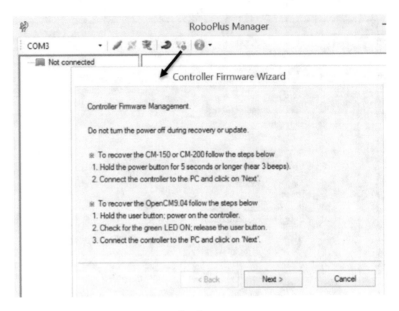

Fig. 5.15 MANAGER V.1 access to controller firmware management

5.2.2 MANAGER V.2 Capabilities

For this section, a CM-150 Probing Car (Fig. 5.16) was chosen to illustrate the use of the MANAGER V.2 tool. This Probing Car used two continuous-turn motors (GM-12A) through Ports 1 and 2, and 2 servo motors (SM-10A) through Ports 3 and 4.

All three modes of communication between the PC and the CM-150 were used successfully with this setup (plain USB cable, LN-101 and BT-210).

The current version (v.2.1.12) of R+MANAGER V.2 can only perform two functions, "Update & Test" and "Firmware Recovery". Some other functions are planned for the near future: "Packet Terminal", "Remote Control", "Self Checklist" and "Dynamixel Calibration" (for more details please visit web link http://support.robotis.com/en/software/roboplus2/r+manager2/quickstart/menu_description.htm).

Figure 5.17 showed the final operational window when the CM-150 was fully connected to MANAGER V.2 via an LN-101 module. If the CM-150 happened to have an out-of-date firmware, MANAGER V.2 would automatically switch to its firmware updating procedure (multi-steps) before the PC user would see the final operational window as shown in Fig. 5.17.

In Fig. 5.17, the third panel from the left (Control Table) showed the accessible parameters:

- Parameters from Addresses 0 through 11 were in the EEPROM area, while those with Addresses from 21 to 129 were in the RAM area.

Fig. 5.16 Probing car
(CM-150) setup for use
with MANAGER V.2

- Furthermore, "grayed-out" parameters were READ-ONLY while the rest were USER-ASSIGNABLE.
- Although the user could actually set the physical speed and direction of rotation to the GM-12A motors attached to Ports 1 and 2, only the mode of operation (wheel or joint) could be set for the SM-10A servos attached to Ports 3 and 4, and not their goal positions. This lack of feature was due to the design of Ports 3 and 4 as GPIO ports (where other actuators/sensors could be attached) and also due to the fact that the SM-10A was not a Dynamixel like the AX-12A.

A more interesting example was to use a hardware combination of an OpenCM9.04-C (blue) and an OpenCM-485-EXP (green) as shown in Fig. 5.18. Please note that the OpenCM9.04-C had to be powered by its own LiPo batteries (top left in Fig. 5.18) while the OpenCM-485-EXP was also powered with its own LiPo battery (bottom right in Fig. 5.18).

In Fig. 5.18, a BIOLOID STEM Hexapod robot (top right) was connected to the 485-EXP board (via its AX/MX-TTL hub) so that its three AX-12A servos (with IDs 3, 4 and 5) would be accessible to MANAGER V.2, while a single XL-320 (bottom left, with ID = 1) was connected to the OpenCM9.04-C's XL-TTL hub.

The Control Table of an OpenCM9.04 controller inside MANAGER V.2 had a new type of parameter named "Dynamixel Channel" at address 16 (see Fig. 5.19).

When this parameter was set to "1", MANAGER V.2 was able to see the Dynamixels physically connected to the 485-EXP board which in this case were the

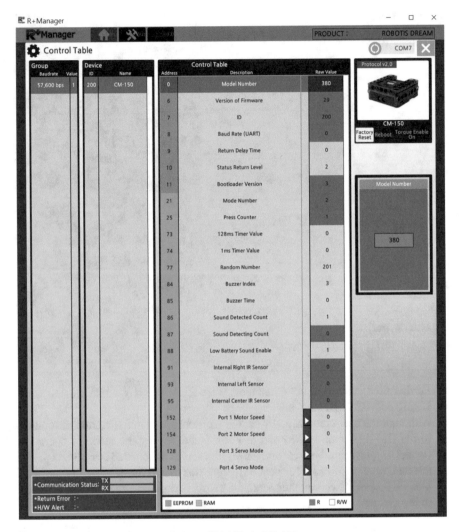

Fig. 5.17 MANAGER V.2 connected to a CM-150 via LN-101

three AX-12As with IDs 3–5 (as shown in Fig. 5.19 inside the "Device" sub-panel). However, MANAGER V.2 would presently (Spring 2017) classify them as "Unknown" as its current version (V. 2.1.12) could only work with the newer XM and XH servos. Thus users of AX and MX servos attached to an 485-EXP board would need to wait perhaps until Summer 2017 to fully utilize MANAGER V.2 for their needs.

On the other hand, when this parameter was set to "0", MANAGER V.2 was able to see the single XL-320 (ID = 1 also) connected to the 9.04-C board as shown in Fig. 5.20 – inside the "Device" and "Control Table" subpanels.

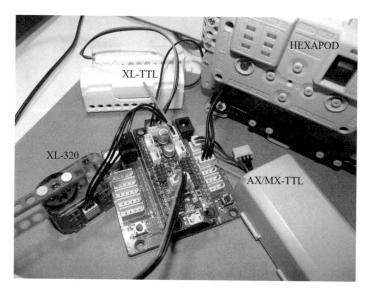

Fig. 5.18 Setup using OpenCM9.04C (*blue*) and OpenCM-485-EXP (*green*) boards

Fig. 5.19 Control table for OpenCM9.04C + 485-EXP boards. (Dynamixel channel set to 1)

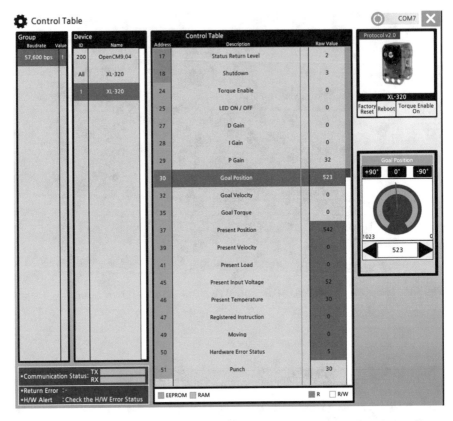

Fig. 5.20 Control table for OpenCM9.04C + 485-EXP boards. (Dynamixel channel set to 0)

Fig. 5.21 Setup to test XM-430 and XL-320 actuators with MANAGER V.2

Figure 5.20 also showed that once the XL-320 Dynamixel was fully recognized inside MANAGER V.2, all its operating parameters were accessible to the users as if they were using the older MANAGER V.1 (see Goal Position subpanel on the right). One inconvenience that the user currently had to put up with was that the system power had to be turned off and back on every time that the Dynamixel Channel got changed.

The last exercise for this section used the setup as shown in Fig. 5.21 whereas an XM-430-W210-R was additionally connected to the 485-EXP.

After setting the OpenCM-904's Dynamixel Channel to 1 and re-power the above setup, MANAGER would then show the connection to the XM-430 actuator and its control table as shown in Fig. 5.22. The user then had full access to the XM-430 actuator.

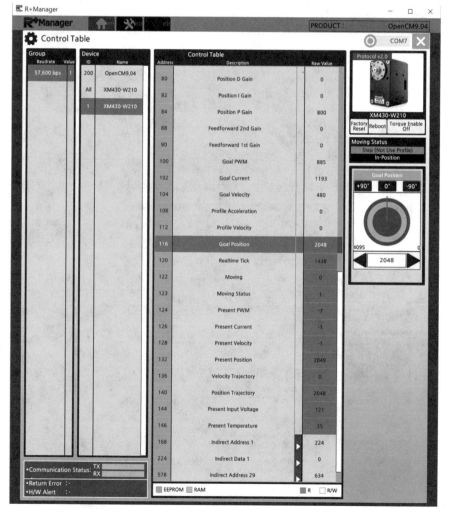

Fig. 5.22 Control table showing access to actuator XM-430

At this point, an advanced user may wonder if this Dynamixel Channel parameter could be set inside a TSKX program (i.e. TASK V.2.1.3 or later) so that the control of Dynamixels attached to either boards (OpenCM9.04-C or 485-EXP) could be achieved "simultaneously"? And the quick answer was "Yes" but not "Simultaneously", and the OpenCM9.04's firmware had to be V.16 or higher. The enclosed "CM904_DynamixelChannel.tskx" code was created to explore this feature (see Fig. 5.23 for an excerpt of this TSKX code).

Essentially, this code was written to allow the user to push Button 1 on the Virtual RC-100 to control an XL-320 (ID = 1) connected to the OpenCM9.04C, and if the user pushed Button 2 then an XM-430 (also ID = 1), connected to the 485-EXP board, would be under user programming control. The author had verified that Statements 16 and 33 did set the OpenCM9.04-C into the proper Dynamixel Channel and thus allowed the TASK program to control the respective Dynamixels attached to **either 9.04C** (i.e. XL type) **or 485-EXP** board (i.e. XM/XH type) **at any one time**. Currently, AX and MX type actuators were not supported on the 485-

13	// Push Button 1 on Virtual RC-100 to set Dynamixel Channel to 0
14	IF (Button == [⬎ 1])
15	{
16	🔵 Dynamixel Channel = 0
17	🔵 ID[1]: ⟳ Goal Position = 5
18	🕐 High-resolution Timer = 2.000sec
19	WAIT WHILE (🕐 High-resolution Timer > 0)
20	📇 Print Screen = 🔵 Dynamixel Channel
21	📇 Print Screen = 🔵 ID[1]: ⟳ Present Position
22	🔵 ID[1]: ⟳ Goal Position = 1000
23	🕐 High-resolution Timer = 2.000sec
24	WAIT WHILE (🕐 High-resolution Timer > 0)
25	📇 Print Screen with Line = 🔵 ID[1]: ⟳ Present Position
26	🕐 High-resolution Timer = 4.000sec
27	WAIT WHILE (🕐 High-resolution Timer > 0)
28	}
29	// Push Button 2 on Virtual RC-100 to set Dynamixel Channel to 1
30	IF (Button == [⬎ 2])
31	{
32	// XM-430 (ID=1) on 485-EXP board
33	🔵 Dynamixel Channel = 1
34	🔵 ID[1]: 🔧 Operating Mode = 3
35	🔵 ID[1]: 🔧 Torque ON/OFF = TRUE (1)
36	🔵 ID[1]: ⟳ Goal Position = 600

Fig. 5.23 Excerpt from program "CM904_DynamixelChannel.tskx" to set Parameter "Dynamixel Channel"

EXP board if TASK V.2 was used (their support may occur at a later time). Thus, at present, if a mixture of XL, AX or MX actuators were used, a mixed-channel/mixed-protocol control procedure could be achieved using the ROBOTIS IDE (see Sect. 9.2 for an example of mixed-channel and mixed-protocol control of XL-320s and AX-12As). Please also note that the development work on the ROBOTIS IDE had stopped at the Dynamixel Pro series, so it is not known at this time if the X series of Dynamixels would ever be ported to the ROBOTIS IDE.

5.2.3 Basic TASK Usage (V.1 and V.2)

If you happen to be an absolute beginner with the TASK tool (V.1 or V.2), please watch the Video 5.5 first before reading on the rest of this section. This video demonstrated the basic steps needed, from creating a TSK or TSKX program from scratch to running it on a CM-510/530 CarBot (Fig. 5.8). It showed how to declare parameters and use printing facilities and also discussed the use of control structures (sequence, loops and conditions). Additionally, it showed how to use functions to modularize coding. Please also refer to the example code files "IR-Sensors-1.tsk/tskx" and "IR-Sensors-2.tsk/tskx".

The next example, "IR-Motor.tsk/tskx", illustrated the closing of the "Sense-Think-Act" loop and was the first example using a Reactive Control approach (well described in Chap. 14 of Matarić (2007)). The Video 5.6 described the main steps for a Reactive Control approach:

1. Determine the "mutually exclusive conditions" that the bot would encounter during its operation in the world.
2. Match up each of those conditions with "appropriate" action(s). Create a table of matching conditions and actions to keep track of one's current thinking about the problem to be solved.
3. From this "Condition|Action" table, generate a matching "Input|Output" or "Sensor|Actuator" table that would correspond to the actual sensors and actuators used in the bot design (see Fig. 5.24a, b).

a

Given Condition >> Appropriate Robot Action	
Conditions	Actions
Condition 1	Action A
Condition 2	Action B
Condition 3	Action C
Condition 4	Action B

b

Given Input Sensor(s) >> Activate Appro. Actuator(s)	
Input Sensors	Output Actuators
NIR	DC MOTORS
TOUCH	SERVO MOTORS
MICROPHONE	BUZZER-SPEAKER
LIGHT LEVEL	LEDs

Fig. 5.24 (**a**) Condition|Action table; (**b**) Sensor|Actuator table

4. Next translate the "Sensor|Actuator" table into coding facilities that were provided with the chosen software development environment (see examples in Figs. 5.3, 5.4, and 5.6).
5. Ones can expect to revisit Steps 1 through 4 several times before achieving a satisfactory solution.

5.3 "Sequence Commander" Project

This project used the CarBot described in Fig. 5.8 and it was a 4-WD adaptation of the 2-WD Sequence Racer robot belonging to the BIOLOID STEM system. The sample code "SequenceCommander_CM-510-530.tsk/tskx" was designed to work on CM-510/530 controllers, while the other version "SequenceCommander_CM-5. tsk/tskx" was made for a CM-5 controller using the Integrated Sensor AX-S1 (ID = 100).

The "Sequence Commander" robot operated in two distinct phases:

1. This robot started out in a "Learning" mode where it would wait for the user to press any of the Up-Down-Left-Right buttons on the controller, in any combination or sequence, but up to five buttons only (see Fig. 5.25 for a code snippet showing how Button 1 was handled, also see Video 5.7 for more details). As soon as the user pressed the sixth button, the robot would sound out an "error" signal, clear all previous inputs and restart on its learning phase (see Fig. 5.26).
2. Once the user had entered a "legal" sequence of button presses (<= five buttons), the user would press on the "START" button to make the robot shift into its "Replay" mode. The robot would then move in the directions as recorded in the sequence made by the user during the learning phase. Each forward or backward step would last 1 s, while the left and right turn maneuvers would last only 0.5 s (see Fig. 5.27).

```
26    ENDLESS LOOP
27    {
28        //  Do nothing if there is no button being pressed
29        WAIT WHILE  (  Controller Button  ==   --  )
30
31        IF  (  Controller Button  ==  U  ||  Controller Button  ==  D  ||  Controller Button  ==  L  ||  Controller Button  ==  R  )
32        {
33            Button  =  Controller Button
34            CALL  Button_Standby
35
36            IF  (  ButtonNumber  ==  1  )
37            {
38                CALL  Button_Init
39                Button1  =  Button
40                CALL  Button_Standby
41                CALL  Buzzer_Button
42                ButtonNumber  =  ButtonNumber  +  1
43            }
```

Fig. 5.25 Start of learning cycle and how button 1 was recorded (from Controller Button)

```
77    ELSE IF  (  ButtonNumber  ==  6  )
78    {
79        CALL  Buzzer_Failure
80        CALL  Button_Init
81        ButtonNumber  =  1
82    }
```

Fig. 5.26 Button 6 triggered an error condition and a restart of learning cycle (resetting ButtonNumber back to 1)

```
94    //  Replay the recorded sequence of maneuvers
95    CALL  Buzzer_Success
96    MoveType  =  Button1
97    CALL  Move
98
99    MoveType  =  Button2
100   CALL  Move
101
102   MoveType  =  Button3
103   CALL  Move
104
105   MoveType  =  Button4
106   CALL  Move
107
108   MoveType  =  Button5
109   CALL  Move
```

Fig. 5.27 Replay of carbot moves according to recorded buttons 1 through 5

There were some other program design features needing to be discussed further:

(a) Each of the four AX-12 motors was provided with its own speed parameter (Speed1 through Speed4). This feature was provided so that the user could fine-tune these parameters in case the robot did not move "straight enough" when commanded to do so. This problem may come from robot construction mis-alignments or from non-uniform performance among the four motors. In practice, it may even come from the non-uniform properties of the surface that the robot would be running on, but that issue would be beyond the user's control (see Fig. 5.28).

149	FUNCTION Forward
150	{
151	**ID[1]:** **Goal Velocity** = CW:0 (0.00%) + Speed1
152	**ID[3]:** **Goal Velocity** = CW:0 (0.00%) + Speed3
153	
154	**ID[2]:** **Goal Velocity** = CCW:0 (0.00%) + Speed2
155	**ID[4]:** **Goal Velocity** = CCW:0 (0.00%) + Speed4
156	
157	WaitTime = ForwardTime
158	CALL Wait
159	CALL Stop
160	}

Fig. 5.28 Function forward with independent parameters Speed1 through Speed4

128	FUNCTION Button_Standby
129	{
130	// Waiting for user to release buttons
131	WAIT WHILE (**Controller Button** != **--**)
132	}

Fig. 5.29 Function to help with "debouncing" button presses

(b) The "Button_Standby" function (see Fig. 5.29) was very short, but it fulfilled an important task of adjusting the extremely fast processing speed of the controller to the rather slow speed of human motion (when the user pressed and released a given button).

Before the execution of this function (essentially statement 131), the controller had already determined and saved the value of the controller's button just pressed by the user into a parameter named "Button" (statement 33 in TASK codes). However, the user's finger might still be hovering over the pressed button as human reaction times are at best in tenths of a second, while the controller can execute its commands in microseconds. This "WAIT WHILE" construct essentially halted the controller's progress as long as it detected that some button(s) were still being pushed. Please note that we had to use "double negation" to achieve this goal, as "NOT (No Button Was Pressed)" was equivalent to "Some Buttons Were Being Pressed". Please also note that it would be "computationally inefficient" to use a WAIT WHILE condition that involved "OR" operators and the status of each of the U-D-L-R buttons as shown below, although it would be logically equivalent.

WAIT WHILE (**Button** == U || **Button** == R || **Button** == D || **Button** == L)

The video file "Video 5.7" described the code in more details and contained webcam recordings of this robot in action.

5.4 "Smart Avoider" Project

This project used a CM-5 CarBot described in Fig. 5.30 and it used an AX-S1 to detect obstacles in front of it.

True to its namesake of "Integrated Sensor", the AX-S1 has many functions as shown in Fig. 5.31:

1. Active NIR sensing via "IR Fire Data" [0–255] (Addresses 26–28 for specific Left-Center-Right sensors). When these values are larger than a threshold value set at Address 52 (defaulting to 32), a 3-bit (RCL) flag would be set at Address 32.
2. Passive NIR sensing via "Light Data" [0–255] (Addresses 29–31 for specific Left-Center-Right sensors). Similarly as for active mode, when these values are larger than a threshold value set at Address 53, a 3-bit (RCL) flag is set at Address 33.
3. Detecting sound claps via parameters at Addresses 35–37.
4. Generating sound via buzzer (addresses 40 and 41).
5. Threshold settings for Active and Passive NIR sensing modes (addresses 52 and 53 respectively). When Address 52 is set to 0, the Active NIR sensors switch to a short-range mode and they can only detect objects within 12 cm of each sensor's opening. But if Address 52 is set to a non-zero number, the Active NIR sensors go to a long-range mode whereas they can detect objects up to 37 cm away from their openings.

In this section, the goal was to illustrate two control approaches (Reactive and Behavior-Based) on the same physical CarBot (Fig. 5.8). The Reactive Control approach yielded a solution represented by the TASK program "ObstacleDetectionCar.tsk/tskx", while the Behavior-Based Control approach yielded "SmartAvoider.tsk/tskx".

As previously discussed in Sect. 5.2.3, a Reactive Control approach required the generation of "mutually-exclusive" conditions for the robot to monitor along with specific and appropriate robot action(s) to be triggered when each of those conditions became true. Figure 5.32 was the Condition-Action table that was generated upon analysis of the "object-detection" capabilities of the AX-S1 (in NIR long-range detection mode).

The translation of this Condition-Action table into a TASK program yielded the code fragment as shown in Fig. 5.33 where the "Object Detected" Flag was used to represent the eight possible conditions to be monitored. Please note that a Reactive Control approach usually lead into an IF-ELSE-IF logical structure. The video clip "Video 5.8" had more detailed explanations of the program design and webcam views of the robot performance in avoiding obstacles.

Upon viewing the video "Video 5.8", the reader would notice that the behavior for dealing with a front obstacle was not very satisfactory as it kept on repeating seemingly un-intelligent actions. A possible remedy was to add a slight left or right turn to the MovFrd function (this is left to the reader to do as an exercise).

Fig. 5.30 CM-5 CarBot used

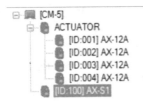

	Address	Description	Value
	26	IR Left Fire Data	0
	27	IR Center Fire Data	0
	28	IR Right Fire Data	0
	29	Light Left Data	0
	30	Light Center Data	0
	31	Light Right Data	1
	32	IR Obstacle Detected	0
	33	Light Detected	0
	35	Sound Data	127
	36	Sound Data Max Hold	255
	37	Sound Detected Count	1
	40	Buzzer	0
	41	Buzzer Timer	0
	52	IR Obstacle Detect Compare	32
	53	Light Detect Compare	32

Fig. 5.31 MANAGER V.1's display for a CM-5 with AX-S1

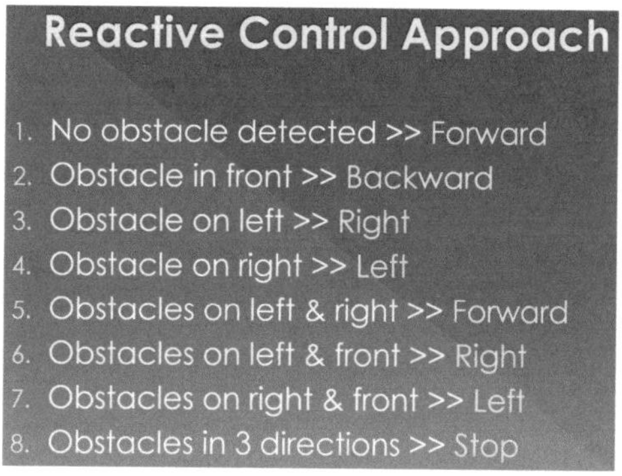

Fig. 5.32 Condition-action table for CM-5/AX-S1 CarBot

Reactive Control Approach

1. No obstacle detected >> Forward
2. Obstacle in front >> Backward
3. Obstacle on left >> Right
4. Obstacle on right >> Left
5. Obstacles on left & right >> Forward
6. Obstacles on left & front >> Right
7. Obstacles on right & front >> Left
8. Obstacles in 3 directions >> Stop

Fig. 5.33 IF-ELSE-IF structure corresponding to condition-action table of Fig. 5.31

```
16   Direct = ID[100]: Object Detected
17   IF ( Direct == 0000 0000 0000 0000 )
18      CALL MovFrd
19   ELSE IF ( Direct == 0000 0000 0000 0010 )
20      CALL MovBrd
21   ELSE IF ( Direct == 0000 0000 0000 0001 )
22      CALL MovR
23   ELSE IF ( Direct == 0000 0000 0000 0100 )
24      CALL MovL
25   ELSE IF ( Direct == 0000 0000 0000 0101 )
26      CALL MovFrd
27   ELSE IF ( Direct == 0000 0000 0000 0011 )
28      CALL MovR
29   ELSE IF ( Direct == 0000 0000 0000 0110 )
30      CALL MovL
31   ELSE IF ( Direct == 0000 0000 0000 0111 )
32      CALL Stop
```

Matarić (2007) presented a compact introduction to the Behavior-Based Control (BBC) approach (Chap. 16) but the seminal work in this area was from Arkin (1998). Matarić described BBC as "the use of behaviors as modules for control" and that "behaviors achieve and/or maintain particular goals". When translated to the SmartAvoider solution, there were three behaviors to emulate:

1. How to go forward.
2. How to avoid obstacles that come in from the side.
3. How to escape from dead-ends.

Figure 5.34 displayed a code fragment representing the main logic of the "SmartAvoider.tsk/tskx" program whereas the robot would go forward as its default behavior, but it would also additionally act appropriately to avoid side obstacles and to escape from cul-de-sacs.

The reader would surely appreciate the coding differences between Figs. 5.33 and 5.34 (i.e. "IF-ELSE-IF" vs. "parallel IFs" with a default function call Go_forward).

```
15      ENDLESS LOOP
16      {
17          IF ( ⬤ ID[100]: 🔵 IR Left  > 200 || ⬤ ID[100]: 🔵IR Right  > 200 )
18              CALL Avoiding_side
19          IF ( ⬤ ID[100]: 🔺 IR Center  >= 252 )
20              CALL Escape
21          CALL Go_forward
22      }
```

Fig. 5.34 Code fragment corresponding to BBC as applied to SmartAvoider

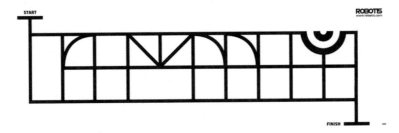

Fig. 5.35 BIOLOID STEM line track

The video clip "Video 5.9" would present more explanations about the coding of those three behaviors and about additional programming decisions taken to improve NIR sensors use. This video clip closed by showing the performance of this "unaltered" SmartAvoider robot as it tried to navigate through a maze, although it was not originally designed for this purpose. This was to illustrate the concept of "emerging behavior" (which was not completely successful in this example as the robot struggled very hard to get out of the cul-de-sac). The interested reader can also visit YouTube for other performances by the same CM-5 CarBot on a different maze (http://www.youtube.com/watch?v=tR0dWVppFFM and http://www.youtube.com/watch?v=iFgVpc5U7Rg).

5.5 "Line Tracer" Project

This project came from the BIOLOID STEM STANDARD kit (CM-530) and used two AX-12W for the motorized wheels and the IR Sensor Array (IRSA) to find its way on a very complex line track (see Fig. 5.35).

It was well designed mechanically to make it nimble in maneuvers and it also illustrated the appropriate uses of timed turns and conditional turns.

5.5.1 Mechanical Design Features

Some well thought out mechanical design features needed to be noticed:

1. The batteries and AX-12Ws (the heaviest components) were laid out such that the robot's center of gravity was just slightly ahead of the wheels contact line (Fig. 5.36). This helped the maneuverability of the Line Tracer.
2. Figure 5.37 showed the subtle ~1 mm clearance in the back, while the front weight rested on the two big LEDs of the IRSA. Thus this robot has four points of contact with respect to the travel surface and the front LEDs offered the minimum friction possible.

As a design exercise, the author adapted the BIOLOID STEM Avoider robot to also have a line-tracing ability using the IRSA (see Fig. 5.38). However as its CG was shifted more forward as compared with the Line Tracer's, this modified Avoider was found to be not as maneuverable as it tended to drag on the "larger" metallic hemispheroidal support piece when a turn was executed.

Fig. 5.36 Line tracer's CG and wheel contact line

Fig. 5.37 Line Tracer's points of contact with the travel surface

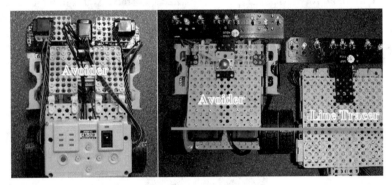

Fig. 5.38 Line Tracer's points of contact with the travel surface

5.5.2   IR Array Sensor (IRSA)

The IRSA had seven NIR LED sets to detect changes in brightness, i.e. whether it was on the black track or not. The LED sets were positioned to match the IRSA with the width of the black track and also for the detection of the circular and diagonal branches (Fig. 5.39).

The video file "Video 5.10" demonstrated how to set up threshold values for each NIR LED set to result in a proper setting of the IRSA's IR Obstacle Detected flag (a 7-bit parameter) which was an important parameter used in programming the maneuvers needed by the robot to navigate a user-defined path within the complex line track as shown in Fig. 5.39 or for any arbitrary single line track as shown in Fig. 5.40 in the next Sect. 5.5.3. One hard lesson learned "from the field" was that all these settings would depend on the "current" power level of the battery pack powering the robot (in other words, remember to refresh your robot's batteries often).

Fig. 5.39 Avoider robot on complex line track

5.5.3 Programming Maneuvers for Arbitrary Single-Line Tracks

The TASK programs used in this section were adapted from the original TSK/TSKX codes provided by ROBOTIS at their web site (http://support.robotis.com/en/product/bioloid/stemkit/download/bio_stem_standard_apps.htm) for their robot named "LineFollower".

The TASK program "LineTracer_ODR.tskx" illustrated two maneuvering algorithms A and B: Algorithm A was to make the carbot follow an arbitrary track created with black electrical tape laid out on brown craft paper (see Fig. 5.40), and Algorithm B was to make the carbot swing around and go the other way (while still following the black track) when an obstacle was encountered in front of it.

But before getting down to the logical details of the algorithms used, let's have a closer look at the possible physical interactions between the IRSA and the black track, to be more exact between the track's width and the layout of the seven IR sensors of the IRSA (see Fig. 5.41).

The reader could see that the black tape width was slightly smaller than the distance between IRSA3 and IRSA5. Thus if the IRSA module was perfectly lined up with the track, then only IRSA4 would see "black". Next, if the carbot was slightly off to the left or right of this perfect alignment, then we would have two mutually exclusive conditions: either (IRSA4 and IRSA5) or (IRSA4 and IRSA3) would see "black". Similarly, if the IRSA was further off-track, then we would have two additional and mutually exclusive conditions: either (IRSA6) or (IRSA2) would see "black". IRSA1 and IRSA7 were not shown in Fig. 5.41, but they would conform to the same trend. These important physical results were used in the development of algorithms A and B.

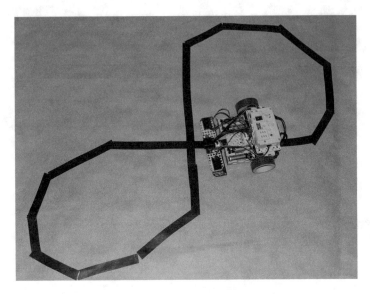

Fig. 5.40 Line Tracer carbot on an arbitrary single line track

Fig. 5.41 A close-up view of the IRSA and a single black track

Algorithm A first relied on a set of seven binary flags, each corresponding to each of the seven IR sensors of the IRSA. The Function "detect_black" showed the computational details involved in extracting these seven individual flags from the global flag labeled "IR Obstacle Detected" (line 166 in code snippet shown in Fig. 5.42). Once the "IR Obstacle Detected" flag was saved into the parameter "black_result" (line 166), the next task was to extract this 7-bit parameter into its individual components "black_1" though "black_7". For example, let's say that the "black_result" parameter happened to be equal to this binary number "0010011", meaning that somehow the individual IR sensors 1, 2 and 5 happened to be "seeing" pieces of the black line(s) (i.e. this binary number should be "read" from right to left).

Upon execution of the next seven statements, 168 through 174, which were logical "AND" operations with specific pre-defined constants, the results would be as follows:

- "black_1" would be equal to "0000001".
- "black_2" would be equal to "0000010".
- "black_3" would be equal to "0000000".

```
162   FUNCTION  detect_black
163   {
164       black_result  =  [ID[100]:  IR Obstacle Detected]
165
166       black_1  =  black_result  &  [ 1 ]
167       black_2  =  black_result  &  [ 2 ]
168       black_3  =  black_result  &  [ 3 ]
169       black_4  =  black_result  &  [ 4 ]
170       black_5  =  black_result  &  [ 5 ]
171       black_6  =  black_result  &  [ 6 ]
172       black_7  =  black_result  &  [ 7 ]
173   }
```

Fig. 5.42 Code for function "detect_black"

- "black_4" would be equal to "0000000".
- "black_5" would be equal to "0010000".
- "black_6" would be equal to "0000000".
- "black_7" would be equal to "0000000".

In short, if any particular "black_N" parameter was greater than zero, the corresponding IR Sensor N was seeing "black".

Thus the first action that Algorithm A, i.e. Function "follow_line", took was to CALL Function "detect_black" to get the current status of these seven parameters in line 86 (see Fig. 5.43). Once this knowledge was obtained, the carbot could then attempt to figure out how to run the Left motor (ID = 1) faster or slower than the Right motor (ID = 2), i.e. to maneuver itself so that the carbot would stay WITH the black track through its changing straight and curving sections (see Fig. 5.40). In order to accomplish this "driving" task, the carbot had to monitor for five "mutually exclusive" main conditions:

1. Condition 1: was IRSA4 (located at the mid-line of IRSA) seeing "black"? If yes, then the carbot was pretty much on track or perhaps just off track slightly, so the wheel speed can be at 100 % forward (statements 103–104 in Fig. 5.43) or at 75 % of maximum for the inner turning wheel – to slightly turn left or right as needed (see statements 93 and 99 in Fig. 5.43).
2. If Condition 1 was false (i.e. Condition 2 was true), IRSA4 was not seeing 'black", therefore the carbot was off-track some, then the carbot needed to check whether IRSA3 or IRSA5 were still seeing "black"? (see Fig. 5.44) and should slow down to 62% of maximum speed for the inner turning wheel (statements 113 and 119 of Fig. 5.44).
3. If Condition 2 was false (i.e. Condition 3 was true), then the carbot was off-track some more, and it needed to check further out whether IRSA2 or IRSA6 were

```
84    FUNCTION  follow_line
85    {
86        CALL  detect_black
87        //  Check for Condition 1 - Carbot is pretty much on track
88        IF  (  black_4  >  0  )
89        {
90            //  If Carbot is OFF the track a little bit but IRSA_4 is still seeing black
91            IF  (  black_4  >  0  &&  black_3  >  0  )
92            {
93                ID[1]:  Goal Velocity  =  CCW:0 (0.00%)  +  l_wheel_speed_3_4
94                ID[2]:  Goal Velocity  =  CW:0 (0.00%)  +  r_wheel_max_speed
95            }
96            ELSE IF  (  black_4  >  0  &&  black_5  >  0  )
97            {
98                ID[1]:  Goal Velocity  =  CCW:0 (0.00%)  +  l_wheel_max_speed
99                ID[2]:  Goal Velocity  =  CW:0 (0.00%)  +  r_wheel_speed_4_5
100           }
101           ELSE
102           {
103               ID[1]:  Goal Velocity  =  CCW:0 (0.00%)  +  l_wheel_max_speed
104               ID[2]:  Goal Velocity  =  CW:0 (0.00%)  +  r_wheel_max_speed
105           }
106       }
```

Fig. 5.43 Code snippet for Condition 1 of Algorithm A (max speed on both wheels or 75% of max on the "inner" turning wheel)

```
108       //  Condition 2 - If Carbot is OFF track some more - only IRSA_3 or IRSA_5 are still seeing black
109       ELSE IF  (  black_3  >  0  ||  black_5  >  0  )
110       {
111           IF  (  black_3  >  0  )
112           {
113               ID[1]:  Goal Velocity  =  CCW:0 (0.00%)  +  l_wheel_speed_3
114               ID[2]:  Goal Velocity  =  CW:0 (0.00%)  +  r_wheel_max_speed
115           }
116           ELSE
117           {
118               ID[1]:  Goal Velocity  =  CCW:0 (0.00%)  +  l_wheel_max_speed
119               ID[2]:  Goal Velocity  =  CW:0 (0.00%)  +  r_wheel_speed_5
120           }
121       }
```

Fig. 5.44 Code snippet for Condition 2 of Algorithm A. (62% of max speed on the inner turning wheel)

still seeing "black"? (see Fig. 5.45) and should slow down to 50% of maximum speed for the inner turning wheel (statements 128 and 133 of Fig. 5.45).
4. If Condition 3 was false (i.e. Condition 4 was true), then the carbot was off-track badly now, and it needed to check whether IRSA1 or IRSA7 were still seeing "black"? (see Fig. 5.46) and the inner turning wheel should now be locked so that a tighter turn could be achieved, using that locked wheel as a pivot point (statements 143 and 149 of Fig. 5.46).
5. Lastly, if Condition 4 was false also (i.e. Condition 5 was true), then the carbot had lost track completely and it needed to go forward at 75% of maximum speed and "hoped" that it might encounter the black track again "somewhere" at a later time (see Fig. 5.47 and statements 156–157).

Algorithm B was simpler to explain than Algorithm A. Whenever the carbot saw an obstacle in front of it, via its IRSS-10 sensor at Port 6, it would trigger the Function "SwingAround" (see Line 24 of Fig. 5.48).

The Function "SwingAround" (Figs. 5.49 and 5.50) illustrated a combination of "timed" and "conditional" right turns to accommodate for the fact that the "obstacle" could be set/found at any possible spot on the black track such as the one shown in Fig. 5.40. A 0.125 s timed right-turn was first used (Fig. 5.49) because at this point in time, the carbot was already lined up with the black track and we just needed to nudge the carbot a bit to the right to make sure that IRSA4 did not see black anymore.

In the next section of code as shown in Fig. 5.50, the reader had to remember that the previous commands for the carbot to turn right (Lines 229–230) were still in effect. The four WHILE loops (Lines 235–253) were "conditional" loops so that the carbot could adjust to "any" section of the curved track where the "SwingAround" maneuver may be called upon. These four loops were used to ensure that an orderly

```
123      // Condition 3 - If Carbot is OFF track even more - only IRSA_2 or IRSA_6 are still seeing black
124      ELSE IF ( black_2 > 0 || black_6 > 0 )
125      {
126          IF ( black_2 > 0 )
127          {
128              ID[1]:  Goal Velocity = CCW:0 (0.00%) + l_wheel_speed_2
129              ID[2]:  Goal Velocity = CW:0 (0.00%) + r_wheel_max_speed
130          }
131          ELSE
132          {
133              ID[2]:  Goal Velocity = CW:0 (0.00%) + r_wheel_speed_6
134              ID[1]:  Goal Velocity = CCW:0 (0.00%) + l_wheel_max_speed
135          }
136      }
```

Fig. 5.45 Code snippet for Condition 3 of Algorithm A (50% of max speed on the inner turning wheel)

```
138   //  Condition 4 - If Carbot is OFF track badly - only IRSA_1 or IRSA_7 are still seeing black
139   ELSE IF  (  black_1  >  0  ||  black_7  >  0  )
140   {
141       IF  (  black_1  >  0  )
142       {
143           [●] ID[1]: [⬜] Goal Velocity  =  CCW:0 (0.00%)  +  l_wheel_speed_1
144           [●] ID[2]: [⬜] Goal Velocity  =  CW:0 (0.00%)  +  r_wheel_max_speed
145       }
146       ELSE
147       {
148           [●] ID[1]: [⬜] Goal Velocity  =  CCW:0 (0.00%)  +  l_wheel_max_speed
149           [●] ID[2]: [⬜] Goal Velocity  =  CW:0 (0.00%)  +  r_wheel_speed_7
150       }
151   }
```

Fig. 5.46 Code snippet for Condition 4 of Algorithm A. (Inner turning wheel locked – i.e. speed = 0).

```
153   //  Condition 5 - Bot has lost track completely - Just go forward at 75% of max speed
154   ELSE
155   {
156       [●] ID[1]: [⬜] Goal Velocity  =  CCW:0 (0.00%)  +  l_wheel_speed_3_4
157       [●] ID[2]: [⬜] Goal Velocity  =  CW:0 (0.00%)  +  r_wheel_speed_4_5
158   }
```

Fig. 5.47 Code snippet for Condition 5 of Algorithm A

process was followed to detect the current status of the IRSA sensors seeing "white" or "black", from the outermost sensor IRSA7 to the central sensor IRSA4:

- The first loop (Lines 235–238) would be active as long as IRSA7 saw "white", i.e. did not yet cross into the black track.
- Next the second loop (Lines 240–243) achieved the previous maneuver but using IRSA6.
- And then similarly for IRSA5 and IRSA4 (Lines 245–253). At this point in time, the carbot ought to be mostly lined up with the black track (but in the reverse direction), thus the next function to be called was "FollowLine" (Line 25 of Fig. 5.48).

It was left as an exercise to the reader to figure out whether the three loops using parameters "black_7", "black_6" and "black_5" were really necessary.

Video 5.11 showed the actual performances of these two algorithms for two tracks: one already shown in Fig. 5.40, the other was like a misshapen O. Video 5.11 showed that Algorithm A, as applied to these particular tracks, eventually got

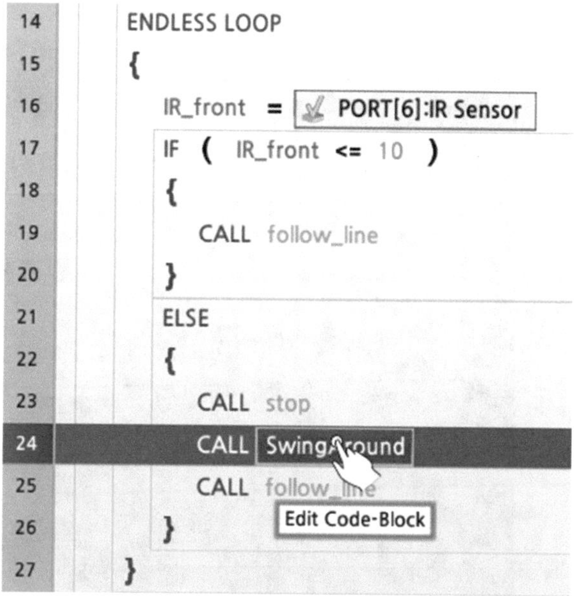

Fig. 5.48 Code snippet for calling "SwingAround" when "obstacle" detected by IRSS-10 Sensor at Port 6

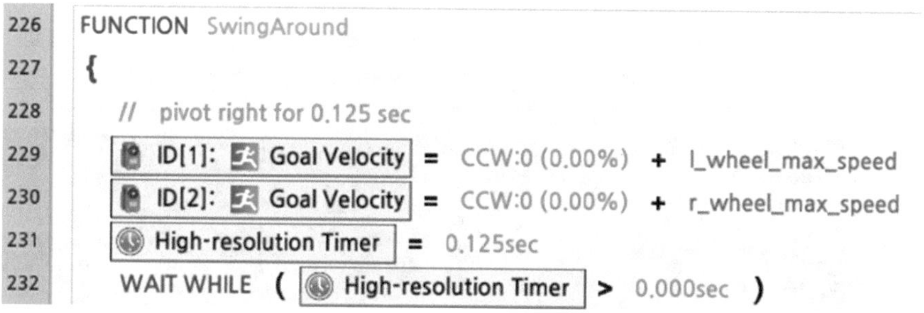

```
14    ENDLESS LOOP
15    {
16        IR_front  =  [✓] PORT[6]:IR Sensor
17        IF  (  IR_front  <=  10  )
18        {
19            CALL  follow_line
20        }
21        ELSE
22        {
23            CALL  stop
24            CALL  SwingAround
25            CALL  follow_line
                              Edit Code-Block
26        }
27    }
```

```
226    FUNCTION  SwingAround
227    {
228        //  pivot right for 0.125 sec
229        [🔋] ID[1]: [⚙] Goal Velocity  =  CCW:0 (0.00%)  +  l_wheel_max_speed
230        [🔋] ID[2]: [⚙] Goal Velocity  =  CCW:0 (0.00%)  +  r_wheel_max_speed
231        [🕐] High-resolution Timer  =  0.125sec
232        WAIT WHILE  (  [🕐] High-resolution Timer  >  0.000sec  )
```

Fig. 5.49 Code snippet for initial timed right turn to clear black track

"stuck" in using "only" either IRSA Sensors 1 or 7 (i.e. the outer most sensors), also that it could not handle sharp curves like a 90-degrees one very well.

The TASK program "LineTracer_ODR_C.tskx" was created to address this shortcoming by adding at appropriate steps of Algorithm A the "Centering" function (Lines 122 and 138 of Fig. 5.51).

The actual code for the Function "Centering" was shown in Fig. 5.52 and the critical logic was implemented in the WHILE loops based on "black_4" (Lines 186–191 and Lines 195–200) whereas essentially the carbot was not allowed to see "white" through its IRSA4 sensor if the "near-by" IRSA sensors such as 2–3 or 5–6 were still seeing "black". In other words, the maneuvering strategy was not to let the "track deviation" situation develop into an involvement of the outermost IRSA sensors 1 or 7 as it would then be too late to do any correction. Video 5.12 showed a much more "satisfactory" performance for the same carbot with the same test tracks

```
233    CALL  detect_black
234    //  Keep on turning right as long as IRSA_7 sees white
235    LOOP WHILE  (   black_7  ==  0  )
236    {
237       CALL  detect_black
238    }
239    //  Keep on turning right as long as IRSA_6 sees white
240    LOOP WHILE  (   black_6  ==  0  )
241    {
242       CALL  detect_black
243    }
244    //  Keep on turning right as long as IRSA_5 sees white
245    LOOP WHILE  (   black_5  ==  0  )
246    {
247       CALL  detect_black
248    }
249    //  Keep on turning right as long as IRSA_4 sees white
250    LOOP WHILE  (   black_4  ==  0  )
251    {
252       CALL  detect_black
253    }
```

Fig. 5.50 Code snippet for the four conditional loops to make the carbot line up again with the black track, but in the reverse direction

using this "Centering" function (also see Exercise 11 at the end of this chapter for another possible approach).

It should also be pointed out that the above algorithms were designed for "arbitrary" but "single-line" tracks like the ones shown in Fig. 5.40. For tracks that had multiple lines crossing each other like shown in Fig. 5.39, different approaches using the concept of "nodes" would have to be used as described in the next section.

```
109    //  Condition 2 - If Carbot is OFF track some more - only IRSA_3 or IRSA_5 are still seeing black
110    ELSE IF  (   black_3  >  0  ||  black_5  >  0  )
111    {
112        IF  (  black_3  >  0  )
113        {
114            [ID[1]: Goal Velocity]  =  CCW:0 (0.00%)  +  l_wheel_speed_3
115            [ID[2]: Goal Velocity]  =  CW:0 (0.00%)  +  r_wheel_max_speed
116        }
117        ELSE
118        {
119            [ID[1]: Goal Velocity]  =  CCW:0 (0.00%)  +  l_wheel_max_speed
120            [ID[2]: Goal Velocity]  =  CW:0 (0.00%)  +  r_wheel_speed_5
121        }
122        CALL  centering
123    }
124
125    //  Condition 3 - If Carbot is OFF track even more - only IRSA_2 or IRSA_6 are still seeing black
126    ELSE IF  (   black_2  >  0  ||  black_6  >  0  )
127    {
128        IF  (  black_2  >  0  )
129        {
130            [ID[1]: Goal Velocity]  =  CCW:0 (0.00%)  +  l_wheel_speed_2
131            [ID[2]: Goal Velocity]  =  CW:0 (0.00%)  +  r_wheel_max_speed
132        }
133        ELSE
134        {
135            [ID[2]: Goal Velocity]  =  CW:0 (0.00%)  +  r_wheel_speed_6
136            [ID[1]: Goal Velocity]  =  CCW:0 (0.00%)  +  l_wheel_max_speed
137        }
138        CALL  centering
139    }
```

Fig. 5.51 Adding appropriate calls of the "Centering" function to algorithm A

5.5.4 Programming Maneuvers for a Multi-crossing Line Track

In this section, the "LineTracer1.tsk/tskx" program took this robot on a path that had only straight line runs and 90 degree turns (left and right). Figure 5.53 showed this path (in red) and the corresponding function calls for the first seven maneuvers. Video 5.13 explained these programming aspects in more details (which were more complex and harder to explain when restricted to using "text-only" as in Sect. 5.5.3 – hopefully the reader agrees with the author here).

```
178    // adjust direction so that sensor 4 (middle) is on the black line

179    FUNCTION centering

180    {

181        CALL detect_black

182        IF ( black_4 == 0 )

183        {

184            IF ( black_2 > 0 || black_3 > 0 )

185            {

186                LOOP WHILE ( black_4 == 0 )

187                {

188                    ID[1]: Goal Velocity = CW:0 (0.00%) + l_wheel_max_speed

189                    ID[2]: Goal Velocity = CW:0 (0.00%) + r_wheel_max_speed

190                    CALL detect_black

191                }

192            }

193            ELSE IF ( black_5 > 0 || black_6 > 0 )

194            {

195                LOOP WHILE ( black_4 == 0 )

196                {

197                    ID[1]: Goal Velocity = CCW:0 (0.00%) + l_wheel_max_speed

198                    ID[2]: Goal Velocity = CCW:0 (0.00%) + r_wheel_max_speed

199                    CALL detect_black

200                }

201            }

202        }

203    }
```

Fig. 5.52 Function "Centering" code details

The "LineTracer2.tsk/tskx" illustrated how the robot could detect and handle diagonal and curve "nodes" – see Fig. 5.54 and Video 5.14 for more programming explanations.

ROBOTIS also provided a WMV video file of LineTracer2 program in action at this web site (http://www.robotis.com/video/BIO_STEM_LineFollower.wmv).

In "LineTracer3.tsk/tskx", two tasks/missions were added to the "LineTracer2" code using light and audio activations (Fig. 5.55).

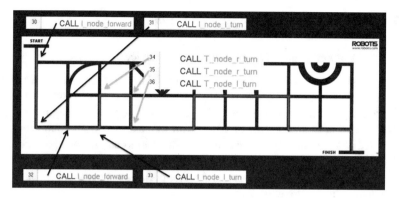

Fig. 5.53 Path (*red*) taken by robot via LineTracer1.tsk

Fig. 5.54 New maneuver
types for "LineTracer2.tsk"

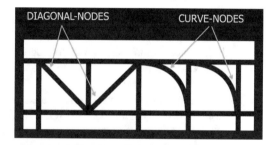

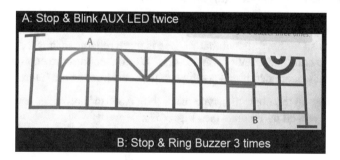

Fig. 5.55 Added missions for "LineTracer3.tsk/tskx"

5.6   Basic Remote Control Concepts

This project demonstrated basic concepts in wireless remote control (RC) programming of a CarBot and applied them into a mixed "human-control" and "autonomous-behavior" situation. We will also assess the performance of ROBOTIS ZigBee and BlueTooth communication devices and software tools such as TASK and the "R+SCRATCH/SCRATCH2" tool chain. We will have a "sneak peek" at the ROBOTIS SMART structure by using a cell phone's Tilt Sensor to achieve the standard U-D-L-R controls of a robot. More advanced communication programming topics such as bot-to-bot and multi-bot controls will be demonstrated in Chap. 7.

5.6.1   Remote Control Using TASK Tool

ROBOTIS supports quite a few communications devices:

- The RC-100 device is the main means to control ROBOTIS' robots, http://www.
  robotis-shop-en.com/?act=shop_en.goods_view&GS=1487&GC=GD080302)
  (currently at version B, see Fig. 5.56). Its default communication means is via
  NIR which is limited to line of sight, short range and low communication rate.
  The RC-100 can also be combined with ZIG-100 to use ZigBee protocols (any
  RC-100 version), or with BT-100 (versions A and B) or with BT-210 (version B
  only) to use BlueTooth protocols (http://support.robotis.com/en/product/auxde-
  vice/communication_main.htm).
- For wireless communications and controls from a Personal Computer, the user
  has two options:

  (a) If the user's PC has BlueTooth (BT) capabilities, then use BT-110 or BT-210
      on the robots, and just access the appropriate COM ports created by the PC's
      BT services upon pairing and connection. Using BT protocols would allow
      TASK to perform "program download" during editing time and "remote
      control" at run time.
  (b) If the user's PC has no BT capabilities, then the user needs to use a combina-
      tion of USB2DYNAMIXEL-ZIG2SERIAL modules and also ZIG-100 or
      BT-100 modules for the protocol needed. Option (b) is obviously more
      expensive than option (a), but the USB2DYNAMIXEL and ZIG2SERIAL
      can perform other functions such as direct access to actuators or changing
      the baud rates. BT is potentially faster than ZigBee, but my personal experi-
      ences so far have shown that ZigBee is more reliable than BT. Also the user
      has to consider whether "one-to-one" BT connections are enough or that he
      or she would need broadcast-type communications in the future.

Depending on the controller type used, the TASK tool provides up to six communications related parameters (see Fig. 5.57):

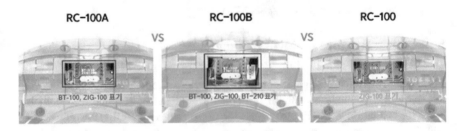

Fig. 5.56 Different versions of the RC-100

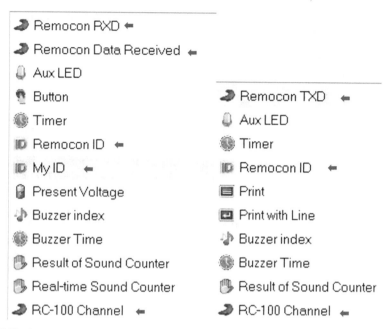

Fig. 5.57 Communication parameters available in TASK tool

1. "Remocon Data Received" is a logical flag which is set to TRUE when the controller received a new message, otherwise it is set to FALSE. This parameter is available on all controllers (CM-5, CM-50, CM-510, CM-530 and OpenCM-9.04).
2. "Remocon RXD" is a 16-bit parameter corresponding to the message received (available on all controllers).
3. "Remocon TXD" is used to send out a 16-bit message (available on all controllers). This procedure will be demonstrated in Chap. 8.
4. "My ID" is available on both TASK V.1 and V.2 tools for the CM-5XX controllers (i.e. Firmware 1.0) and corresponded to the ZigBee ID of either the ZIG-100 or ZIG-110 module used on the robot being programmed. It is not available on TASK V.2 for "Firmware 2.0" type of controllers, however the ZIG-100/110 modules would just work fine with them, but the user would have to use MANAGER V.1 and a "Firmware 1.0" controller such as the CM-530 to manage certain functions of the ZIG-100/110.
5. "Remocon ID" corresponds to the ZigBee ID for the "other" ZIG-100 or ZIG-110 modules (on the "other" robot or on the RC-100 being used).
6. "RC-100 Channel" only pertains to the NIR communication option.

More information can be found at the ROBOTIS e-Manual web site (http://support.robotis.com/en/software/roboplus/roboplus_task/programming/parameter/controller(roboplus_task).htm).

In this section, only the use of "Remocon Data Received", "Remocon RXD" and "Remocon TXD" would be demonstrated. "TestComm.tsk/tskx" was made for an

OpenCM-9.04-C using 4 XL-320s (ID = [1,2,3,4]). Figure 5.58 listed a code fragment to illustrate the basic algorithm:

- After initializing the XL-320s to wheel mode and setting their speeds to a value of 512 (CALL Initialize), the controller went into an endless loop whereas:

  (a) It waited for new data to come in, i.e. stayed at Statement 7 as long as "Remocon Data Received" was FALSE.

  (b) When this flag became TRUE, because new data just came in, it would save this data in parameter "Data" (Statement 8).

  (c) Then it would go through a procedure to determine the digit that corresponded to the "thousand" position in "Data" (i.e. parameter "Thousands" in Statement 10). If "Thousands" is positive, it would turn Motor 1 as many times as the actual value found in "Thousands". If not, it would flash on the LED of Motor 1 once for 0.5 s.

  (d) Then it would proceed similarly to determine and handle the digits corresponding to the "hundred" (Motor 2), "ten" (Motor 3) and "unit" (Motor 4) positions in "Data".

  (e) At the end of the endless loop, the robot would send back, via "Remocon TXD", to the PC the original "Data" that it received (Statement 75). The robot would also "Print Line" back to the PC this original "Data" (Statement

```
3       CALL Initialize
4
5       ENDLESS LOOP
6       {
7           WAIT WHILE ( ⮑Remocon Data Received   == FALSE )
8           Data = ⮑Remocon RXD
9
10          Thousands = Data / 1000
11          ▤Print = Thousands
12          ID = 1
13          IF ( Thousands > 0 )
14          {
15              LOOP FOR ( I = 1 ~ Thousands )
16              {
17                  CALL Motors
18              }
19          }
20          ELSE
21          {
22              CALL LED
23          }
```

Fig. 5.58 Algorithm to parse out the "Thousand" digit from "Data"

76). This was to show that the "Remocon TXD" command sent data in a different format, i.e. in a 6 byte packet that the "Program Output Monitor" window could not unpack and decode properly, thus it could only display "U" instead of the actual data. On the hand, because the "Print Line" command sent data in the clear, the "Program Output Monitor" could display the data properly (see Video 5.15).

Video 5.15 also explained more details for this program and showed how to use the "Zig2Serial Management" feature of the RoboPlus Manager V.1 tool to provide inputs from the PC to this program wirelessly via ZigBee (ZIG-100/110A) and BlueTooth (BT-110/210) devices.

The next three TASK programs were created to illustrate the application of RC concepts to a CarBot of a type as shown in Fig. 5.30:

1. "RC100_Carbot_1.tsk/tskx" demonstrated the use of the "&" (i.e. AND) operator to filter out unwanted user button pushes unless they were the Up-Down-Left-Right buttons (Statement 18). It also showed the use of the IF-ELSE-IF structure to enable the activation of one and only one condition, in other words allowed the use of only 1 button on the RC-100 at any one time.
2. "RC-100_Carbot_2.tsk/tskx" showed the use of parallel IFs so as to enable the user to activate several buttons at the same time, for example "Up" and "Left" together to perform a left turn with a wider arc instead of a pivot turn when "Left" is used alone.
3. "RC-100_Carbot_3.tsk/tskx" built on "RC-100_Carbot_2.tsk/tskx" and added a new functionality by allowing the user to change the robot speed "on the fly" using the buttons "1-2-3-4".

Video 5.16 had more details about these three TASK programs and showed how they performed at run time.

5.6.2   Remote Control Using the R+SCRATCH/SCRATCH2® Tool Chain

In late 2016, ROBOTIS started support for using the MIT's SCRATCH 2 Offline Editor with their products called PLAY 700 (CM-50) and IoT (OpenCM7.00) and the author believed that ROBOTIS would eventually expand this capability to the CM-150/200 in the future as they were designed for the younger children market.

Although SCRATCH 2 was designed for the younger beginner programmers, it had very advanced event-driven concepts (with codes running at the MS Windows OS level) that were worthwhile to be contrasted to the remote control concepts and coding approaches described in the previous Sect. 5.6.1 for the TASK tool which ultimately yielded machine codes that ran at the controller level of the test robot (i.e. two very different run-time environments).

In this section, we will be using the DOG robot from the PLAY 700 system as the test bed for illustrating the similarities and differences between the TASK and SCRATCH 2 approaches. Let's first start with a simple 1-button-only Up-Down-Left-Right control of the DOG robot using the TASK tool (see resulting code in Fig. 5.59 and the program "P700_RC_Basics1.tskx").

The program "P700_RC_Basics1.tskx" used a familiar endless loop whereas, at run-time, the button input from the Virtual RC-100 controller was parsed and acted upon, based on whether the pushed button was U, D, L or R, using another familiar IF-ELSE-IF structure. Please see Video 5.17 for the actual performance of this TASK solution which could not handle multiple U-D-L-R buttons being pushed.

7	ENDLESS LOOP
8	{
9	WAIT WHILE ([≫ Remocon Data Received] == FALSE (0))
10	Button = [≫ Remocon RXD]
11	IF (Button == [≫ --])
12	{
13	CALL Stop
14	}
15	IF (Button == [≫ U])
16	{
17	CALL Forward
18	}
19	IF (Button == [≫ D])
20	{
21	CALL Backward
22	}
23	IF (Button == [≫ L])
24	{
25	CALL LeftTurn
26	}
27	IF (Button == [≫ R])
28	{
29	CALL RightTurn
30	}
31	}

Fig. 5.59 TASK solution to a simple U-D-L-R control scheme of the DOG robot (P700_RC_Basics1.tskx)

Converting the "P700_RC_Basics1.tskx" program to an equivalent event-driven SCRATCH 2 code yielded two possible solutions: "P700_RC_Basics_1.sb2" and "P700_RC_Basics2.sb2". Both versions used keyboard event handling facilities provided via the Windows OS/SCRATCH 2 environment.

Figure 5.60 described Version 1 whereas the approach was to handle each keyboard event "completely" from when a particular arrow key was pushed until it was released. It used a SCRATCH 2 Event Block named "when ___ key pressed" for each of the arrow keys (Up, Down, Left, Right) found on a typical ASCII computer keyboard, thus Fig. 5.60 showed four independent set of blocks ("threads" if you will) corresponding to each of the U-D-L-R arrow keys.

After each "arrow key" Event Block, the next two blocks were actual R+SCRATCH Extension Blocks to "set" the geared motors on Port 1 and Port 2 to the appropriate turning direction (CW or CCW) and power level (500 or 0) to make the robot go in the direction(s) wanted by the user. These two blocks are "write" blocks in the sense that they are commands sent from the PC (i.e. from SCRATCH 2 IDE) to the robot for execution on its hardware. Via personal communications, ROBOTIS shared that each "write" block would take around 50 ms to execute due to its processing steps through the helper application R+SCRATCH (which was the software layer between SCRATCH 2 and the PLAY 700 controller – see Sect. 4.6). Due to these "delays", Video 5.17 showed that, for any "Go Forward" execution, Port 1 motor would "start" about a fraction of a second before Port 2 motor would start, thus creating a left jerk motion before the robot could drive forward evenly as normal. This delay-generated left jerk motion was non-existent (or at least not perceptible to the human eyes) for the TASK version "P700_RC_Basics1.tskx" as its compiled code ran directly on the CM-50 controller, i.e. "much closer" to the actual motors than the SCRATCH 2 commands (which had to be interpreted at the Windows OS level and then sent via R+SCRATCH to the robot). For the author, this was an interesting side effect to observe.

Next the WAIT block used a NOT logic to check whether the user had released the arrow key that had been pushed earlier. If this key was released, then the next two commands were "write" commands designed to turn off the two motors. This step was definitely necessary to prevent the robot from continuing to do what it was doing previously even though the user had released his or hers previously chosen arrow key! In other words, robotics programmers must remember to put the robot into a known state (usually a "stop" or "home" configuration) whenever the robot received no commands, for obvious safety reasons.

The second part of Video 5.17 showed how the Version 1 solution (P700_RC_Basics_1.sb2) actually behaved and interestingly this solution would also allow the user to push several buttons at the same time, but with "unexpected" results. For example, if the user pushed the "UP" and "RIGHT" arrow keys at the same time, then usually the user meant for the robot to make a wider turn to the right. However Video 5.17 showed that the right-turn maneuver dominated over the go-forward maneuver (for reasons not yet investigated thoroughly as a "debug" version of SCRATCH 2 was not available to the author).

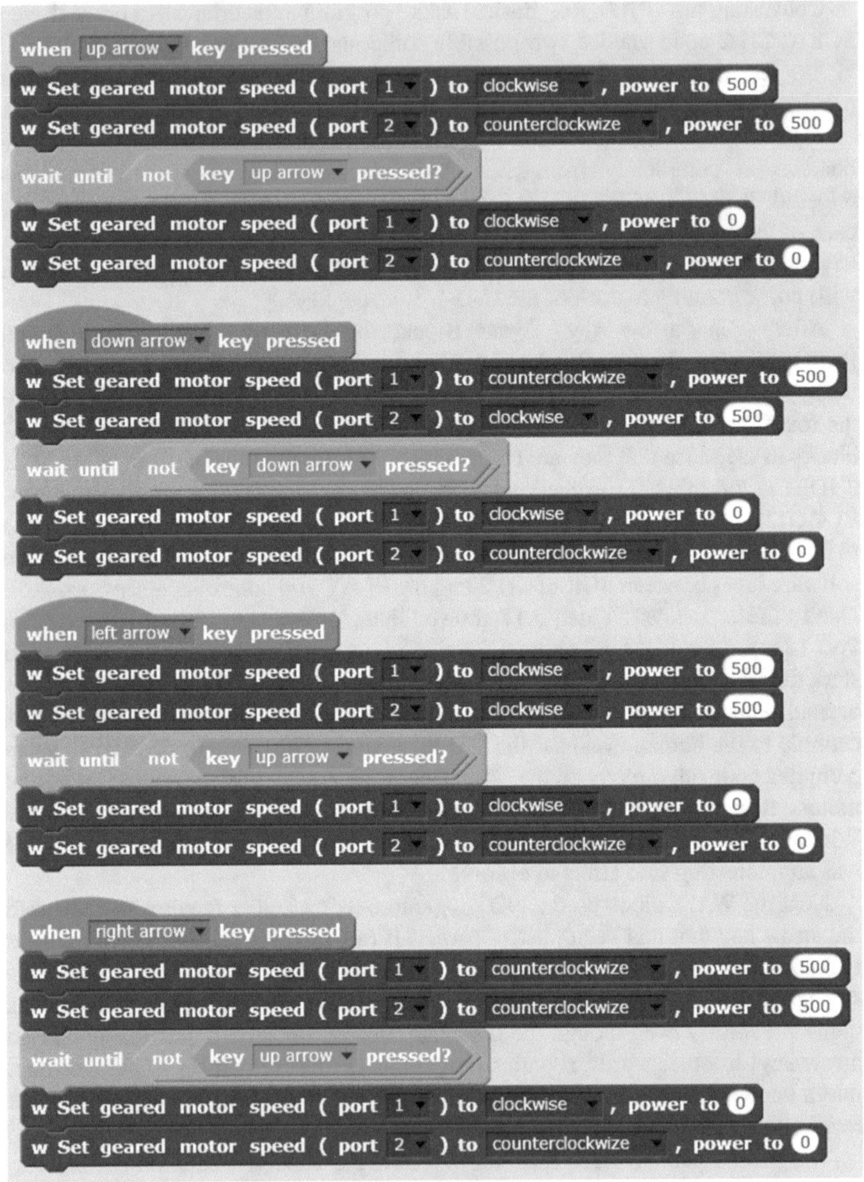

Fig. 5.60 SCRATCH 2 solution to a simple U-D-L-R control scheme of the DOG robot (Version 1)

The Version 2 solution (P700_RC_Basics_2.sb2) was shown in Fig. 5.61 where essentially the default "all-keys-released" state (i.e. "stop" state) was designed much more explicitly by its own "thread" using the FLAG Event Block in combination with a FOREVER loop (i.e. a sort of a "permanent" event!). Version 2 would

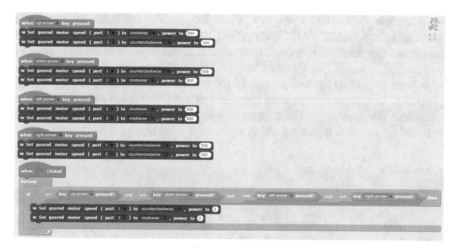

Fig. 5.61 SCRATCH 2 solution to a simple U-D-L-R control scheme of the DOG robot (Version 2)

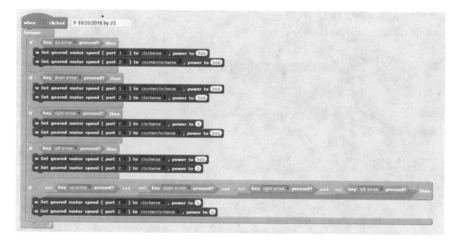

Fig. 5.62 SCRATCH 2 solution to a combo key control scheme of the DOG robot (Version 3)

probably be more computationally demanding on the PC but it did reduce the number of SCRATCH 2 blocks used by 30% (from 24 to 17). The third part of Video 5.17 did not show any outwardly better or worse performance for Version 2 as compared to Version 1.

The Version 3 solution (P700_RC_Basics_3.sb2), shown in Fig. 5.62, was successful in allowing the user to combine arrow keys (such as Up and Right for example – see Part 4 of Video 5.17). It used a FOREVER loop combined with four "parallel" IFs to allow each pushed arrow key to have a little bit of "action time" during each cycle of the FOREVER loop.

When this SCRATCH 2 program "P700_RC_Basics3.sb2" was converted to a seemingly "equivalent" TASK code "P700_RC_Basics3a.tskx" (see Fig. 5.63), it was found that this TASK code did not work at all when more than one arrow key was pushed by a user (see Part 5 of Video 5.17). In other words, although the arrow keys were pushed by the user in exactly the same way for the SB2 and the TSKX programs, why were the resulting behaviors of the DOG robot different for each situation? The reason for the difference was about how each type of programs received information regarding which arrow keys were being pushed by the user:

5	ENDLESS LOOP
6	{
7	WAIT WHILE (🔊 Remocon Data Received == FALSE (0))
8	Button = 🔊 Remocon RXD
9	
10	IF (Button == 🔊 U)
11	{
12	CALL Forward
13	}
14	ELSE IF (Button == 🔊 D)
15	{
16	CALL Backward
17	}
18	ELSE IF (Button == 🔊 L)
19	{
20	CALL LeftTurn
21	}
22	ELSE IF (Button == 🔊 R)
23	{
24	CALL RightTurn
25	}
26	ELSE IF (Button == 🔊 --)
27	{
28	CALL Stop
29	}
30	}

Fig. 5.63 "P700_RC_Basics3a.tskx" – "equivalent" TASK version of "P700_RC_Basics3.sb2"

1. In the case of the SCRATCH 2 code, via Windows OS it received "separate" information regarding which keys were being pushed, thus multiple "parallel" IF conditions could be "true" and therefore acted upon independently, but sequentially within a little bit of time slice due to the inherent "interpreting" and "communication" delays of SCRATCH 2 and R+SCRATCH (when its FOREVER loop was cycled through). So for example, when "UP" and "RIGHT" arrow keys were pushed, the robot was going forward for a few milliseconds and then right for a few milliseconds, therefore performing a wider right curve maneuver as wanted by the user.

2. In the case of the TASK code, ones had to realize that the RC-100 Remote Controller (both physical and virtual versions) first combined all the status (0 for not-pushed or 1 for pushed) for each of the ten "buttons" U-D-L-R-1-2-3-4-5-6 into a single 10-bit message which was then sent to the robot which had to decode it into its separate components first (and this was not done) before this information can be acted on properly. Please see the ROBOTIS e-Manual web link   (http://support.robotis.com/en/product/auxdevice/communication/rc100_ manual.htm) for more information.

Thus the program "P700_RC_Basics3b.tskx" (Fig. 5.64) was created to first parse the 10-bit message and then add about a 250 ms time delay for each "robot action", because the TASK code ran directly on the CM-50 controller so it did not have "interpretation/communication" delays to contend with as for the SCRATCH 2 environment, in order to attain similar "physical" results on the robot (see last part of Video 5.17). Please note that this was the same procedure used to separate the seven IR Sensors of the IRSA module in Sect. 5.5.3.

5.6.3   Remote Control Using Tilt Sensor from Smart Phone

In December 2016, ROBOTIS released the R+m.PLAY700 Smart Phone App for Android and iOS devices, originally intended only for users of the PLAY700 kit (i.e. controller CM-50). However this R+m.PLAY700 resource turned out to be applicable to all Firmware 2.0 controllers (see Fig. 5.65 – i.e. CM-150, CM-200 (not shown), OpenCM-9.04 and OpenCM-7.00).

This App offered many Multimedia features that would be described in depth in Chap. 11, but in this Sect. 5.6.3, the use of its tilt sensor would be showcased as a Remote Control device for the little CM-150 Avoider as shown in Fig. 5.65 at the bottom left corner.

The corresponding TASK code was named "RC_Tilt_Sensor.tskx" and a basic sensing and controlling algorithm was reported in Figs. 5.66, 5.67, 5.68, and 5.69. The phone's tilt sensor was used as a joy-stick controller, thus the nomenclature of the sensing parameters and their use in various conditional statements would read, at places, like a complete reversal of the logic used with the RC-100 controller:

```
9      ENDLESS LOOP
10     {
11         WAIT WHILE  (  [ 🖱 Remocon Data Received ]  ==  FALSE (0)  )
12         Data  =  [ 🖱 Remocon RXD ]
13         Up_button  =  Data  &  [ 🖱 U ]
14         Down_button  =  Data  &  [ 🖱 D ]
15         Left_button  =  Data  &  [ 🖱 L ]
16         Right_button  =  Data  &  [ 🖱 R ]
17
18         IF  (  Data  ==  [ 🖱 -- ]  )
19         {
20             CALL  Stop
21         }
22
23         IF  (  Up_button  >  0  )
24         {
25             CALL  Forward
26             CALL  Delay
27         }
28
29         IF  (  Down_button  >  0  )
30         {
31             CALL  Backward
32             CALL  Delay
33         }
```

Fig. 5.64 "P700_RC_Basics3b.tskx" – truly "equivalent" TASK version of "P700_RC_Basics3.
sb2"

1. Figure 5.66 showed a familiar Endless Loop containing the sensing and controlling algorithm which was started by fixing the phone into a landscape display
   mode (Statement 7). This action was necessary to prevent the phone from switching, unintentionally, to a portrait mode later during the actual demonstration phase
   when the phone was manipulated as a joy-stick, as this would confuse the user
   considerably (i.e. a Human-Machine-Interface issue more than a Robotics one).
2. The phone's reference plane is the horizontal plane. Thus all four "Gradients or
   Accelerations" parameters (in degrees) would be reported as 0 degree when the

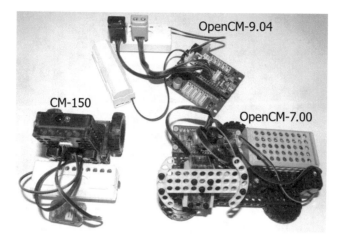

Fig. 5.65 Selected controllers compatible with the R+m.PLAY700 Smart App

5	ENDLESS LOOP
6	{
7	☐ SMART: 🔄 Screen Rotation = Landscape Mode (2)
8	Gradient_Up = ☐ SMART: 📐 Acceleration Sensor (U)
9	Gradient_Down = ☐ SMART: 📐 Acceleration Sensor (D)
10	Gradient_Left = ☐ SMART: 📐 Acceleration Sensor (L)
11	Gradient_Right = ☐ SMART: 📐 Acceleration Sensor (R)
12	☐ SMART: 🔢 Number Display = [Position:(3,2)],[Item:Gradient_Up],[Size:100],[Color:White]
13	☐ SMART: 🔢 Number Display = [Position:(2,3)],[Item:Gradient_Left],[Size:100],[Color:White]
14	☐ SMART: 🔢 Number Display = [Position:(4,3)],[Item:Gradient_Right],[Size:100],[Color:White]
15	☐ SMART: 🔢 Number Display = [Position:(3,4)],[Item:Gradient_Down],[Size:100],[Color:White]

Fig. 5.66 Part 1 of sensing and controlling algorithm for using tilt sensor

16	IF (Gradient_Up < 5 && Gradient_Down < 5 && Gradient_Left < 5 && Gradient_Right < 5)
17	{
18	CALL Stop
19	}

Fig. 5.67 Part 2 of sensing and controlling algorithm for using tilt sensor

phone was positioned flat in a horizontal manner (which was not easily achieved in practice in a user's hands). Furthermore the phone would only provide POSITIVE Gradient values when they were ABOVE the horizontal plane. For example, if the Front edge of the phone was above the horizon, the Gradient_Up parameter would show a positive value while the Gradient_Down parameter would show a "zero" value (although logically it should have been a "negative" value but equal in magnitude to Gradient_Up). Statements 8–11 recorded the

20	ELSE
21	{
22	IF (Gradient_Down >= 5)
23	{
24	CALL Forward
25	}
26	IF (Gradient_Up >= 5)
27	{
28	CALL Backward
29	}
30	IF (Gradient_Right >= 5)
31	{
32	CALL LeftTurn
33	}
34	IF (Gradient_Left >= 5)
35	{
36	CALL RightTurn
37	}
38	}

Fig. 5.68 Part 3 of sensing and controlling algorithm for using tilt sensor

U-D-L-R gradients and Statements 12–15 displayed them on the phone display in the standard cross configuration centered on the screen.

3. Next, the author wanted the robot to be still when the phone was held "mostly" in a horizontal plane thus the use of the limits of 5° in Statement 16 of Fig. 5.67.

4. Thus, when any of the Gradients U-D-L-R were equal or above 5°, the ELSE branch would be taken by the algorithm (Statement 20 of Fig. 5.68). Four parallel IFs constructs (Statements 22–37) were used in this ELSE branch as the author wanted the robot to be able to move in more than just one direction (U-D-L-R) at any one time to achieve smoother curved trajectories.

5. In Fig. 5.68, the various motion functions called were designed with a fixed Speed setting of 256 on the robot's motors. But the reader would be encouraged to also try out the "gradient-proportional speed" motion functions (with suffix "_P") which were also provided in the file "RC_Tilt_Sensor.tskx" (see Fig. 5.69 and Review Question 12).

42	FUNCTION Forward
43	{
44	⊙ PORT[1]:Geared Motor = CCW:0 (0.00%) + Speed
45	⊙ PORT[2]:Geared Motor = CW:0 (0.00%) + Speed
46	}
47	
48	FUNCTION Forward_P
49	{
50	Speed_P = Gradient_Down / 5
51	Speed_P = Speed_P * Speed
52	⊙ PORT[1]:Geared Motor = CCW:0 (0.00%) + Speed_P
53	⊙ PORT[2]:Geared Motor = CW:0 (0.00%) + Speed_P
54	}

Fig. 5.69 Part 4 of sensing and controlling algorithm for using tilt sensor

Video 5.18 showed the performance on this remote controlling approach on a CM-150 Avoider CarBot.

Recapping Chap. 5, although the "Sense-Think-Act" paradigm was key for problem-solving and software-generation in robotics, the side effects of communication characteristics should also be recognized as they could have important influences on the overall behaviors of the robot at run times.

5.7 Review Questions for Chap. 5

1. What is the name of the sensor module in the Bioloid robotics system?
2. What is the name of the servo motor module in the Bioloid robotics system?
3. How many UART ports does the Bioloid Controller CM-5 provide?
4. What are the two ways that we can use the AX-12+ module for?
5. How many symbols are used in a hexadecimal numbering system?
6. What is the numerical range on decimal values that a 10-bit number can represent?
7. What is the UART_1 communication port used for on the CM-5?
8. What are the name and default speed (how many bits per second) for the communication protocol between the CM-5 and the various servo motor and sensor modules?
9. Please name two possible control architectures that can be used in robotics.

10. Which tool of the RoboPlus Suite to use when programming several servo motors in concert?
11. What is the type of communication protocol used between the PC and the typical CM-5XX controller.
12. Draw up the diagram for the World-Computer Interactions.
13. In the TASK tool, what are the two types of commands that can be used to set the rotation direction and speed for the AX-12 module?
14. In the TASK tool, when should we use the "CUSTOM" option?
15. What is the actual time elapsed when the standard timer counter changed by three counts?
16. The Bioloid system uses a Full-Duplex communication protocol. (T-F)
17. Inside the TASK tool, which parameter should you set in order to change the range mode of the AX-S1, from long range to short range, and vice-versa?
18. Which parameter should we set in the Task tool to make the AX-S1 NIR sensors go into the Short Range detection mode? Parameter name and correct value to set?
19. What is the minimum time period between sound claps for the AX-S1 to discern them as distinct sound claps?
20. How long does it take before a sound count detected in the AX-S1 becomes accessible to a program running on the CM-5?
21. The NIR sensors on the AX-S1 modules can be set to be used in a passive mode. (T-F)
22. Which parameter should we set in the Task tool to make the AX-12+ modules go into the Wheel Mode? i.e. Address number and correct value to set?
23. What is the "duplex" type of the communication protocol used by the Bioloid system to communicate between different devices?
24. To control the accuracy of the Goal Position on an AX-12, we need to reduce its SLOPE value to the minimum value of 1. (T-F)
25. What is the purpose of an IF-ELSE-IF structure?
26. What is the purpose of a parallel IFs structure?
27. Which communication protocol does the Virtual RC-100 Controller use to communicate with the CM-5XX controller?
28. Describe procedure to ignore all inputs from the RC-100 except for buttons "2" and "3".

5.8   Review Exercises for Chap. 5

1. To practice setting an AX-12 into Continuous Turn mode and using basic timed operations for it with the LOAD or COMPUTE commands, please use the enclosed programs "AX12ContinuousTurn_1.tsk" and "AX12ContinuousTurn_2.tsk".
2. To practice reading from the DMS and NIR sensors, please use the enclosed program "DMS-NIR_sensing.tsk".

3. To practice using the AX-S1 as a light sensor, please use the enclosed program "AX-S1-LightSensing.tsk".

4. To practice using the AX-S1 as a NIR sensor in a direct mode, please use the enclosed program "AX-S1-IR-SensingDirect.tsk". This program also shows how to get the AX-S1 into its long-range and short-range mode.

5. To practice using the AX-S1 as a NIR sensor using its Obstacle Detected Flag and its related Threshold setting, please use the enclosed program "AX-S1-IR-SensingFlag.tsk". Please modify this program to make it work similarly but in its Light Sensing mode.

6. The enclosed program "CM5Music.tsk" demonstrates how to get the AX-S1 to play music using a CM-5. Please modify this program to make it work on a CM-530 controller.

7. Expand enclosed "Sequence Commander" code so that the bot can handle up to 8 maneuvers in a given sequence.

8. Modify "Sequence Commander" code so that each left and right turn corresponds to a 90-degree turn. Does your solution work for all running surfaces? And at all battery levels? i.e. these "timed" maneuvers have their limitations and more flexible "sensor-based" approaches are used in the "Line Tracer" project in Sect. 5.5.

9. Please start from the sample program "RC100_Carbot_2.tsk" and implement the following features:

   (a) If no button is pushed on the RC-100, the carbot should stop.
   (b) Buttons U-D-L-R are still used in the usual manner.
   (c) Buttons 1-2-3-4 will be used to adjust the nominal speed level of the carbot:

      • Button 1 will set the speed to 128 (initial default speed).
      • Button 2 will set the speed to 256.
      • Button 3 will set the speed to 512.
      • Button 4 will set the speed to 1023.
      • The user should have to push any of the buttons (1 through 4) only ONCE to set the "new" speed (i.e. the user SHOULD NOT have to keep pushing down on the "speed" buttons to affect a speed change).
      • Please make sure that the user can change the speed setting (1,2,3,4) while performing a car maneuver (U,D,L,R) AT THE SAME TIMER.
      • Remember that you can use the Virtual RC-100 (i.e. using Mouse click or keyboard from inside the Task Tool – the View Print of Program sub-window to be exact) to test out your solution before using the Physical RC-100.

10. Practice in combining Remote Control and Autonomous Behavior programming in one application:

      • Simulation of a tail-gating situation between two carbots (both under remote control by the RC-100's), when the front carbot suddenly stops (an equivalent static frontal obstacle could also be used).

- Rear carbot using its NIR sensors will trigger an autonomous response (i.e. ignoring commands from RC-100) to help it avoid colliding into front car).
- The Spring 2013 solutions from the students at National Taiwan University can be viewed at this link https://www.youtube.com/playlist?list= PLVHBjRDK0kAJA_GC3UMreuKVeWnuASw7P&feature=view_all.

11. Another possible variation for Algorithm A in Sect. 5.5.3 was to use "hard" maneuvers no matter what the condition was (Conditions 1 through 5), i.e. set the speed of the inner turn wheel to ZERO. How do you like the performance of the CarBot then?
12. In Sect. 5.6.3, the reader was encouraged to try out the "gradient-proportional" speed setting functions also. What were the results on the robot motions? Did we need the big IF of Statement 16 any more?

References

Arkin RC (1998) Behavior-based robotics. The MIT Press, Cambridge

Campion G, Chung W (2008) Wheeled robots. In: Siciliano B, Khatib O (eds) Springer handbook of robotics. Springer, Heidelberg, pp 391–410

Cook G (2011) Mobile robots. Wiley, Hoboken

Dudek G, Jenkin M (2010) Computational principles of mobile robotics. Cambridge University Press, Cambridge

Ford JL Jr (2014) Scratch 2.0 programming for teens. CENGAGE Learning PTR, Boston

Matarić MJ (2007) The robotics primer. The MIT Press, Cambridge

Nourbakhsh IR (2013) Robot futures. The MIT Press, Cambridge

Warner TL (2015) Teach yourself Scratch 2.0 in 24 hours. SAMS, Indianapolis

Winfield A (2012) Robotics. Oxford University Press, Oxford

Chapter 6
Actuator Position Control Applications

In Chap. 3, the hardware characteristics for representative actuators such as the AX-12 and MX-28 were described and the key information was on the rotational encoders that were used for the AX-12/18 (a variable potentiometer, restricted to a 300° range) and for the MX-28/64 (a magnetic field detector allowing a full 360° range). In this chapter, the programming aspects for controlling such actuators via the TASK and MOTION EDITOR (V.1) tools are described using concepts/parameters such as Present Position, Goal Position, Goal Speed, Torque Limit, Present Load, Motion Page and Joint Offset. The capabilities of the newer R+MOTION (V.2) tool will be discussed in Chap. 10 using the DARWIN-MINI and the XL-320.

This chapter's main topics are listed below:

- How does an actuator reach and maintain a given position via TASK?
- Concept of Motion Page in MOTION EDITOR (V.1) and its integration with the TASK tool using two applications: a robotic arm and a biped robot.
- Walking robots – Forms and Functions: BUGFIGHTER, HEXAPOD, DROID, GERWALK, BIPED and HUMANOID.
- Remote Control of Rover Robot with Arm.
- Interactions between various actuator parameters: Goal Position, Slopes, Margins, Punch, Present Position, Present Load, Torque Limit and Join Offset.
- Use of "CALLBACK" function inside the Task tool.
- Illustrations of above concepts to a "load-sensing" gripper.

For more advanced topics on legged and humanoid robots and also on mobile manipulators, the reader is referred to Kajita and Espiau (2008), Kemp et al. (2008), Chevallereau et al. (2009), Abdel-Malek and Arora (2013), Kajita et al. (2014), Cook (2011), Li and Ge (2013), to Jazar (2010), Niku (2011) and Burdet et al. (2013).

© Springer International Publishing Switzerland 2017
C.N. Thai, *Exploring Robotics with ROBOTIS Systems*,
DOI 10.1007/978-3-319-59831-4_6

6.1   AX-12/18 Position Control with TASK

In Position Control mode, the AX-12/18 was restricted to a range of 300°, with 0°
(i.e. Goal Position = **0**) located in the lower right corner of the actuator front view
(see Fig. 6.1), and 300° (i.e. Goal Position = **1023**, i.e. a 10 bit parameter) located
in its lower left corner (Fig. 6.1).

First, ones must take care in setting all actuators to the Position Control mode
before programming them in TASK for this mode. This action could be performed
inside the RoboPlus Manager tool by setting the "CCW Angle Limit" parameter to
a value of 1023 or by using a LOAD command inside a TSK/TSKX program with a
CUSTOM WRITE of a WORD-sized value of 1023 into Address 8 of appropriate
Dynamixel IDs (preferred by author, see Fig. 6.2).

Fig. 6.1 Front and back
views of AX-12

Fig. 6.2 TASK function to
initialize AX-12s to
position control mode

```
FUNCTION Initialize_motor
  {
      // Set_Position Control mode to all servos
      ID[1]: ADDR[8(w)]   = 1023
      ID[2]: ADDR[8(w)]   = 1023
      ID[3]: ADDR[8(w)]   = 1023
      ID[4]: ADDR[8(w)]   = 1023
      ID[5]: ADDR[8(w)]   = 1023
      ID[6]: ADDR[8(w)]   = 1023
      ID[7]: ADDR[8(w)]   = 1023
      RETURN
  }
```

Fig. 6.3 Example of a parameter naming error in TASK tool

The Control Table for all operational parameters for the AX-12 (as an example) could be accessed at this link http://support.robotis.com/en/product/actuator/dyna-mixel/ax_series/dxl_ax_actuator.htm. Please note that most parameters had Read/Write properties, but some were set to Read-Only. The user also should be aware that the **name** for a particular parameter as shown in this **Control Table** may not match with the **name** for the same parameter as shown in the **TASK** program pop-up window, so it is highly recommended for users to rely on the **address** of such parameter instead. For example, Address 34 in the Control Table was named "Torque Limit" (allowable range [0–1023], but the same parameter in TASK tool was named "Goal Torque" (see Fig. 6.3). **This book would only use the parameter names as provided in the Control Table web links**.

Figure 6.3 also listed often-used parameters:

1. "Goal Position" [0–1023] stood for the location where the user wanted the actuator to go next.
2. The CW/CCW "Margin" and "Slope" parameters worked in concert to provide the "mechanical behavior" of the actuator when it traveled from a "Present Position" to the already set "Goal Position" (more details later in this chapter).
3. "Torque Limit" [0–1023] denoted the maximum load (electrical current) allowed on the actuator. In practical terms, when the "Torque Limit" was set to 0 for any actuator, the user could rotate such actuator by hand (as no electrical current was allowed into the motor, thus no resisting torque to maintain a given "Goal Position"). This effect would be demonstrated in Chap. 7 for the implementation of a Bilateral Control of two GERWALKS robots. "Torque Limit" worked in concert with "Moving Velocity" to provide the actual (final) rotational speed of the actuator.

4. "Moving Velocity" should be considered as a relative setting [0–1023], whereas 0 meant to stop the actuator, and 1023 meant the maximum speed allowable by the current setting of "Torque Limit". For example, a setting combination of "Torque Limit" (= 512) and "Moving Velocity" (= 1023) would only result in half the rotational speed that the actuator was capable of doing physically. Users also would not need to set the rotation directions – CW or CCW when actuators were in Position Control mode (in contrast to when they were in wheel mode for Chap. 5).

For actuators, there was another important parameter called "Punch" that users could adjust by using a CUSTOM WRITE of a WORD-sized value (default = 32) into Address 48 of appropriate Dynamixel IDs. It corresponded to the minimum electrical current used to drive the motor from a full-stop, i.e. to overcome static friction and inertia (parameter E in Fig. 6.5).

Another issue to be aware of when using TASK was that just because a parameter showed up in the pop-up menu did not mean that it really existed and was operational for the actuator being used for the user's particular robot. In Fig. 6.4, the

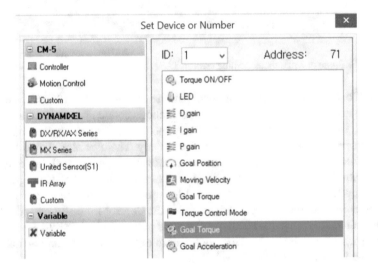

Fig. 6.4 "Goal Torque" (address 71) operational only for MX-64 and not for MX-28

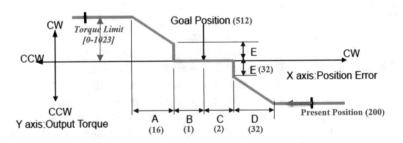

Fig. 6.5 Output torque vs. actuator position far and near a goal position

"Goal Torque" parameter (address 71) was really operational for the MX-64 actuator only and did not exist at all on the Control Table of the MX-28, please see links to both Control Tables below:

• http://support.robotis.com/en/product/actuator/dynamixel/mx_series/mx-28at_ar.htm.
• http://support.robotis.com/en/product/actuator/dynamixel/mx_series/mx-64at_ar.htm.

Figure 6.5 is key to the understanding of how a typical AX actuator got controlled by its own embedded controller (Atmel AVR – 8 KB – 16 MHz) to go from a Present Position (let's say 200) to a Goal Position of 512. Furthermore let's assume that Parameter D (CCW Slope) is set to 32, Parameter C (CCW Margin) set to 2, Torque Limit (TL) set to 1023, Moving Velocity (MV) set to 1023, and E (Punch) set to 32:

1. As soon as this LOAD command **ID[1]: Goal Position** = 512 was issued to Actuator[1], its embedded controller would move the horn CCW towards the Goal Position of 512 at the highest output torque/speed possible (because TL=1023 and MV=1023) while monitoring its "Present Position" (PP).
2. When its PP reached the value of 478 (=512-2-32), the controller would start to reduce its output torque "linearly" so that it could reach an output torque of 32 (=E) by the time the actuator's PP reached its value of 510 (=512-2).
3. When the PP reached 510, the controller would cut off power to the motor to let it "coast" towards 512. If from inertial effects, the PP overshot to a value higher than 513 (=512+1), the controller would generate a CW output torque to bring the horn back towards 512. Please note that the user did not have to set the CW/CCW Margin and Slope values in a symmetrical manner as the plot in Fig. 6.5 would have suggested. In the previous discussion, B was set to 1, and A set to 16, so the constraints could be set to be tighter for the CW approach to 512 than for the CCW approach to 512, or vice-versa, as needed.
4. This cycle of monitoring the PP, comparing it to the GP and turning the motor CW or CCW as needed, was repeated every 7.8 ms by the embedded controller. This hardware/software refresh cycle is fundamental to the working of all ROBOTIS actuators, and Chap. 10 would show how this timing was used to generate motion frames inside the R+Motion tool (V.2) to enable its synchronous mode.
5. Lastly, this mode of controlling the AX actuators could be classified as Proportional Control. The control scheme for other actuators such as the MX-28 or the XL-320 would be of the Proportional-Integral-Derivative (PID) type (see Sect. 6.2).

Video 6.1 demonstrated more details for these characteristics using the following example TASK programs on a CM-5/510/530:

1. AX-12-Position.tsk/tskx
2. AX-12-MoveNoHold.tsk/tskx
3. AX-12-MoveMargin.tsk/tskx
4. AX-12-MoveSmooth.tsk/tskx
5. AX-12MonitorPositionSpeed.tsk/tskx

6.2   Using MOTION and TASK Tools (V.1)

Currently, ROBOTIS is supporting two motion-editing tools, one bundled with the RoboPlus suite (Motion V.1) and the other, self-standing, called R+Motion (currently at V.2.4.1). Motion V.1 had been around since 2009 and is currently at v1.0.29.0. Motion V.1 still supports some CM-5 based robots (BIOLOID COMPREHENSIVE kit), so it may still be around for a few years but its days are numbered. R+Motion V.2 only supports the BIOLOID PREMIUM kit and later kits, such as STEM, SMART and ROBOTIS-MINI and this tool will be discussed in more details in Chap. 10.

As pointed out in Sect. 4.4, conceptually these Motion tools were based on cartoons (or frames) animation which was essentially the re-playing of a sequence of poses of the considered character (robot) within a time line, with slight variations in each pose to create the illusion (perception) of motion. The key idea is to use MOTION tools to create motion **data sets** (saved as *.MTN/MTNX files) while using the TASK tool to create the **logical flow** among these motion data sets (saved as *.TSK/TSKX files). In other words, the two tools, MOTION and TASK, have to work together in order to create robot moves that are both mechanically realizable and logically coherent.

The MOTION V.1 tool is quite a sophisticated tool and the user can purchase a copy of the User's Guide at http://www.robotis-shop-en.com/?act=shop_en.goods_view&GS=1486&GC=GD080400. This manual is highly recommended as it provides extensive procedures on how to use specific features of the tool which won't be repeated in this book. Instead, this book will strive to provide an integrative description of key concepts and practical demonstrations of selected procedures via video recordings.

6.2.1   Characteristics of a Motion Page in Motion V.1

Motion V.1 was designed to operate with an actual robot connected via a serial communication port in real-time to the PC running this tool. Figure 6.6 showed the complete window for Motion V.1.0.29.0 when started up on a PC. It contained two main panels:

1. The right panel was designed to do pose editing, i.e. to interact directly with the robot by turning selected actuators on and off, and thus enabling the setting of the robot into various "poses" which were then saved as sequential "steps" on the left panel.
2. The left panel was designed to allow the grouping of these "sequential steps" into "motion pages" which could be left self-standing or further linked together for more extensive robot moves.

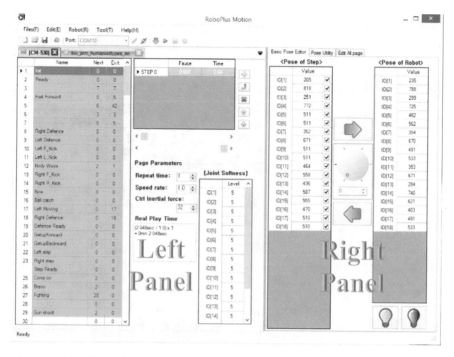

Fig. 6.6 Motion V. 1 tool (full view)

Conceptually, the user would need to consider the Motion Pages, located on the left edge of the Left Panel, as "macro" lines of code, and as such, they could be copied and pasted somewhere else for further editing. They could be renamed as needed, or named if a blank page was created.

Each Motion Page could have up to seven steps (i.e. robot poses) labeled from Step 0 to Step 6. A **Pose** was essentially a data set of "n" actuator positions [0–1023] where "n" was the number of actuators used for generating robot moves (thus not all actuators making up the robot needed to be listed here). A **Step** was then a Pose along with a play **Time** period (0.072–2.04 s) allocated to the robot for it to reach this Pose, from whatever its actuator positions were at the start of this Step. A **Pause** time (0.000–2.04 s) was optional and specifies the time period for the controller to wait out (doing nothing) before it can start on the next Step.

Each Motion Page could be connected to another Motion Page by specifying that Page Number in the "**Next**" column (see Left Panel), i.e. after all the steps in the current Motion Page were executed, the controller would play the "Next" Page Number as specified in its own "Next" cell value. A "0" in this "Next" cell meant for the controller to stay at the same Motion Page when the last Step of the current Motion Page was finished. For example, as shown in Fig. 6.6, let's trace the route that the robot would take if Motion Page 4 (Fast Forward) was initiated. After Page 4 was done, it would go to Page 5, then Page 6, then Page 3, then Page 7, then back to Page 5, i.e. the robot would "Fast Forward" continuously until its power is turned off or the controller gets reset.

Fig. 6.7 Exit page 42, eventually ending at page 43

42	F_E_R		43	43
▶ 43			0	0

Fig. 6.8 Page parameters that affect the final execution of motion pages

Page Parameters

Repeat time: 1

Speed rate: 1,0

Ctrl Inertial force:
32

Real Play Time

(3.984sec / 1.0) x 1
= 0min 3.984sec

[Joint Softness]

	Level
ID[1]	5
ID[2]	5
ID[3]	5
ID[4]	5
ID[5]	5
ID[6]	5
ID[7]	5
ID[8]	5
ID[9]	5
ID[10]	5
ID[11]	5
ID[12]	5
ID[13]	5
ID[14]	5

It was also possible that a command to stop a Motion Page needed to be executed while the robot was still going through a particular Motion Page, and as a consequence of this command, the robot could be put into an unstable situation/pose. To alleviate this problem, the user could specify a defined Motion Page as an "**Exit**" Page where the robot could go to, after an emergency stop command had been issued. For example, if the robot was doing Page 5 when it was ordered to stop, it would "exit" to Page 42 which would lead to Page 43 for a final stop, setting the robot into a ready position (Fig. 6.7).

Found in the middle of the Left Panel, the "Page Parameters" section (Fig. 6.8) could be used to specify four parameters for each Motion Page, independently of each other:

- "Repeat Time" and "Speed Rate" were rather self-explanatory parameters (see the formula used in the Real Play Time box).
- The "Control Inertial Force" (CIF) parameter (0–127 range, default value = 32) represented the level of acceleration and deceleration used to move the robot **from Pose to Pose**. It was inversely proportional to the actual acceleration/deceleration level. Thus a small CIF translated to higher speed increases from the starting Pose and also higher speed decreases when near the ending Pose (like for

a "punch"), so outwardly the robot would move fast but might shake and get unstable. Conversely, a large CIF would be equivalent to a "caress" then.

- The "Joint Softness" (JS) parameter (1–7 range, default value is 5) was related to the compliance (Margin and Slope) settings for the actuator, so it applied only when near the starting and ending Poses of a Motion Step. When JS was large, resulting movements would be smooth but would not be suitable for legs that needed much support. So the robot could collapse at the Ready Pose already. A small JS would provide steadiness of motion but resulting movements might look too rigid.

Motion V.1 also offered a Calibration tool which could be used to define the individual actuator's MOTION OFFSET values that were particular to a given robot, as no robot could really be built exactly (in a mechanical sense) like another of the same model. However, we would expect them to be able to use the same Motion files and Task programs (for more details visit web link http://support. robotis.com/en/software/roboplus/roboplus_motion/etc/roboplus_motion_offset. htm). This particular way of using actuator offset values could be considered as "static" and it would be demonstrated in Sects. 6.2.2 and 6.2.3 respectively for a robotic arm and a Gerwalk bipedal robot. A more "dynamic" application of these offsets values would be demonstrated in Sect. 6.6 for a "load sensing" gripper.

6.2.2 Application to a PhantomX REACTOR Robotic Arm

The "barebones" kit for this robotic arm could be purchased from Trossen Robotics (http://www.trossenrobotics.com/p/phantomx-ax-12-reactor-robot-arm.aspx)   and the reader could use the existing controller (CM-5/510/530), AX-12As and sensors from a BIOLOID COMPREHENSIVE or PREMIUM kit to construct and program this robotic arm.

The PhantomX REACTOR arm used in this section had a wrist and a parallel gripper with a total of eight AX-12As, along with added IR sensors (see Fig. 6.9). Actuator 1 was set as the panning servo (beneath pan/slew-bearing plates). Actuators 2 and 3 formed the shoulder joint, while Actuators 4 and 5 formed the elbow joint. Actuator 6 took care of the Up-Down motion of the wrist, while Actuator 7 took care of the wrist rotation function. Actuator 8 was used to open and close the parallel gripper. One IR sensor (Port 4) was installed at the gripper and aimed forward to detect frontal objects. Two other IR sensors were positioned at the elbow joint and were aimed left (Port 1) and right (Port 2) respectively.

TASK and MOTION V.1 would be used in this section (but converted TSKX and MTNX files were also included in the Springer's Extra Material files for the reader's reference). Video 6.2 showed the procedure for calibrating the AX-12As using the REACTOR_Arm.mtn/mtnx file and its Motion Page No. 1 named "Calibrate". The reader would learn how to adjust the MOTION OFFSET values of individual actuators so as to correspond to a given Calibration Pose from inside the Motion Editor

Fig. 6.9 PhantomX
REACTOR robotic arm
with added IR sensors

tool. The normal range of these "static" offset values were between −255 and 255, and in practice they would be usually less than 20 in magnitude, unless serious robot building errors had occurred!

6.2.2.1 Joint Offsets and Exit Pages

The first REACTOR arm project was about how to use a TASK function named JOINT OFFSET to isolate a specific actuator from the Motion Pages' control, and also how to use the TASK's function named MOTION INDEX NUMBER (also known as MOTION PAGE on the ROBOTIS e-Manual web site) for emergency exits from a Motion Page still in progress (see http://support.robotis.com/en/software/roboplus/roboplus_task/programming/parameter/motion/roboplus_task_jointoffset.htm and http://support.robotis.com/en/software/roboplus/roboplus_task/programming/parameter/motion/roboplus_task_motionpage.htm).

This project used the files REACTOR_Moves.tsk/tskx and REACTOR_Arm.mtn/mtnx, specifically the Motion Pages No. 2 (Init) and No. 3 (Move1). Figure 6.10 showed Part 1 of the algorithm:

- Statement 6 set the Gripper's JOINT OFFSET to 1024 which effectively isolated this Dynamixel from the effects of all Motion Pages where it may be included when the user created the "REACTOR_Arm.mtn/mtnx" file previously.

```
3    START PROGRAM
4    {
5        // Isolate Servo 8 from Motion Page being played
6        ⚙ ID[8]:Joint Offset = 1024
7
8    Start :
9        // Get into Init Pose
10       ▶ Motion Index Number = 2
11       🔊 Buzzer Time = Play Melody
12       ♪ Buzzer index = Melody3
13       WAIT WHILE ( ▶ Motion Status == TRUE )
14       WAIT WHILE ( 🔊 Buzzer Time > 0.000sec )
15
16       // Endless loop for RC input from user
17       ENDLESS LOOP
18       {
19           IF ( ▶ Motion Index Number == 2 )
20           {
21               ▶ Motion Index Number = 3
22           }
23           ELSE
24           {
25               ▶ Motion Index Number = 2
26           }
```

Fig. 6.10 Part 1 of "REACTOR_Moves.tsk/tskx"

This setting also allowed "direct" GOAL POSITION commands to be sent to Dynamixel 8, at a later time in the algorithm (and while a motion page was being played).

- Statement 10–14 commanded the robot arm to play Motion Page 2 and its CM-510/530 Controller to play Melody 3 at the same time.
- Statement 17 started the main loop, and Statements 19–26 commanded the arm to keep on cycling between Motion Pages 2 and 3 at all times (even if the user did intervene via the RC-100's buttons in Part 2 of the algorithm).

Figure 6.11 showed Part 2 of the algorithm whereas:

- At any time after the cycling of Motion Pages 2 and 3, if the user had pushed Button "U" on the RC-100 Remote Controller, the Gripper would go to Goal Position 512 and thus opened the Gripper (Statement 30). Otherwise if Button "D" was pushed, Dynamixel 8 would go to its Goal Position of 172, i.e. closed the Gripper (Statement 37).
- However if the user had pushed Button "1", the robot would execute Motion Page = 0 (Statement 41), i.e. stopped all eight Dynamixels wherever they happened to be at this point in time. It next played Melody 4 (Statements 42–44) and then exited the Endless Loop (Statement 45) to go to Part 3 of the algorithm.

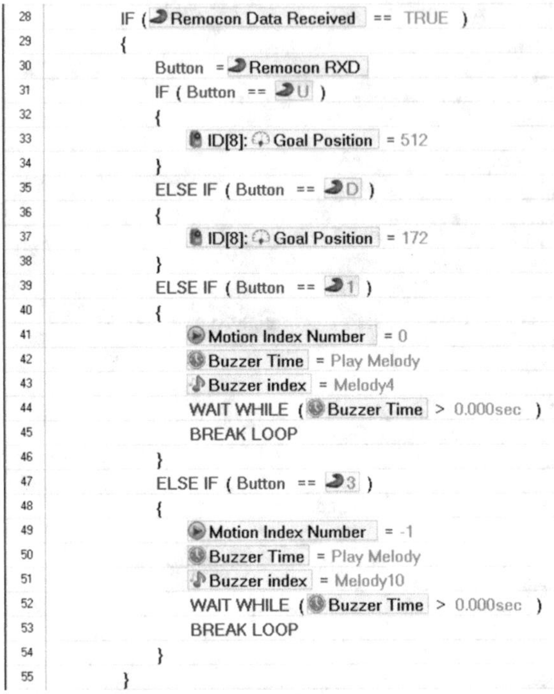

```
28          IF ( ➲Remocon Data Received  == TRUE )
29          {
30              Button = ➲Remocon RXD
31              IF ( Button == ➲U )
32              {
33                  🔋ID[8]: ⊕Goal Position  = 512
34              }
35              ELSE IF ( Button == ➲D )
36              {
37                  🔋ID[8]: ⊕Goal Position  = 172
38              }
39              ELSE IF ( Button == ➲1 )
40              {
41                  ▶Motion Index Number  = 0
42                  🔊Buzzer Time = Play Melody
43                  ♪Buzzer index = Melody4
44                  WAIT WHILE ( 🔊Buzzer Time > 0.000sec )
45                  BREAK LOOP
46              }
47              ELSE IF ( Button == ➲3 )
48              {
49                  ▶Motion Index Number  = -1
50                  🔊Buzzer Time = Play Melody
51                  ♪Buzzer index = Melody10
52                  WAIT WHILE ( 🔊Buzzer Time > 0.000sec )
53                  BREAK LOOP
54              }
55          }
```

Fig. 6.11 Part 2 of "REACTOR_Moves.tsk/tskx"

- Or if the user had pushed Button "3" instead, the robot would execute Motion Page $=-1$ (Statement 49), i.e. the robot arm would finish its current Motion Page which could be either Motion Page 2 or 3 and then stopped all Dynamixels at the end of that page (2 or 3). It next played Melody 10 (Statements 50–52) and then exited the Endless Loop (Statement 53) to go to Part 3 of the algorithm.

Figure 6.12 showed Part 3 of the algorithm whereas:

- Statement 59 cleared the RxD buffer of the "0" value that was sent by the RC-100, when the user released any button at the end of Part 2 of the algorithm. Then the robot just waited for a new command from the user (Statement 60).
- If the new command was the Button "R", then the controller played Melody 16 and then jumped to START (i.e. Statement 8) to restart the algorithm all over again. Else it would set the Joint Offset of Dynamixel 8 back to 0 and then exited the TASK program completely.

The reader should recognize that MOTION OFFSET (used in MOTION tools) and JOINT OFFSET (used in TASK tools) referred to the same parameter inside the firmware of typical ROBOTIS controllers and actuators. Similarly, MOTION PAGE (used in MOTION V.1 tool) and MOTION INDEX NUMBER (used in MOTION V.2 tool) pointed to the same parameter in ROBOTIS controller firmware. Video 6.3 demonstrated the performance of this first REACTOR project.

```
58   // To clean out of the RxD buffer the "0" sent when a button is released
59   Button  = ⤵ Remocon RXD
60   WAIT WHILE ( ⤵ Remocon Data Received  == FALSE )
61
62   Button  = ⤵ Remocon RXD
63   IF ( Button  == ⤵ R )
64   {
65        🔵 Buzzer Time  = Play Melody
66        ♪ Buzzer index  = Melody16
67        WAIT WHILE ( 🔵 Buzzer Time > 0.000sec )
68        JUMP Start
69   }
70
71   // Reset Joint Offset to Servo 8
72   🔵 ID[8]:Joint Offset  = 0
```

Fig. 6.12 Part 3 of "REACTOR_Moves.tsk/tskx"

6.2.2.2   Avoider Arm

The second REACTOR project showed how to use IR sensors to help the arm avoid obstacles appropriately while performing prescribed motion pages. It was another application of the TASK's JOINT OFFSET function but this time using the normal range of values from −255 to 255.

This project used the files REACTOR_Arm.mtn/mtnx and CM530-AvoiderArm_1.tsk/tskx (file CM5-AvoiderArm_1.tsk also enclosed for users of CM-5s). Figure 6.13 showed Part 1 of the algorithm used:

- Statements 7–14 initialized the parameters used:

   - The three "Steps" (statements 7–9) were discrete change values to be applied to the respective joints "Shoulder", "Elbow" and "Wrist" whenever there was a need to modify their current motions to avoid an obstacle.
   - Statements 10–13 initialized the four Joint-Offset parameters that would need to be computed and set to the appropriate Dynamixels so as to override the current Motion Page being played whenever an obstacle got within the alarm ranges as monitored by the three IR sensors (see Fig. 6.9).
   - "StopServo" was a flag to be set when an obstacle was detected and it was used to stop/restart Servo 1 (i.e. the panning actuator).

- Statement 18 set the arm into its ready pose (i.e. Motion Page 4) and Melody 2 was then played (Statements 19–21). Finally Motion Page 5 was executed (Statement 23).

The main algorithm was set inside an Endless Loop starting at Statement 26 (see Fig. 6.14 for Part 2):

Fig. 6.13 Part 1 of
"CM530-AvoiderArm_1.
tsk/tskx"

```
4    START PROGRAM
5    {
6        // Initialize parameters
7        ShoulderStep  = 10
8        ElbowStep  = 5
9        WristStep  = 5
10       ShoulderFB  = 0
11       ShoulderLR  = 0
12       ElbowUD  = 0
13       WristUD  = 0
14       StopServo  = FALSE
15
16   Start  :
17       // Get into Ready Pose
18       ▶ Motion Index Number  = 4
19       🔊 Buzzer Time  = Play Melody
20       ♪ Buzzer index  = Melody2
21       WAIT WHILE ( 🔊 Buzzer Time  > 0.000sec  )
22       // Move arm to left (Motion Page 5)
23       ▶ Motion Index Number  = 5
```

```
26   ENDLESS LOOP
27   {
28       // Check for obstacle in Front
29       IF ( 📶 PORT[4]:IR Sensor  > 75  )
30       {
31           ShoulderFB = ShoulderFB + ShoulderStep
32           IF ( ShoulderFB > 255  )
33           {
34               ShoulderFB = 255
35           }
36           // Move shoulder servos backwards quickly to avoid obstacle
37           CALL AvoidFront
38           CALL Delay2
39       }
40       ELSE
41       {
42           ShoulderFB = ShoulderFB - ShoulderStep
43           IF ( ShoulderFB < 0  )
44           {
45               ShoulderFB = 0
46           }
47           // Move shoulder servos forward slowly
48           CALL AvoidFront
49           CALL Delay1
50       }
```

Fig. 6.14 Part 2 of "CM530-AvoiderArm_1.tsk/tskx"

Fig. 6.15 Functions
"AvoidFront/Left/Right"
used in "CM530-
AvoiderArm_1.tsk/tskx"

```
102   FUNCTION AvoidFront
103   {
104        ID[2]:Joint Offset   = 0  -  ShoulderFB
105        ID[3]:Joint Offset   = 0  +  ShoulderFB
106        RETURN
107   }
108
109   FUNCTION AvoidLeft
110   {
111        ID[1]:Joint Offset   = 0  -  ShoulderLR
112        RETURN
113   }
114
115   FUNCTION AvoidRight
116   {
117        ID[1]:Joint Offset   = 0  +  ShoulderLR
118        RETURN
119   }
```

- Statements 29–38 – If the IR Sensor at Port 4 detected a "frontal" object (i.e. IR sensor value greater than 75), the algorithm computed a new "increasing" value for "ShoulderFB" (which however could not go beyond 255), and then called upon the Function AvoidFront (see Fig. 6.15) to set the JOINT OFFSETs of Actuators 2 and 3 with the appropriate values of "ShoulderFB", as Actuators 2 and 3 were mounted as mirror image of each other (Statements 104–105). Function "Delay2" was a 0.128 s. delay (Statement 38).
- Else, i.e. no "frontal" obstacle, the algorithm computed a new "decreasing" value for "ShoulderFB" (which however could not go negative), and then similarly called upon the Function AvoidFront (see Fig. 6.15) to set the JOINT OFFSETs of Actuators 2 and 3 with the appropriate values of "ShoulderFB". Function "Delay1" was a 0.512 s. delay (Statement 49). In other words the arm would back away from the frontal obstacle quickly but then would return to its Ready pose at a slower rate.
- In Part 3 of the algorithm (Statements 52–78), it independently checked on IR Left (Port 1) and IR Right (Port 2) for any obstacle present within their detected ranges (i.e. greater than 100). For example, in Statements 52–63 in Fig. 6.16, if an obstacle were present on the left (Port 1), the algorithm first would set a value for "ShoulderLR" (but no greater than 255). Next, it called Function AvoidLeft (see Fig. 6.15) which moved the arm to the right for 0.128 s using Actuator 1 (Statements 60–61), then stopped Actuator 1 for 2.048 s (i.e. Functions "StopServo1" and "Delay3" – see Fig. 6.17). Figure 6.17 also showed how the flag "StopServo" was used and how the Joint Offset of Servo 1 was set to "1024" to get it off the influence to the motion pages being played (see Sect. 6.2.2.1).

```
51        // Check for obstacle on Left
52        IF (  PORT[1]:IR Sensor  > 100  )
53        {
54            ShoulderLR  = 2 * ShoulderStep
55            IF ( ShoulderLR  > 255  )
56            {
57                ShoulderLR  = 255
58            }
59            // Move shoulder servo 1 right quickly to avoid obstacle
60            CALL AvoidLeft
61            CALL Delay2
62            CALL StopServo1
63            CALL Delay3
64        }
```

Fig. 6.16 Case when obstacle found on the arm's left side in "CM530-AvoiderArm_1.tsk/tskx" (Part 3 of algorithm)

Fig. 6.17 Functions to stop and run Servo 1 in "CM530-AvoiderArm_1. tsk/tskx"

```
143   FUNCTION StopServo1
144   {
145        ID[1]:Joint Offset   = 1024
146        ID[1]:  Moving Velocity  = 0
147        StopServo  = TRUE
148        Buzzer Time  = Play Melody
149        Buzzer index  = Melody6
150        WAIT WHILE ( Buzzer Time  > 0.000sec  )
151        RETURN
152   }
153
154   FUNCTION RunServo1
155   {
156        ID[1]:Joint Offset   = 0
157        ID[1]:  Moving Velocity  = 1023
158        StopServo  = FALSE
159        RETURN
160   }
```

- In the case of no obstacle found on the left or right of the arm (Statements 80–83 – see Fig. 6.18 for Part 4 of algorithm), the Actuator 1 would be allowed to resume its activity by calling "RunServo1" (see Fig. 6.17 also).
- Figure 6.19 showed Part 5 of the algorithm where if ("StopServo!= TRUE), i.e. no left or right obstacle were present, it was OK for the arm to alternate between its Motion Pages 5 and 6.

Video 6.4 demonstrated the performance of this "Avoider Arm" project.

```
65    // Check for obstacle on Right
66    ELSE IF ( ⚙PORT[2]:IR Sensor  > 100  )
67    {
68        ShoulderLR  = 2  *  ShoulderStep
69        IF ( ShoulderLR  > 255  )
70        {
71            ShoulderLR  = 255
72        }
73        // Move shoulder servo 1 left quickly to avoid obstacle
74        CALL AvoidRight
75        CALL Delay2
76        CALL StopServo1
77        CALL Delay3
78    }
79    // Clear on Left & Right - No Offset needed
80    ELSE
81    {
82        CALL RunServo1
83    }
```

Fig. 6.18 Part 4 of "CM530-AvoiderArm_1.tsk/tskx"

```
85    IF ( StopServo  !=  TRUE  )
86    {
87        // If arm was moving left (Motion Page 5) then move it right (Motion Page 6)
88        IF ( ⏵Motion Index Number  == 5  )
89        {
90            ⏵Motion Index Number  = 6
91        }
92        // Otherwise move it left (Motion Page 5)
93        ELSE
94        {
95            ⏵Motion Index Number  = 5
96        }
97    }
```

Fig. 6.19 Part 5 of "CM530-AvoiderArm_1.tsk/tskx"

6.2.2.3   Pick and Place Dowels

The third REACTOR arm project showed how to use an IR sensor to search for, pick up and move 1 wooden dowel to a fixed location. Please note that the IR Sensor at the Gripper's location got repositioned to better detect a "vertical" object such as a wooden dowel (see Fig. 6.20 and review Fig. 6.9).

New MOTION files "Reactor_ArmDowel.mtn/mtnx" (see Fig. 6.21) and corresponding TASK files "REACTOR_DowelSearch_n_Grab.tsk/tskx" were created for this project (they were derived from the work done by Mr. Erik Lockhart as a project for the course ENGR-4310 in Fall Semester 2015 at UGA). The basic strategy was to swing the arm back and forth while using the gripper's IR Sensor first to detect the wooden dowel, then to use it to put the dowel within the grasping zone of

Fig. 6.20 IR sensor
repositioned for better
detection of "vertical"
wooden dowel

Fig. 6.21 Motion pages no. 2, 7 and 8 were used in the Dowel search project

the gripper. As the distance between the dowel and the IR sensor was variable, the
Joint Offset feature was once again used to terminate early the behavior of a critical
Motion Page (No. 7) so that the arm/gripper would stop at a proper distance at all
times in order to grab the dowel.

The TASK file "REACTOR_DowelSearch_n_Grab.tsk" started out with some
parameter definitions (see Fig. 6.22) and a call to Function Init (see Fig. 6.23).

Fig. 6.22 Starting code
for "REACTOR_
DowelSearch_n_Grab.tsk"

```
4    RightPosition  = 282
5    LowRight   = RightPosition   - 2
6    HighRight  = RightPosition  + 2
7    LeftPosition  = 722
8    LowLeft  = LeftPosition   - 2
9    HiLeft  = LeftPosition  + 2
10   CALL Init
```

```
54  FUNCTION Init
55  {
56      // Isolate Servo 8 from Motion Page being played & make sure that gripper is opened
57      ⚙ ID[8]:Joint Offset  = 1024
58      🔋 ID[8]: ⊕ Goal Position  = 512
59      // Get into Init Pose
60      ▶ Motion Index Number  = 2
61      WAIT WHILE ( ⦿ Motion Status  ==  TRUE  )
62      // Isolate Servo 1 from Motion Page being played & move it to Far Left position
63      ⚙ ID[1]:Joint Offset  = 1024
64      🔋 ID[1]: 🎚 Moving Velocity  = CCW:75
65      🔋 ID[1]: 🔵 Goal Torque  = 400
66      🔋 ID[1]: ⊕ Goal Position  = LeftPosition
67      WAIT WHILE ( 🔋 ID[1]: 🏁 Is Moving  ==  TRUE  )
68  }
```

Fig. 6.23 Function Init

The main purpose of Function Init was to uncouple Servo 8 (i.e. Gripper) and Servo 1 (Slew Motor) from the effects of the Motion Pages to be activated in later code sections:

- Statement 57 set the Joint Offset of Servo 8 to 1024 and thus isolated it from the effects of Motion Pages. Statement 58 made sure that the gripper was open.
- Statements 60–61 put the REACTOR Arm into its "Init" pose (see Fig. 6.9) by activating Motion Page No. 2.
- Once the arm was in its "Init" pose, Servo 1 got "uncoupled" from Motion Pages with Statement 63.
- Statements 64–67 moved Servo 1 and consequently the whole arm to its "left-most" Goal Position (Statement 66) with a Torque Limit of 400 and at a CCW Speed Setting of 75.

As usual, the main algorithm was contained with an Endless Loop starting at Statement 12 (see Fig. 6.24):

- Upon the user pushing down the UP button on the CM-5XX controller (Statement 15), the Scan Function would be called (Statement 18 and see Fig. 6.25).
- Function "Scan" essentially slewed Servo 1 between its "RightPosition" and "LeftPosition" as long as the IR Sensor at Port 4 (i.e. Gripper) was not detecting

```
12    ENDLESS LOOP
13    {
14        // Arm searchs and grabs dowel
15        IF ( 🔘 Button  ==  🔘 U  )
16        {
17            // Scan for dowel
18            CALL  Scan
19            // Motion Page 7 extends the arm towards the dowel and "beyond"
20            ▶ Motion Index Number  = 7
21            WAIT WHILE ( 🔧 PORT[4]:IR Sensor   <=  400  )
22            // Uncouple remaining servos from Motion Page 7 & Stop all servos
23            ⚙ ID[2]:Joint Offset   = 1024
24            ⚙ ID[3]:Joint Offset   = 1024
25            ⚙ ID[4]:Joint Offset   = 1024
26            ⚙ ID[5]:Joint Offset   = 1024
27            ⚙ ID[6]:Joint Offset   = 1024
28            ⚙ ID[7]:Joint Offset   = 1024
29            🔋 ID[All]: 🖼 Moving Velocity  = 0
30            WAIT WHILE ( ▶ Motion Status   ==  TRUE  )
```

Fig. 6.24 Part 1 of main algorithm of Dowel search project

```
70    FUNCTION  Scan
71    {
72        // Slew arm back and forth until dowel is found
73        LOOP WHILE ( 🔧 PORT[4]:IR Sensor  <=  0  )
74        {
75            IF ( 🔋 ID[1]: 🎯 Present Position  >=  LowLeft  )
76            {
77                🔋 ID[1]: 🎯 Goal Position  = RightPosition
78                Adjust  = -15
79            }
80            ELSE IF ( 🔋 ID[1]: 🎯 Present Position  <=  HighRight  )
81            {
82                🔋 ID[1]: 🎯 Goal Position  = LeftPosition
83                Adjust  = 15
84            }
85        }
86        // Dowel found, adjust to center it to Gripper
87        Final_Position  = 🔋 ID[1]: 🎯 Present Position  +  Adjust
88        🔋 ID[1]: 🎯 Goal Position  = Final_Position
89
90        🎵 Buzzer index  = Melody6
91        🔊 Buzzer Time  = Play Melody
92        WAIT WHILE ( 🔊 Buzzer Time  >  0.000sec  )
93        RETURN
94    }
```

Fig. 6.25 Function "Scan" of Dowel search project

any wooden dowel (Statements 73–85). The "Adjust" parameter was used so as
to correct for position overshoots due to unavoidable hardware and software
delays for the arm.

- Once a dowel was found, an "adjusted" Final_Position value was computed and
  set as the Goal Position for Servo 1, so as to line up the dowel with the gripper's
  opening (Statements 87–88).
- Lastly, Melody 6 was played as audio feedback to the user.
- At this point, the algorithm returned to Statement 20 (see Fig. 6.24) and activated
  Motion Page No. 7 which extended the Arm forward quite far (see Fig. 6.26). As
  a consequence of this move, the Gripper's IR Sensor reading on Port 4 would
  keep on increasing towards the 400 value.
- When this "400" value was reached, the dowel would be found within the grasping
  zone of the Gripper and thus the whole arm needed to be stopped, so to speak, "in its
  track". However, at this point in time, the Motion Page No. 7 would most likely have
  not yet reached its endpoint poses for all its servos, and thus the only way to stop all
  the servos was to set their Joint Offset parameters to 1024 (Statements 23–28) and
  then set all their Moving Velocities to zero (Statement 29). Please note that Servos 1
  and 8 were already uncoupled from the Motion Pages earlier in the algorithm.
- Statement 30 was quite important and critical to use as the Motion Page No. 7 needed
  "time" to "time itself out", even though all the servo speeds were already set to zero.
- Next Melody 3 was played (Statements 32–34 in Fig. 6.27), and the Gripper was
  closed onto the dowel by issuing a Goal Position command in Statements 36–37.

Fig. 6.26 Motion page no.
7 extended the arm forward

Fig. 6.27 Part 2 of main algorithm of Dowel search project

```
39              // Return control to motion file and move dowel to final position
40              ⬡ ID[1]:Joint Offset  = 0
41              ⬡ ID[2]:Joint Offset  = 0
42              ⬡ ID[3]:Joint Offset  = 0
43              ⬡ ID[4]:Joint Offset  = 0
44              ⬡ ID[5]:Joint Offset  = 0
45              ⬡ ID[6]:Joint Offset  = 0
46              ⬡ ID[7]:Joint Offset  = 0
47              ⬡ ID[8]:Joint Offset  = 0
48              ▶ Motion Index Number  = 8
49              WAIT WHILE  ( ▶ Motion Status  ==  TRUE )
50          }
51      }
```

Fig. 6.28 Part 3 of main algorithm of Dowel search project

- Figure 6.28 showed Part 3 of the main algorithm whereas all the servos would need to be put back under the influence of the Motion Page 8 which was created to move the dowel to a designated unloading location. Thus Joint Offsets for Servos 1–7 needed to be reset to zero first (Statements 40–46), while the Joint Offset for the Gripper (Servo 8) was reset to zero in the "last possible millisecond" so as to prevent it from dropping the dowel too early (Statements 47 and 48).

Video 6.5 demonstrated this project performance with a square wooden dowel.

6.2.2.4 Remote Control of Mobile Manipulator

Figure 6.29 illustrated a mobile manipulator robot by mounting the REACTOR Arm onto a chassis with four AX-12W acting as wheels (with Dynamixel IDs set from 11 to 14). This project's goal was to contrast the Remote Control approach used for a carbot (wheel control) vs. the one used for a manipulator arm (position control). The complete algorithm was described in the files "MobileReactor_RC_Demo.tsk/tskx" and "REACTOR_Arm.mtn/mtnx").

The "Car Control" component was adapted from the file "RC100_Carbot_2. tsk" (see Sect. 5.6.1) whereas each car maneuver was user-specified by a combination of the U-D-L-R buttons of the RC-100 and a specific function was written for each specific RC-100's button (i.e. a wanted car maneuver). For example, the "Up" button would be matched to the Function "Go_forward" and any/all Button Release would correspond to the "Stop" Function. On the other hand, the "Manipulator Control" component used a common Function "Execute" issuing "Goal Position" commands to all seven Dynamixels used for the REACTOR arm (see Fig. 6.30), whereas the appropriate values for parameters "Pos_1" to "Pos_8" were determined in the Main program depending on the user's inputs via the RC-100's buttons.

Fig. 6.29 Mobile
manipulator robot based on
the REACTOR arm

Fig. 6.30 Common
function "Execute" used
for mobile manipulator
robot

```
172   FUNCTION Execute
173   {
174       ID[1]:  Goal Position   = Pos_1
175       ID[2]:  Goal Position   = Pos_2
176       ID[3]:  Goal Position   = Pos_3
177       ID[4]:  Goal Position   = Pos_4
178       ID[5]:  Goal Position   = Pos_5
179       ID[6]:  Goal Position   = Pos_6
180       ID[7]:  Goal Position   = Pos_7
181       ID[8]:  Goal Position   = Pos_8
182   }
```

The "Manipulator" algorithm started by setting all Joint Offsets to 1024 so that "Goal Position" commands could be used effectively later (Statements 15–22 in Fig. 6.31). It also saved all "Present Positions" i.e. each actuator's position as commanded by Motion Page No. 2 (Statements 24–31). Please note that Step size for each actuator adjustment (as wanted by the user) was set to 4 (Statement 33). It next used the usual "Endless Loop" construct to read in RC-100 inputs and processed them accordingly.

Using an analogy to the human arm, the REACTOR arm could be considered as having three joints plus hand grasp (i.e. gripper):

• The shoulder joint was represented by Actuators 1, 2 and 3.
• The elbow joint was represented by Actuators 4 and 5.

```
11      // Get into Init Pose
12      ▶ Motion Index Number  = 2
13      WAIT WHILE  ( ▶ Motion Status  ==  TRUE  )
14      // Switch to Position Control mode
15      ID[1]:Joint Offset   = 1024
16      ID[2]:Joint Offset   = 1024
17      ID[3]:Joint Offset   = 1024
18      ID[4]:Joint Offset   = 1024
19      ID[5]:Joint Offset   = 1024
20      ID[6]:Joint Offset   = 1024
21      ID[7]:Joint Offset   = 1024
22      ID[8]:Joint Offset   = 1024
23      // Pick up Init Positions
24      Pos_1  =  ID[1]:  Present Position
25      Pos_2  =  ID[2]:  Present Position
26      Pos_3  =  ID[3]:  Present Position
27      Pos_4  =  ID[4]:  Present Position
28      Pos_5  =  ID[5]:  Present Position
29      Pos_6  =  ID[6]:  Present Position
30      Pos_7  =  ID[7]:  Present Position
31      Pos_8  =  ID[8]:  Present Position
32      // Step size to adjusting servos at each RC command received
33      Step  = 4
```

Fig. 6.31 Start of "Manipulator" algorithm

- The wrist joint was represented by Actuators 6 and 7.
- The hand grasp/gripper was represented by Actuator 8.

Thus the shoulder joint's actions were designed with a combination of the "U-D-L-R" buttons with the "1" button:

- "L+1" and "R+1" would slew the arm left and right respectively and only Actuator 1 was involved, thus only Parameter "Pos_1" was determined (see Fig. 6.32).
- "U+1" and "D+1" would move Actuators 2 and 3 "up" or "down" respectively (see Fig. 6.33). Please note that Actuators 2 and 3 were mounted as mirror images of each other on the REACTOR arm. Thus Parameter "Step" was used as "negative" or "positive" values for each Actuator 2 or 3 depending on the motion wanted (Statements 61–62 or 70–71).

As the elbow joint had only one axis of rotation controlled by Actuators 4 and 5, "U-D" buttons were used in combination with Button "2" to activate this joint (see Fig. 6.34).

The wrist's flicking movement was controlled by Actuator 6, thus "U-D" buttons were used in combination with Button "3" for this motion (see Fig. 6.35).

Fig. 6.32 "Shoulder L/R" component of "Manipulator" algorithm

```
43    IF ( Button  ==  ⟳L+1 )
44    {
45            Pos_1  = Pos_1  +  Step
46            IF ( Pos_1  >=  0  &&  Pos_1  <=  1023  )
47            {
48                    CALL Execute
49            }
50    }
51    ELSE IF ( Button  ==  ⟳R+1 )
52    {
53            Pos_1  = Pos_1  -  Step
54            IF ( Pos_1  >=  0  &&  Pos_1  <=  1023  )
55            {
56                    CALL Execute
57            }
58    }
```

```
59    ELSE IF ( Button  ==  ⟳U+1 )
60    {
61            Pos_2  = Pos_2  -  Step
62            Pos_3  = Pos_3  +  Step
63            IF ( Pos_2  >=  0  &&  Pos_2  <=  1023  &&  Pos_3  >=  0  &&  Pos_4  <=  1023  )
64            {
65                    CALL Execute
66            }
67    }
68    ELSE IF ( Button  ==  ⟳D+1 )
69    {
70            Pos_2  = Pos_2  +  Step
71            Pos_3  = Pos_3  -  Step
72            IF ( Pos_2  >=  0  &&  Pos_2  <=  1023  &&  Pos_3  >=  0  &&  Pos_4  <=  1023  )
73            {
74                    CALL Execute
75            }
76    }
```

Fig. 6.33 "Shoulder U/D" component of "Manipulator" algorithm

The wrist's rotating movement was controlled by Actuator 7, thus "L-R" buttons were used in combination with Button "3" for this motion (see Fig. 6.36).

Lastly, Button "5" was used for closing the gripper by a step of four positional values for each button push (Actuator 8) and Button "6" was used for opening it up all the way in one button push (see Fig. 6.37).

Video 6.6 demonstrated this performance of this mobile manipulator. Another RC approach was suggested in the Exercise section of this chapter.

```
77      // Use Button 2 in combination with U-D for elbow servos IDs=4,5 for U-D
78      ELSE IF ( Button == [U+2] )
79      {
80          Pos_4 = Pos_4 + Step
81          Pos_5 = Pos_5 - Step
82          IF ( Pos_4 >= 0 && Pos_4 <= 1023 && Pos_5 >= 0 && Pos_5 <= 1023 )
83          {
84              CALL Execute
85          }
86      }
87      ELSE IF ( Button == [D+2] )
88      {
89          Pos_4 = Pos_4 - Step
90          Pos_5 = Pos_5 + Step
91          IF ( Pos_4 >= 0 && Pos_4 <= 1023 && Pos_5 >= 0 && Pos_5 <= 1023 )
92          {
93              CALL Execute
94          }
95      }
```

Fig. 6.34 "Elbow U/D" component of "Manipulator" algorithm

Fig. 6.35 "Wrist's Flick" component of "Manipulator" algorithm

```
97      ELSE IF ( Button == [U+3] )
98      {
99          Pos_6 = Pos_6 + Step
100         IF ( Pos_6 >= 0 && Pos_6 <= 1023 )
101         {
102             CALL Execute
103         }
104     }
105     ELSE IF ( Button == [D+3] )
106     {
107         Pos_6 = Pos_6 - Step
108         IF ( Pos_6 >= 0 && Pos_6 <= 1023 )
109         {
110             CALL Execute
111         }
112     }
```

6.2.3   Application to a GERWALK Robot

To further illustrate the main concepts described in Sect. 6.2.1, a CM-510 GERWALK robot (Fig. 6.38) was used to create the following videos:

- Video 6.7 demonstrated the concept of a Calibration Pose and showed how to use the Calibration sub-tool and to do basic motion editing.

Fig. 6.36 "Wrist's Rotate" component of "Manipulator" algorithm

```
113    ELSE IF ( Button  ==  ➎L+3 )
114    {
115        Pos_7  = Pos_7  -  Step
116        IF ( Pos_7  >=  0  &&  Pos_7  <=  1023  )
117        {
118            CALL Execute
119        }
120    }
121    ELSE IF ( Button  ==  ➎R+3 )
122    {
123        Pos_7  = Pos_7  +  Step
124        IF ( Pos_7  >=  0  &&  Pos_7  <=  1023  )
125        {
126            CALL Execute
127        }
128    }
```

Fig. 6.37 "Gripper's Open/Close" component of "Manipulator" algorithm

```
129    // Opening gripper using Button 6
130    ELSE IF ( Button  ==  ➎6 )
131    {
132        Pos_8  = 512
133        CALL Execute
134    }
135    // Closing gripper by STEP=4 using Button 5
136    ELSE IF ( Button  ==  ➎5 )
137    {
138        Pos_8  = Pos_8  -  Step
139        IF ( Pos_8  >=  150  )
140        {
141            CALL Execute
142        }
143    }
```

- Video 6.8 used the sample Motion and Task files provided by ROBOTIS to illustrate the planning of motion design inside MOTION V.1 and the integration of Motion Pages with the logic flow design provided inside TASK.
- For users of the CM-530 along with the MOTION V.2 and TASK V.2 tools, the corresponding files "PRM_GerwalkDemo.mtnx" and "PRM_GerwalkDemo.tskx" were also included in the Springer Extra Materials for this book.
- An RC_Gerwalk project was also suggested in the Exercise section of this chapter.

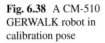

Fig. 6.38 A CM-510
GERWALK robot in
calibration pose

6.3   Form and Function of Walking Robots

In this section, the goal was to illustrate how Form influences Function in the design and operation of walking robots, going from a design with only two actuators (BugFighter) to a Humanoid design with 18 actuators.

The author could be wrong on this issue, but so far the author could not find any legged robot that could move around on only one actuator. Even the Planar One-Leg Hopper from MIT (c. 1982) needed two actuators (http://www.ai.mit.edu/projects/leglab/robots/2D_hopper/2D_hopper.html).

The BIOLOID STEM STANDARD kit offered a 2-servo BugFighter robot using two AX-12W in wheel mode. The walking motion of its 6 legs was based on the conversion of the rotary motion of the AX-12W into a back-and-forth motion for its legs by using a dual set of four-bar linkages for each side (Fig. 6.39). Please note that the four-bar linkages worked in a vertical plane for this robot. Thanks to this mechanical conversion, the control scheme for this robot was the same for the various carbots used in Chap. 5. ROBOTIS has a video of this robot in action at http://www.robotis.com/video/BIO_STEM_BugFighter.wmv.

In the BIOLOID STEM EXPANSION kit, the Hexapod robot used three AX-12A in Position Control mode to achieve its walking ability: one actuator to shift its weight left and right using its middle leg in a seesaw motion (Fig. 6.40), and one actuator on each side to provide the forward-backward and steering motions. The hexapod also used a four-bar linkage but as a parallel linkage between its front and back leg, and it worked in a horizontal plane instead (Fig. 6.41).

Fig. 6.39 Dual set of
four-bar linkages used for
BugFighter

Fig. 6.40 ROBOTIS
hexapod middle leg (3rd
actuator) is used to shift its
weight left and right

Fig. 6.41 ROBOTIS
hexapod front actuator's
activation of a parallel
four-bar linkage to obtain
its walking motion

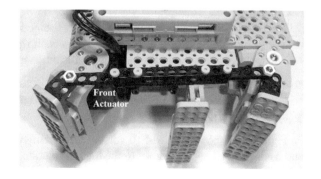

Fig. 6.42 ROBOTIS
PREMIUM GERWALK

ROBOTIS has a video of its walking motion at http://www.robotis.com/video/
BIOLOID_STEM_14.Hexapod.wmv.

TWITCH was also another three-actuator Hexapod using a different design for
its legs (http://forums.trossenrobotics.com/robots.php?project_id=7#ad-image-0).

There are great web resources for mechanical linkage design such as:

- http://www.mekanizmalar.com/.
- Kinematic Models for Design Digital Library from Cornell University (http://
  kmoddl.library.cornell.edu/).

Figure 6.42 showed a CM-510 version of the GERWALK using seven actuators
and having a walking gait closer to the human solution. It swung its upper body left
and right in order to shift its center of gravity towards the supporting leg before making

Fig. 6.43 ROBOTIS BIPED

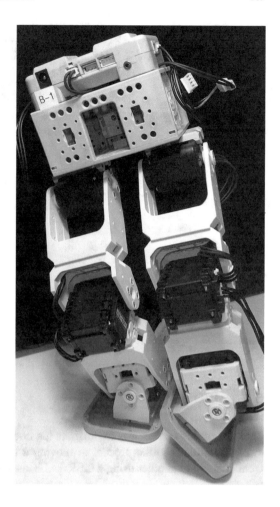

a step with the other leg. As it had no hip joint, it had to use a "moon-walk" like maneuver to generate enough friction forces from the ground surface to allow it to turn sideways (see video at http://www.robotis.com/video/BIO_PRM_GerWalk.wmv).

Figure 6.43 showed a CM-510 version of the BIPED robot using eight actuators and having a walking gait even closer to the human solution. It did have ankle joints but it still had no hip joint, so it had to rotate its ankle and to use another "moon-walk" like solution to generate enough friction forces from the ground surface to turn sideways (see video at http://www.robotis.com/video/BIO_PRM_BipedWalkingRobot.wmv).

Figure 6.44 showed a CM-530 version of a Humanoid-A robot using 12 actuators for its lower body, i.e. it had hip and ankle joints and thus had a walking gait closer to a human gait (see Video 6.9).

The reader is also recommended to read and practice the materials found in Sect. 4.4 (Advanced Learning) of the Software Programming User's Guide that came with the BIOLOID PREMIUM kit. This section has some advanced instructional materials for humanoid gait generation using the Mirror Exchange function. If the

Fig. 6.44 ROBOTIS
HUMANOID-A (lower
body)

reader had bought the PREMIUM kit before 2013, it would not contain this manual and it has to be bought separately at present (http://www.robotis-shop-en. com/?act=shop_en.goods_view&GS=1486&GC=GD080400).

6.4 "Torque" Effects

The goal of this section was to illustrate the interactions between the many parameters that influenced the process of moving a typical actuator (AX-12) from one Goal Position to another using a GERWALK as the test robot.

6.4.1 Torque Limit, Present Position and Present Load

Figure 6.45 was essential to the understanding of the interplay between the three parameters "Torque Limit" (Address 34, also labeled as Goal Torque in TASK), "Present Position" (Address 36) and "Present Load" (Address 40):

1. As previously described in Sect. 6.1, "Torque Limit" [0–1023] denoted the maximum load (electrical current) allowed on the actuator. When it was set high (512–1023), the user would need quite substantial physical effort to manually move the actuator from its current Goal Position. When it was set low (1–10), the user would be able to manually shift the actuator away from its Goal Position, and the actuator would not be able to get back to its Goal Position at all (this feature would be used later in Chap. 7 to illustrate the concept of "Bilateral Control" between two GERWALK robots).

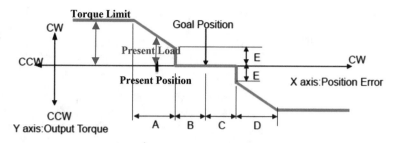

Fig. 6.45 Concept of present load

2. Despite its given name of "Present Load", this parameter should not be understood as the true external load on an actuator. More accurately, it would correspond to the "present electrical current" that was supplied to the motor by the actuator's circuitry as long as the controller realized that the actuator's Present Position was not where it was supposed to be (i.e. Goal Position ± Margins B and C).

Video 6.10 and the TASK file "Gerwalk_TorqueEffects.tsk/tskx" provided a more thorough demonstration for these characteristics by monitoring the Present Position and Present Load during a manually forced resetting of the Goal Position(s) of the GERWALK actuators.

6.4.2 Adjusting Torque Limit Dynamically

In this subsection, the TASK files "Gerwalk_LoadAdjust.tsk/tskx" and "Gerwalk_LoadAdjustFast.tsk/tskx" were used to demonstrate the external behavior of actuators 4 and 5 of a GERWALK robot (Fig. 6.46) when they were manually forced out of their given Goal Position of 512. This would be the first step towards the understanding of a feedback control approach for actuators such as the AX-12 (see Sect. 6.6 for a load-sensing application).

"Gerwalk_LoadAdjust.tsk/tskx" first initialized all 7 AX-12s to a GOAL POSITION of 512 with a TORQUE LIMIT of 1023 during the first 2 s of its run, then it set the TORQUE LIMIT to 1 for Actuators 4 and 5 (to enable the user to manually move these two actuators out of Position 512 at his or her discretion). Next, it got into an endless loop where it printed out the following parameters: actuator's ID, Torque Limit (Address 34), Present Position (Address 36) and Present Load (Address 40). Also within this endless loop, if the controller found that the Present Position of either actuator (4 or 5) was outside of a High-Low boundary around the initial position, it would increase the corresponding Torque Limit by a constant value saved in a parameter called TorqueStep. The net physical result was that one would see the actuators 4 and 5 go back to Position 512 at a constant rate, once the user "disturbed" their initial positions (see Video 6.11).

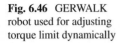

Fig. 6.46 GERWALK
robot used for adjusting
torque limit dynamically

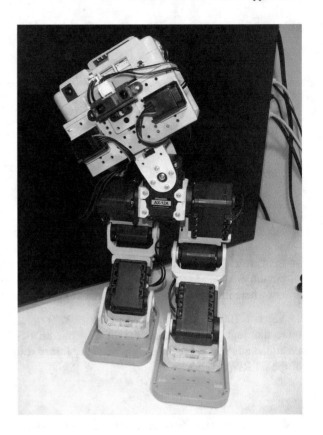

"Gerwalk_LoadAdjustFast.tsk/tskx" was designed to do the same job as "Gerwalk_LoadAdjust.tsk/tskx", except that its TorqueStep value was proportional to the Gap between the actuator's Present Position and either High or Low boundary values set around the Goal Position (see Video 6.12).

6.5   More "Joint Offset" Effects

In Sect. 6.2.3, Video 6.7 showed the reader how to adjust the "Motion Offset" values of individual actuators so as to correspond to a given Calibration Pose from inside the Motion Editor tool. In this Sect. 6.5, this JOINT OFFSET parameter would be re-visited as an advanced programming technique from inside the TASK tool and it would applied to a single-actuator gripper (Fig. 6.47).

Fundamentally, the JOINT OFFSET parameter of a given actuator was designed to work only in conjunction with a MOTION PAGE (involving that particular actuator) being "played". Please note that the GOAL POSITION parameter (Address 30) would not be affected in any way by the JOINT OFFSET parameter.

Fig. 6.47 A single-actuator gripper

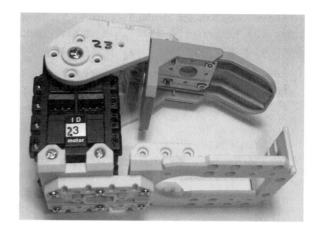

The possible range of values for JOINT OFFSET would be from −255 to +255 and essentially it was "added" to a given GOAL POSITION **as specified in a STEP** of a particular MOTION PAGE (and **not** by a regular LOAD command into Address 30, please make a note of this). For example, if JOINT OFFSET was set as −10 for a particular actuator, and a given STEP or MOTION PAGE commanded this actuator to go to Position 200, the actuator would actually go to a physical Actuator Position equal to (200−10 = 190).

The "BasicGripper.mtn/mtnx" and "AX-12JointOffset.tsk/skx" files, along the video file "Video 6.13" demonstrated the above interactions for a single-actuator gripper. The reader could see that the invoked PAGE 3 (inside the endless loop) commanded the actuator to go to Position 512, but the actuator was actually decrementing its Goal Position by −10 every time that the program went through this loop (i.e. it kept on "opening up" the gripper).

Finally, the reader should note that there was a special value of 1024 that could be set to the JOINT OFFSET parameter of a particular actuator, when the programmer wanted to **exclude** this particular actuator from the effects of a MOTION PAGE being executed. The ROBOTIS e-Manual web site has an application of this feature to the arms and grippers of a humanoid (Type B) robot (http://support.robotis.com/en/product/bioloid/premiumkit/tutorial/bioloid_prem_tutorial_gripper.htm). Please note that Dynamixel IDs 9 and 10 are assigned to the grippers in this particular example, and refer to the sample program files "PRM_HumanoidTypeB.mtn/mtnx" and "PRM_GripperExam.tsk/tskx".

6.6   A Load Sensing Gripper

In this section, the goal was to program a simple gripper that could adjust to different sizes of the object to be held by its fingers.

The software solution involved using the same "BasicGripper.mtn/mtnx" file but a new TSK file called "AX-12Gripper.tsk/tskx" whereas the basic algorithm was described below:

1. Initialization to put the gripper at the neutral position by playing Motion Page 3, i.e. its Goal Position was set to 512.
2. Next the program entered an endless loop whereas:

   (a) If the user pushed on Button U of the Virtual RC-100, the controller played MOTION PAGE 2 (i.e. Actuator Position set to 430 – open gripper position).
   (b) Else if the user pushed on Button D of the Virtual RC-100, the controller played MOTION PAGE 1 (i.e. Actuator Position set to 600 – closed gripper position).
   (c) Else if the user pushed on Button 1 of the Virtual RC-100, the controller played MOTION PAGE 3 (i.e. Actuator Position set to 512 – neutral position).
   (d) Also within this loop the program printed out the actuator Present Position, Joint Offset and Present Load for monitoring purposes.

3. The "size-adjusting" algorithm was implemented by a CALLBACK function which was executed by the controller every 8 ms. The interested reader should check the ROBOTIS e-Manual for more details about CALLBACK at http://support.robotis.com/en/software/roboplus/roboplus_task/programming/command/roboplus_task_cmd_callback.htm. This algorithm essentially read out the Present Load (Address 40) and compared this value:

   (a) To "128". If it was less or equal to 128, parameter Adjustment was set to 0 (i.e. the gripper had not encountered the object yet, or it was only starting to squeeze on the object).
   (b) To parameter MaxSqueezeTorque (=750), if it was greater than MaxSqueezeTorque, meaning that it was squeezing "hard" on the object then parameter Adjustment was decremented by 1 (i.e. the gripper was made to "open-up").
   (c) The final value for Adjustment (whether 0 or a negative value) was assigned to JOINT OFFSET to affect a controlled "opening and closing" of the gripper so as to maintain a PRESENT LOAD "around" the value set by MaxSqueezeTorque.

The video file "Video 6.14" illustrated how this application translated out in practice.

The reader should note that this section was a very basic attempt at "Force Control", Villani and De Schutter (2008) have a more thorough treatment of the issues involved.

6.7   Review Questions for Chap. 6

1. A MOTION file can be considered as DATA, while a TASK file can be considered as LOGIC. (T-F)
2. The Play Motion flag is set to 0 when a given Motion page is being executed. (T-F)

3. To control the accuracy of the Goal Position on an AX-12, we need to reduce its SLOPE value to the minimum value of 1. (T-F)

4. A Joint Offset command using a value within the normal range [−255, 255] could take effect without invoking a Play Motion Page command. (T-F).

5. A Joint Offset of 1024 on a specific actuator will allow the use of "Goal Position" command on this actuator at later times. (T-F).

6. What does parameter PUNCH correspond to in functional terms for the motor component of a Dynamixel actuator?

7. What does parameter PRESENT LOAD correspond to in functional terms for the motor component of a Dynamixel actuator?

8. What does parameter SLOPE correspond to in functional terms for a Dynamixel actuator?

9. What are the ranges of possible values for the "Torque Limit" and "Present Load"?

10. What is the physical unit that parameter SLOPE can be mapped to?

11. What is the physical unit that parameter MARGIN can be mapped to?

12. How does one program the AX-12 to go to its "position control" mode from inside a TASK program?

13. Please match typical servo parameters to their correct addresses:

- Present Load   Address 30
- Present Position   Address 34
- Goal Position   Address 36
- Torque Limit (Goal Torque)   Address 40

14. What is the allowable range of angular positions in degrees for the AX-12 when it is set into Position Control mode?

15. What is the allowable range of angular positions in degrees for the XL-320 when it is set into Position Control mode?

16. When using the Motion Editor V.1 tool, what is the maximum number of steps/poses that the user can create per Motion Page, for each type of CM-5XX controller?

17. How many Motion Pages are allowed for each type of CM-5XX controller?

18. Describe in your own terms what is a POSE?

19. Describe in your own terms what is a STEP?

20. Describe in your own terms what is a MOTION PAGE?

21. How can a robot flow from one MOTION PAGE to another MOTION PAGE?

22. What is the EXIT PAGE used for?

23. To control the acceleration/deceleration patterns for a robot to go from Pose to Pose, we need to adjust the "Joint Softness" parameter of the servos involved in the robot's moves. (T-F)

24. What is the effect of a small CONTROL INERTIAL FORCE setting?

25. What is the effect of a large JOINT SOFTNESS setting?

26. When using the Motion Editor V.1, for each specific step within a Motion Page, there is a value that ones can set for the "PAUSE" field, does that "PAUSE" happen "before" or "after" the execution of that specific step?
27. Describe the procedure to calibrate a multi-link robot using a specially designed pose for this task.
28. What is the common linkage system used to convert a rotational motion into a back-and-forth motion?
29. What is the cycle time for execution of the CALLBACK function?
30. How many CALLBACK functions can be used in a TASK program?
31. What are the two steps needed to exclude a Dynamixel actuator from the effects of MOTION PAGES?
32. If the PRESENT POSITION of an actuator is at 512 and considering that its JOINT-OFFSET is set at −100, what is its final position when a TASK command is issued for it to go to GOAL POSITION 800?
33. What is the procedure to use to allow a manual adjustment of a typical Dynamixel actuator when it is powered?
34. The parameter "Present Load" can tell us about the current mechanical loading on a given actuator. (T-F)

6.8   Review Exercises for Chap. 6

1. The enclosed program "AX-12ChangingModes.tsk" demonstrates how to change an AX-12 to its Continuous Turn mode (set 0 to Address 8) and then to its Position Control mode (set 1023 to Address 8).
2. The enclosed program "AX-12MonitorPositionSpeed.tsk" is a basic data acquisition program that prints data onto the TASK tool's Output Window that the user can cut and paste into a data plotting application such as MS Excel.
3. The enclosed programs "PRM_GerwalkDemo.mtn" and "PRM_GerwalkDemo.tsk" work together to make a PREMIUM GERWALK (see Fig. 6.42) walk and avoid obstacles autonomously. Please modify the TASK code so that the user can control its movement via the RC-100, while keeping intact its obstacle avoiding functions.
4. Create a new Motion file (*.MTN) that would allow a GERWALK to kick a tennis ball. See enclosed video clip "Video_6_15.mp4" for several student solutions to a soccer penalty-kick simulation.
5. Create a new Motion file (*.MTN) that would allow a GERWALK or a BIPED to go up and down stairs steps:

   • Practice in generating appropriate motion pages combining CG shifting forward and up, backward and down as well as sideways recovery.
   • Alex Fouraker's 2008 GERWALK solutions up and down six stair steps (see enclosed video files "Video_6_16.wmv" and "Video_6_17.wmv") – this one used the AX-S1 to check for a step existence first before stepping up.

- Matthew Paulishen's 2011 solution for a BiPed going up six stair steps (see enclosed video file "Video_6_18.wmv").
- Also enclosed is a sample MTN file "BipedUpStairs.mtn" that the user could start from.

6. Compare the servo configurations between the BIOLOID HUMANOID type A and the ROBOTIS-MINI.
7. Contrast similarities and differences about how the three actuators were used between the two designs: BIOLOID HEXAPOD and TWITCH?
8. Ones could use the "independent" button approach as shown in "RC100_Carbot_3.tsk/tskx" to activate the manipulator algorithm, i.e. 3 RC-100 buttons would have to be pushed at the same time.
9. Starting from the files PRM_GerwalkDemo.mtn/mtnx and PRM-GerwalkDemo.tsk/tskx, please create a new TSK file named PRM_GerwalkRC.tsk/tskx that should allow the user to use the U-D-L-R buttons of the RC-100 controller to remotely control the normal motions of the bot (i.e. U for forward motion, D for backward motion, and similarly for R and L). Please keep "active" the autonomous obstacle avoiding features of the PRM_GerwalkDemo.tsk/tskx, i.e. if the user "drives" the bot into a wall, it should be able to avoid the wall regardless of which buttons are being pushed by the user during that time.
10. Start from the program "Gerwalk_LoadAdjust.tsk/tskx", modify it so that:

    (a) It can also print a TIMER value along with the existing parameters. Use the High-Resolution Timer if you happen to work with a CM-530.
    (b) Try different combinations of SLOPE, MARGIN and PUNCH values to see changes in performance of the robot.
    (c) Capture the data streams and plot the data with respect to the Timer counts.

11. Start from the program "Gerwalk_LoadAdjust_Fast.tsk/tskx", modify it so that:

    (a) It can also print a TIMER value along with the existing parameters. Use the High-Resolution Timer if you happen to work with a CM-530.
    (b) Try different MARGIN values to see changes in performance of the robot. Also try different values for the Gap divisor (currently set at 2).
    (c) Capture the data streams and plot the data with respect to the Timer counts.

12. Use the program "AX-12Gripper.tsk/tskx" with different value combinations of the parameters "GrippingPower" and "GrippingSoftness" and let the gripper "chomp" on your finger. Note the differences in feeling on your finger.

References

Abdel-Malek KA, Arora JS (2013) Human motion simulation. Academic, Waltham
Burdet et al (2013) Human robotics. The MIT Press, Cambridge
Chevallereau C et al (2009) Bipedal robots. Wiley, Hoboken
Cooke G (2011) Mobile robots. Wiley, Hoboken

Jazar RN (2010) Theory of applied robotics. Springer, Heidelberg

Kajita S, Espiau B (2008) Legged robots. In: Siciliano B, Khatib O (eds) Springer handbook of robotics. Springer, Heidelberg, pp 361–389

Kajita et al (2014) Introduction to humanoid robotics. Springer, Heidelberg

Kemp CC et al (2008) Humanoids. In: Siciliano B, Khatib O (eds) Springer handbook of robotics. Springer, Heidelberg, pp 1307–1333

Li Z, Ge SS (2013) Fundamentals in modeling and control of mobile manipulators. CRC Press, Boca Raton

Niku SB (2011) Introduction to robotics. Wiley, Hoboken

Villani L, Schutter D (2008) Force control. In: Siciliano B, Khatib O (eds) Springer handbook of robotics. Springer, Heidelberg, pp 161–185

Chapter 7
Communication Programming with "Remocon" Packets

For its robotics systems, ROBOTIS had devised two types of communication packets:

1. "Remocon" packets were first devised for the RC-100 controller, but they could be used for general user-programmable messages between various CM and OpenCM controllers, along with the PC or Smart Phone. This Chap. 7 described applications of these "Remocon" packets using ZigBee communications hardware which could be extended to NIR and BlueTooth hardware.
2. "Dynamixel" packets were designed to communicate with and to control ROBOTIS actuators, sensors, hardware controllers and external devices such as Smart Phones, thus these "Dynamixel" packets had a more complex structure than the "Remocon" packets. These "Dynamixel" packets were further classified as:

   (a) DXL 1.0 for Firmware 1.0 systems (for example, AX-12, AX-S1 and CM-5/510/530/700).
   (b) DXL 2.0 for Firmware 2.0 systems (for example, XM-430, XL-320, Dynamixel-Pro, CM-50, OpenCM-9.04/7.00 and OLLOBOT).

Applications of "Dynamixel" packets will be presented in Chaps. 11 and 12.

ROBOTIS offered three types of wireless communication hardware: NIR, ZigBee and BlueTooth:

- NIR was the default mode for its remote controller RC-100/A/B (http://www.robotis.us/rc-100b/) using the NIR Receiver OIR-10. The user could add the ZIG-100 module to it to get ZigBee capabilities (1-to-1, many-to-one, or broadcast), or the BT-100/BT-210 to get BlueTooth capabilities (1-to-1).
- The CM-5 system could only use the ZIG-100 module (or NIR via its AX-S1 module), while the CM-510, CM-530 and OpenCM-904 systems could interface with all three types of communication hardware.
- For ZigBee interfacing from Windows PCs to ROBOTIS systems, the user needed to use the hardware combination of USB2DYNAMIXEL/ZIG2SERIAL/ZIG-100/ZIG-110A. For BlueTooth interfacing, the user could use any PC

© Springer International Publishing Switzerland 2017
C.N. Thai, *Exploring Robotics with ROBOTIS Systems*,
DOI 10.1007/978-3-319-59831-4_7

BlueTooth devices (V. 2.0 and above) and either BT-110A, BT-210 or BT-410 (available after Summer 2015 and see Sect. 3.2.5).

Although ROBOTIS' NIR, ZigBee and BlueTooth communications hardware were different, applications using "Remocon" packets were standardized and based on a 16-bit user-designed message wrapped up in a 6-byte packet (http://support. robotis.com/en/product/auxdevice/communication/rc100_manual.htm):

- The first two bytes are "FF" and "55" represented the header information and would be fixed.
- A typical 16-bit user-defined message "Data" would be split into a lower byte "Data_L" fragment and a higher byte "Data_H" fragment by the ROBOTIS controller firmware which then reassembled them as the last 4 bytes of the Remocon packet "Data_L", "~Data_L", "Data_H" and "~Data_H", where for example "~Data_L" would represent the Inverse (1's Complement) of the "Data_L" byte.
- This scheme was used so that the controller's firmware could check on the integrity of a particular Remocon packet that it received.

Section 5.6 presented the basic communication concepts of transmitting and receiving data from the RC-100 (Virtual and Physical) on a 1-to-1 basis. Later in this chapter, we will take on more advanced concepts such as broadcast programming and message shaping.

This chapter's main topics are listed below:

- Hardware channel differences between ZIG-100 and ZIG-110A.
- Embedding signals to control multiple robots with a single RC-100 (NIR or ZigBee).
- Message shaping concepts as applied to Leader/Follower grippers and robots (bilateral control), and to multiple-user situations.
- PC to robot communications via Manager tool and via C/C++ programming tools.
- Comparing ZigBee and BlueTooth performances.

7.1 ZigBee® Broadcast Channel Differences

ROBOTIS designed their ZigBee modules to operate on four possible broadcast channels, one through four. ZIG-100 is defaulted to Channel 1, but could be changed to the other channels by modifying the status of Pins 7 and 8 (http://support.robotis. com/en/product/auxdevice/communication/zigbee_manual.htm). On the other hand, ZIG-110A was defaulted to Channel 4 and it was unchangeable. A further complication arose if the user needed to have ZigBee communications on the PC side by using the combination USB2Dynamixel/ZIG2Serial/ZIG-100 through an USB port, as the ZIG2Serial could also be set into 4 possible broadcast channels by opening or shorting 3 resistors R5, R6 and R7 (http://support.robotis.com/en/product/auxdevice/communication/zig2serial_manual.htm and see Fig. 7.1).

When using the ZigBee modules, either ZIG-100 or ZIG-110A, in a 1–1 mode such as in this example:

• ZIG-100 on an RC-100 or on a combo as shown in Fig. 7.1 on a PC.
• ZIG-110A on a CM-510 or CM-530.

The user only had to make sure that the ZigBee IDs matched via the MANAGER tool or by programming them directly inside a TASK program (http://support.robotis.com/en/product/auxdevice/communication/zigbee_manual.htm).

Also when the user needed to have broadcast ZigBee from the PC to other ZIG-100 modules, there would be no need for the user to do anything special at all ZigBee modules involved were defaulted to Channel 1 already.

However when the user needed to have ZigBee broadcast capability from the PC to a set of ZIG-110A modules, the user needed to modify a ZIG2Serial module "permanently" in such a way the ZIG2Serial from then on would broadcast only on Channel 4 (see Fig. 7.2).

Fig. 7.1 USB2Dynamixel/ZIG2Serial/ZIG-100 combo

Fig. 7.2 ZIG2Serial
module with R7 removed

7.2   Broadcast Use of RC-100 (NIR and ZigBee)

The goal of this section was to illustrate the extended use of a single RC-100 in the selective control of multiple robots. Please note that this project would work properly only when NIR or ZigBee communications hardware are used, as the BT-100/110A/210/410 hardware sets could only be used on a 1-to-1 basis (http://support.robotis.com/en/product/auxdevice/communication/bt100_110.htm).

A typical BIOLOID kit would come with an RC-100/A/B (http://www.robotis-shop-en.com/?act=shop_en.goods_view&GS=1487&GC=GD080302) and an NIR receiver   OIR-10   (http://www.robotis-shop-en.com/?act=shop_en.goods_view&GS=1291&GC=GD080301). The user could set the RC-100/A/B in NIR mode to 8 distinct 1-to-1 channels (http://support.robotis.com/en/product/auxdevice/communication/rc100_manual.htm). To complete the "communication loop", the user would also need to set the matching channel inside the TASK program (for example, ➿RC-100 Channel = CH3 in case Channel 3 was used). However the special Channel 0 (➿RC-100 Channel = CH0) would make the robot "listen" to any channel, effectively allowing several robots to receive commands from a single RC-100 remote controller (and so, in a way, achieved broadcast control via NIR). The user should also take note that NIR communications was constrained physically to a "line-of-sight", thus the group of robots would need to stay "close to one another".

If ones used the ZigBee route, i.e. using ZIG-100/110A, the "line-of-sight" constraint would be removed (http://support.robotis.com/en/product/auxdevice/communication/zigbee_manual.htm). However, the setting of the "broadcast" mode (Remote ID=65535) on the ZIG-100/110A hardware would be more involved. The ZIG-100 module would need to be mounted on a ZIG2SERIAL+USB2DYNAMIXEL hardware combination (as modified in Sect. 7.1) and set to "broadcast ID" from inside MANAGER using its ZIG2SERIAL Management tool (see Fig. 7.3 and review the video file "Video 7.1" and the "TestZigBee.tsk" program).

Furthermore a matching ZIG-110A module would need to be mounted on a typical CM-510/530 controller, and its "broadcast ID" (65535) would need to be set from the "Controller" panel inside the MANAGER tool (see Fig. 7.4). Please note that currently the MANAGER and TASK tools (V.1 and V.2) do not support such operation with the OpenCM-904-C controller.

The RC-100 had ten buttons: U-D-L-R and numerical 1 through 6. In this application, the intent was to keep the U-D-L-R buttons for directional control of the movements of a typical robot, while the numerical buttons (1,2,3) were used to select a specific robot among three BIOLOID STEM robots (see Fig. 7.5): BugFighter (i.e. button 1), Droid (i.e. button 2) and Hexapod (i.e. button 3).

Figure 7.6 was an excerpt from the file "RC_1_BugFighter.tsk" which described this basic algorithm as applied to the BugFighter:

- Statement 17 saved the "information" in the Remocon packet received into Parameter Message.
- Statements 19–22 separated out further information about the status of Parameters Button1, Button2, Button3 and Direction.

Fig. 7.3 Setting ZIG-100 to Broadcast mode (Remote ID=65535)

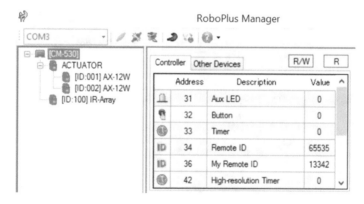

Fig. 7.4 Setting ZIG-110 to Broadcast mode (Remote ID=65535)

- Next, if Button 1 was set, only the BugFighter was moved around (Statements 24–28).
- Else If all three Buttons were not set (Statements 29–33), the BugFighter would also be moved around (and the other bots also).
- In any other situation, the BugFighter stayed still (Statements 34–37).

During run time, when the user used only the U-D-L-R buttons, all three robots would perform the same commanded maneuver, but if the user added one, two or all three (1,2,3) buttons, then only the selected robot(s) would perform the commanded maneuvers. Please review the 4 TASK programs (RC_1_BugFighter.tsk, RC_1_BugFighter_NIRReceiver.tsk, RC_2_Droid.tsk, RC_3_Hexapod.tsk) and also the video file "Video 7.2" for more details.

Fig. 7.5 Remote control
of three BIOLOID STEM
robots

```
16      // Save RC-100 signal received into variable called Message
17      Message  =  Remocon RXD
18      // Save status of Button1, Button2, Button3 and Direction
19      Button1  = Message  &  1
20      Button2  = Message  &  2
21      Button3  = Message  &  3
22      Direction  = Message  &  U+D+L+R
23
24      IF ( Button1  ==  1 )
25      {
26          // Only Bugbot moves around
27          CALL MoveAround
28      }
29      ELSE IF ( Button1  != 1 && Button2  != 2 && Button3  != 3 )
30      {
31          // All 3 bots move around
32          CALL MoveAround
33      }
34      ELSE
35      {
36          CALL Stop
37      }
```

Fig. 7.6 Basic control algorithm as applied to the BugFighter robot

7.3 Message "Shaping" Concepts

When using the RC-100 as previously described, we were using only 10 bits out of 16 bits possible that were transmitted or received via ROBOTIS communication protocols (NIR, ZigBee or BlueTooth) within a **single** "Remocon" communication packet. In this section, several applications would be described, exploiting this capability:

- Mimicking grippers.
- Leader and Follower GERWALKs.
- Multiple users and multiple robots (ZigBee only).

7.3.1 Mimicking Grippers

In this sub-section, two applications of 1-to-1 ZigBee message-shaping techniques would be described using two single-actuator grippers, each controlled by its own CM-5XX controllers as shown in Fig. 7.7.

The first application was of an "Open Loop" system as the command structure went only one way – from Leader to Follower. Please review the TASK programs ("CM5XX-MimicGripperL_OpenLoop.tsk" and "CM5XX-MimicGripperF_OpenLoop.tsk"), MOTION files ("BasicGripper16.mtn" and "BasicGripper17. mtn") and watch the video file "Video 7.3" for demonstration purposes, but the key concepts would be explained in the following paragraphs.

First, the only difference between the two MTN files mentioned earlier was that the Leader gripper had ID=16, while the Follower gripper had ID=17. The actual Motion Pages inside these MTN files had the same information:

- Page 1 (Close Grip) corresponded to Actuator Position 600 on the used AX-12s (ID=16 or ID=17).

Fig. 7.7 Leader and Follower Grippers using 1-to-1 ZigBee

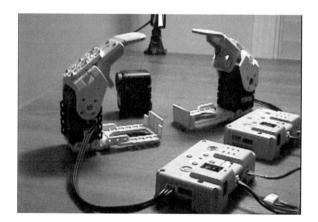

- Page 2 (Open Grip) corresponded to Actuator Position 430 on either AX-12.
- And Page 3 (Neutral Pose) corresponded to Actuator Position 512 on the same AX-12s.

Next the feature called "Torque Limit" (or "Goal Torque"), first described in Sect. 6.4.1, was applied to both grippers but in a differential manner:

- The Leader gripper had its "Torque Limit" set to 1, so as to allow the user to manipulate this gripper as needed (Statement 26 in "CM5XX-MimicGripperL_OpenLoop.tsk").
- The Follower gripper had its "Torque Limit" set to 1023 so as to allow it to do the real work (Statement 11 in "CM5XX-MimicGripperF_OpenLoop.tsk").

The most important procedure was how to shape the 16-bit message so that a single communication packet could contain as many pieces of information that were needed to accomplish a given job. For this "Open Loop" application, the need was for the Leader to send its Actuator ID (=16) and its Present Position (variable value) over to the Follower which would then mapped the Leader's Actuator ID (=16) to match with its own corresponding ID (=17) and to set its Goal Position to the same value as the Leader's Present Position as received in the 16-bit message:

- On the Leader side ("CM5XX-MimicGripperL_OpenLoop.tsk"), the goal was to configure the 16-bit message so that it contained the Actuator ID in the highest 6 bits and the Actuator's Present Position (Address 36, and numerical range [0–1,023]) in the lowest 10 bits (see Fig. 7.8). This meant that the ID value needed to be shifted 10 bits to the left of the 16-bit message which was accomplished by "multiplying" ID with the binary constant 0000 0100 0000 0000 (statement 31). Next the current Present Position value was added to the same 16-bit message (statement 33). This final combined message was then sent to the Follower with the "Remocon_TXD" function (statement 35). In the rest of this TSK file, minor procedures were written to cycle through all actuators used if more than one were used (statements 38–46), and to include some do-nothing code so as not to overhelm the ZigBee channel (statements 49–53).
- On the Follower side ("CM5XX-MimicGripperF_OpenLoop.tsk"), Fig. 7.9 showed how to unpack the received 16-bit message so that it contained the Actuator ID in the highest 6 bits, by first bit-wise ANDing it with "1111 1100 0000 0000" (statement 23) and then by shifting right this temporary result by 10 bits with its division by "0000 0100 0000 0000" (statement 25). Statement 27 was used to adjust for the difference between the IDs used for the Leader and

```
30    // Move current Leader ServoID to highest 6 bits to create ServoID part of message, i.e. shift left 10 bits
31    MessageSent = 0000 0100 0000 0000 * ID
32    // Combine previous ServoID part and Current Position part (address 36 and 10-bit data) into one single 16-bit message
33    MessageSent = MessageSent +  ID[ID]: ADDR[36(w)]
34    // and send it to Follower
35     Remocon TXD  = MessageSent
```

Fig. 7.8 Preparing message to be sent by leader bot

```
21    MessageReceived  = ▶ Remocon RXD
22    // Filtering out the highest 6 bits to find ServoID part of MessageReceived
23    ServoID  = MessageReceived  &  1111 1100 0000 0000
24    // Map the highest 6 bits to bits 6, 5, 4, 3, 2, 1; i.e. shift right 10 bits
25    ServoID  = ServoID  /  0000 0100 0000 0000
26    // Map Leader ServoID (16 for my case) to corresponding Follower ServoID (17 for my case - adjust to suit your situation
27    ServoID  = ServoID  +  1
28    // Filtering out the lowest 10 bits to get the Servo Position part of the message
29    TargetPosition  = MessageReceived  &  0000 0011 1111 1111
```

Fig. 7.9   Processing message received by follower bot

Follower grippers (the user may not have to perform this step if he or she chose the same ID for the actuators). Next the lowest 10 bits was filtered out into the parameter TargetPosition by ANDing the 16-bit message with "0000 0011 1111 1111" (statement 29). Lastly, a precautionary step was taken to check the TargetPosition's value to see if it was less or equal to 1,023 before actually assigning it to the Follower's Goal Position (i.e. address 30 in statement 34).

The second application would be a "Closed Loop" system whereas information flowed both ways between Leader and Follower. Please review the TASK programs "CM5XX-MimicGripperL_ClosedLoop.tsk" and "CM5XX-MimicGripperF_ClosedLoop.tsk" and the video file "Video 7.4" for demonstration purposes, but the key concepts would be explained further in the following paragraphs.

In addition to the features/concepts already explained for the "Open-Loop" case, there were additional codes created to handle the bilateral flow of messages:

- On the Follower side (CM5XX-MimicGripperF_ClosedLoop.tsk), statements 40–76 (see Fig. 7.10) were added to handle:

  1. Prepare the 16-bit message with the ServoID value in the highest 6 bits (statement 41).
  2. Save the Present Load value (address 40) into parameter LoadData and if LoadData was larger than LoadLimit (=512), i.e. there was some obstruction present, then the Follower needed to open up the gripper (statement 47) and sounded an audio alarm (statements 48–57). Next the Follower added its LastPosition (address 36) into the 16-bit message and sent it away to the Leader (statements 58–61). Finally it went into a While Loop waiting for the user to push the "R" button on the CM-5XX controller (essentially to reset this alarm situation – statements 64–68). When "R" was eventually pushed, the Follower sounded the "All Clear" (statements 69–75) and got back to its Main Loop waiting for messages to come in from the Leader gripper.
  3. If LoadData was less than LoadLimit, i.e. no obstruction encountered, the Follower just got back to its Main Loop and waited for messages to come in from the Leader gripper.

- On the Leader side (CM5XX-MimicGripperL_ClosedLoop.tsk), statements 59–102 (see Fig. 7.11) were created to handle possible messages coming from the Follower (in case of an obstruction occurring to the Follower):

```
41    MessageSent  = 0000 0100 0000 0000   * ServoID
42    LoadData = █ ID[ServoID]: ADDR[40(w)]  & 0000 0011 1111 1111
43    IF ( LoadData  >  LoadLimit )
44    {
45        // Open up gripper to release the Load
46        █ ID[All]: ◔ Goal Torque  = 512
47        ◉ Motion Index Number  = OpenPage
48        LOOP WHILE ( ◉ Motion Status  ==  TRUE )
49        {
50            // Sound the alarm
51            █ ID[100]: ♪ Buzzer index  = La#(1)
52            █ ID[100]: ◉ Buzzer Time  = 0.128sec
53            // WAIT WHILE ( █ ID[100]: ◉ Buzzer Time > 0.000sec )
54            ♪ Buzzer index  = La#(1)
55            ◉ Buzzer Time  = 0.128sec
56            WAIT WHILE ( ◉ Buzzer Time > 0.000sec )
57        }
58        // Servo Current Position (address 36) should be about at the Open Position by now
59        LastPosition  = █ ID[ServoID]: ADDR[36(w)]
60        MessageSent  = MessageSent  + LastPosition
61        ◗ Remocon TXD  = MessageSent
62
63        // Wait for user to push R button to reset the system
64        █ ID[All]: ◔ Goal Torque  = 1023
65        LOOP WHILE ( █ Button  !=  █ R )
66        {
67            █ ID[ServoID]: ADDR[30(w)]  = LastPosition
68        }
69        // Sound the all clear
70        █ ID[100]: ♪ Buzzer index  = Melody0
71        █ ID[100]: ◉ Buzzer Time  = Play Melody
72        // WAIT WHILE ( █ ID[100]: ◉ Buzzer Time > 0.000sec )
73        ♪ Buzzer index  = Melody0
74        ◉ Buzzer Time  = Play Melody
75        WAIT WHILE ( ◉ Buzzer Time > 0.000sec )
```

Fig. 7.10 Actions by follower bot in closed loop situation

1. After parsing the Follower's 16-bit message into the ID and TargetPosition components (statements 61–69), the Leader "stiffened" up by setting its Torque Limit setting to 512 (statement 71) and by going to the "Open" position while sounding an audio alarm (statements 72–84). At this point, the user would no longer be able to manipulate the Leader gripper as before.

2. Next it went into a While Loop waiting for the user to push the "R" button on the CM-5XX controller (essentially to reset this alarm situation – statements 86–90).

3. When "R" was eventually pushed, the Follower went "soft" again by resetting its Torque Limit to 1 (statement 92) and sounded the "All Clear" (statements 95–101) and got back to its Main Loop generating 16-bit messages to send over to the Follower gripper, according to the user's manipulations of the robot.

```
59    IF ( ⤵Remocon Data Received  ==  TRUE  )
60    {
61        MessageReceived  = ⤵Remocon RXD
62        // Filtering out the highest 6 bits to find ServoID part of MessageReceived
63        MotorID  = MessageReceived  &  1111 1100 0000 0000
64        // Map the highest 6 bits back down to bits 6, 5, 4, 3, 2, 1, i.e. shift right 10 bits
65        MotorID  = MotorID  /  0000 0100 0000 0000
66        // Map Leader ServoID (16 for my case) to corresponding Follower Servo ID (17 for my case)
67        MotorID  = MotorID  -  1
68        // Filtering out the lowest 10 bits to get the Servo Position part of the message
69        TargetPosition  = MessageReceived  &  0000 0011 1111 1111
70
71        🔩 ID[All]: ◌ Goal Torque  = 512
72        🔩 ID[All]: 🎚 Moving Velocity  = 256
73        // Opening up Leader gripper
74        ▶ Motion Index Number  = OpenPage
75        LOOP WHILE ( ▶ Motion Status  ==  TRUE  )
76        {
77            // and sound the alarm
78            🔩 ID[100]: ♪ Buzzer index    Do#(4)
79            🔩 ID[100]: 🔔 Buzzer Time    0.5sec
80            WAIT WHILE ( 🔩 ID[100]: 🔔 Buzzer Time  > 0.0sec )
81            ♪ Buzzer index  = Do#(4)
82            🔔 Buzzer Time  = 0.5sec
83            WAIT WHILE ( 🔔 Buzzer Time  > 0.0sec  )
84        }
85
86        // Wait for user to reset the system
87        LOOP WHILE ( 🔘 Button  !=  🔘 R  )
88        {
89            🔩 ID[MotorID]: ADDR[30(w)]  = TargetPosition
90        }
91
92        🔩 ID[All]: ◌ Goal Torque  = 1
93        🔩 ID[All]: 🎚 Moving Velocity  = 1
94
95        // Sound the all clear
```

Fig. 7.11 Actions by leader bot in closed loop situation

7.3.2 Leader-Follower GERWALKS

This sub-section extended the previous message-shaping and bilateral control concepts to a more complete robot such as the GERWALK. In this closed-loop application, both Leader and Follower robots were supposed to be ID'd identically (ID=1 to ID=7 in this particular case).

Similarly, please review the TASK programs (Gerwalk_L_ClosedLoop.tsk) and Gerwalk_F_ClosedLoop.tsk), GerwalkDemoMotion.mtn file and the video file "Video 7.5" for demonstration purposes, and the key concepts would be explained in the following paragraphs.

In this application, as we were dealing with seven actuators on the Leader robot, the overall plan was to send "ID" and "Present Position" information for only one actuator in each 16-bit message as before, and additionally used a scheme to cycle through the IDs from 1 to 7 (see Fig. 7.12):

- On the Leader side (Gerwalk_L_ClosedLoop.tsk):

  1. Parameter ID started at 1 (statement 8) and all "Torque Limits" were set to 1 (statement 19) to allow user to manipulate the Leader robot at will.
  2. The endless loop starts with statement 23 where the ID component was shifted to the highest 4 bits while the corresponding Present Position (Addr. 36) stayed in the lower bits (statement 24). The combined message was then sent to the Follower Gerwalk (statement 25).
  3. Statements 27–34 were used to update Parameter ID and cycle it through from one to seven.
  4. The 50-count delay section (statements 35–40) was critical to the behavior of the Follower. For any shorter delay used, the Follower would loose its

```
23      MessageSent   = 0001 0000 0000 0000    * ID
24      MessageSent   = MessageSent  +  🔋 ID[ID]: ADDR[36(w)]
25      🔌 Remocon TXD   = MessageSent
26
27      IF ( ID  >=  7 )
28      {
29          ID  = 1
30      }
31      ELSE
32      {
33          ID  = ID  +  1
34      }
35      // This 50-count delay is critical to the behavior of the slave
36      Delay  = 50
37      LOOP WHILE ( Delay  != 0 )
38      {
39          Delay  = Delay  -  1
40      }
41
42      IF ( 🔌 Remocon Data Received   ==  TRUE  )
43      {
44          MessageReceived   = 🔌 Remocon RXD
45          MotorID  = MessageReceived  &  1111 0000 0000 0000
46          MotorID  = MotorID  /  0001 0000 0000 0000
47          TargetPosition  = MessageReceived  &  0000 1111 1111 1111
48          🔋 ID[MotorID]: ADDR[34(w)]  = 512
49          🔋 ID[MotorID]: ADDR[30(w)]  = TargetPosition
50          ⏱ Timer  = 0.128sec
51          WAIT WHILE ( ⏱ Timer  != 0.000sec  )
52          🔋 ID[MotorID]: ADDR[34(w)]  = 1
53      }
```

Fig. 7.12 Algorithm used by Leader Gerwalk bot in closed loop situation

synchronization, as it could skip on the message received and consequently could also skip setting up properly one or more of its seven actuators.

5. Statements 42–53 were used to decode potential messages from the Follower (when some or all its actuators detected an overload at a certain TargetPosition). In this situation, the Leader would "stiffen" up the corresponding actuator (defined by MotorID) by raising its Torque Limit value to 512 and by holding it at the TargetPosition sent over by the Follower (statements 48–49), but only for 0.128s, and then by resetting it back to 1 (statements 50–52). The net "physical" result would be that the human user would feel this particular actuator "twitch" or "tremble" in his or her hand.

- On the Follower side (Gerwalk_F_ClosedLoop.tsk) – see Fig. 7.13:

1. Similarly as for the Follower gripper, statements 22–29 decoded the message from the Leader GERWALK and set the corresponding actuators to the TargetPosition values received.

2. Statements 31–46 essentially cycled through the Follower's 7 actuators to see which one had its Present Load value (address 40) higher than the LoadLimit of 512. If such a situation arose, the Follower would transmit its own message to the Leader regarding this overloading actuator along with its ID and Current

```
22        MessageReceived   = ➥ Remocon RXD
23        MotorID = MessageReceived  &  1111 0000 0000 0000
24        MotorID = MotorID  /  0001 0000 0000 0000
25        TargetPosition = MessageReceived  &  0000 1111 1111 1111
26        IF ( TargetPosition  <=  1023 )
27        {
28             ⬛ ID[MotorID]: ADDR[30(w)]   = TargetPosition
29        }
30
31        MessageSent  = 4096  *  ID
32        LoadData = ⬛ ID[ID]: ADDR[40(w)]   &  0000 0011 1111 1111
33        LoadLimit = 512
34        IF ( LoadData  >  LoadLimit )
35        {
36             MessageSent  = MessageSent  +  ⬛ ID[ID]: ADDR[36(w)]
37             ➥ Remocon TXD   = MessageSent
38        }
39        IF ( ID  >=  7 )
40        {
41             ID = 1
42        }
43        ELSE
44        {
45             ID = ID  +  1
46        }
```

Fig. 7.13 Algorithm used by Follower Gerwalk bot in closed loop situation

Position (address 36). Please note that this process was performed independently of the setting of the TargetPosition values as required in the messages sent by the Leader (previous paragraph 1.). In other words, messages were constantly "prepared" but not necessarily "sent" by the Follower to the Leader.

This sub-section was a very basic introduction to the bilateral control concept used often in tele-operation systems designed to work in hazardous environments or in medical robot systems. A few selected references are: Vertut (2012), Minh (2013) and Milne et al. (2013).

7.3.3 Multiple Users and Multiple Robots (ZigBee Only)

This sub-section extended the message-shaping technique developed by Mr. Matthew Paulishen (during his student days at my laboratory) for his Mobile Wireless Sensor Network project (see Thai and Paulishen (2011)).

Figure 7.14 illustrated the targeted situation where we would have several users (i.e. several RC-100s with ZigBee being used in broadcast mode) trying to control their own robots (Twin GERWALK) which also had two independent controllers in their design (Leader and Follower, also set to broadcast mode). Please see enclosed video file "Video 7.6" for a solution by Duke TIP students.

Figure 7.15 described the detailed bit assignment when using the RC-100s to send messages out to Leader robots. Essentially, each user was allowed to use the U-D-L-R buttons for maneuvering and the numerical buttons 1 and 2 for special instructions (i.e. $2^2-1=3$ separate special instructions allowed). Each user used

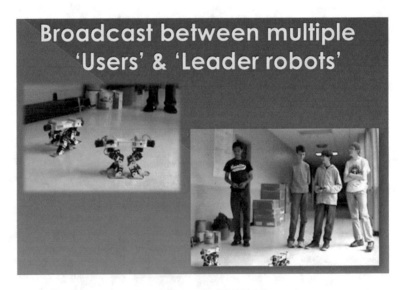

Fig. 7.14 Multiple users controlling multiple twin GERWALKs

Fig. 7.15 Bit configuration for 16-bit messages broadcasted from RC-100s

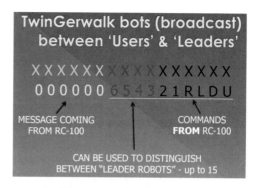

Fig. 7.16 Bit configuration for 16-bit messages broadcasted from robots

buttons 3 through 6 to denote the specific Leader robots that would be assigned to particular users (with 4 bits, that would amount to 15 potential Leaders).

The reader should note that this was a broadcast environment so all the distinctions to be made had to be embedded in the 16-bit messages which were "constantly" broadcasted among the RC-100s and robots at play. Also the reader probably already noted the "honor" system that would have to be followed as any user could "jam" or "misdirect" the other user's robot with the above scheme.

Figure 7.15 also showed that if a message came from the RC-100s, its highest 6 bits would be set to 0 and this feature would be used to distinguish the messages coming out the RC-100s from the ones coming out from either Leader or Follower robots as shown in Fig. 7.16.

Essentially, codes running on the robots would look for the pattern "000000", "101010" or "010101"in the highest 6 bits of every message that they received:

- If this pattern was "000000", then this message came from the RC-100s and would need to be processed further by the Leader robots **only**. Then the Leader robots would investigate the lowest 10 bits according to Fig. 7.15 to decode what actions their assigned users wanted them to do.
- When executing those user-assigned actions, the Leader robots would definitely need to send their own messages to their matching Followers to coordinate the required physical actions between Leader and (matching) Follower to perform a given maneuver. These messages would have the pattern "101010" embedded in the highest 6 bits. The lower 6 bits would contain the "true" message between matching

Leader and Follower, while the mid-4 bits would be used to distinguish which Leader-Follower units (among the 15 possible) that this message pertained to.

* When the Follower robots needed to send their own messages to their matching Leaders, they would use the pattern "010101" in the highest 6 bits.

7.4   PC to Robots Communications via C/C++

If the reader wanted to do some communications programming between the PC and ROBOTIS' robots, but still stayed within the RoboPlus environment, then he or she could use the "Zig2Serial Management" sub-tool inside RoboPlus' Manager tool. As its name indicated, the "Zig2Serial Management" sub-tool was created to help the user with managing the Zig2Serial module when used with a ZIG-100. This sub-tool would allow the user to set the ZIG-100's communication mode (1 to 1, many to 1 and broadcast – as shown in the previous video file "Video 7.1") and also to send numerical data via the associated Windows COM port regardless of which wireless hardware was actually used – ZigBee or BlueTooth (as will be used in Chap. 11 for the ROBOTIS-MINI). As we were using the ROBOTIS' protocol with the standard 6-byte packet, we would also have to stay with the closed-firmware option on the robot side, i.e. operational with all the CM-5XX systems and OpenCM-904-C (closed-firmware mode), but not workable with OpenCM-904-A/B as they used OpenCM IDE.

If the reader wished to use C/C++ programming on the PC side for more flexibility, but still stayed with the closed-firmware option on the robot side to benefit from the TASK and MOTION features of the RoboPlus Suite, then he or she needed to have a closer look at the contents of the enclosed file "WirelessRobot.zip". This ZIP file contained all source codes and workspace files needed when using MicroSoft Visual Studio Express (2010, 2012, 2013 and 2015 editions and even Visual C V.6). This package was created by Mr. Matthew Paulishen with a simple command-line interface with the goal of introducing C/C++ constructs needed to properly integrate with the "ZigBee" and "Dynamixel" SDKs from ROBOTIS (http://support.robotis. com/en/software/zigbee_sdk.htm and http://support.robotis.com/en/software/dyna-mixelsdk.htm). After unzipping this file, the folder "WirelessRobot\code\app\ WirelessRobot DataLogger Binaries" contained the various executables for the DataLogger projects and the TASK files "WirelessCarbot.tsk" and "CM5xxDataLogger.tsk" for the reader to try out (*Note: Depending on what DLLs were installed on the reader's PC, not all executable programs would work properly, but the VC6 version should work for all Windows OS from XP to 10*).

The video file "Video 7.7" illustrated the use of the "DataLogger" executables with a CM-5 CarBot running the "WirelessCarbot.tsk" program via ZigBee hardware. First it demonstrated how the Zig2Serial Management sub-tool (of ROBOPLUS MANAGER) worked with the "WirelessCarbot.tsk" program. Next, it showed how the Visual C/C++ executable "DataLogger.exe" program worked with the same TASK program.

The much longer video file "Video 7.8" illustrated the various C/C++ constructs for understanding ZigBee Communications programming for Visual Studio Express 2012 (but it should applicable to later editions also). This video also referred to the "CM5XXDataLogger.tsk" files.

7.5   ZigBee and BlueTooth® Performances

ROBOTIS ZigBee hardware was of an older design so its maximum baud rate was 115,200 bps, while the latest BT-210 module could potentially go to 400,000 bps (but operated around 250 kbps) and besides BlueTooth devices are widely available in current PCs and mobile devices. Surprisingly, the newer BT-410 series would operate lower at 128 kpbs to save energy on mobile devices. Thus if the reader plans to work with only one robotic system from ROBOTIS, and if he or she already has BlueTooth on their computing platforms, then BlueTooth is the obvious choice for accessibility and cost. However, BlueTooth usually takes rather a long time to connect as compared to ZigBee. BlueTooth also required two COM ports on the PC for each BT connection (one outgoing and one incoming) while ZigBee required only one COM port. This "feature" could become important if the user is interested in controlling teams of robots from a central PC. Hughes et al. (2013) shared some hands-on knowledge for BlueTooth applications to teams of LEGO NXT robots that the reader might find useful. Huang and Rudolph (2007) have essential information for BT programmers in Windows and Linux environments.

In Sect. 4.3, the author had shown that for situations with fast changing communication data, ROBOTIS ZigBee hardware/software outperformed ROBOTIS BlueTooth hardware/software (at least for now).

Lastly, from the point of view of educational flexibility, the option of changing to different ZigBee modes (1 to 1, many to 1 and broadcast) clearly favored ZigBee over BlueTooth. Thus overall, the author still preferred ZigBee over BlueTooth if the development environment can accommodate USB/ZigBee communications.

On mobile devices, all of us would have no choice but to use BlueTooth! The newer BT-410 is supposed to allow one master controlling several slave devices, so another "seesaw" battle!

7.6   Review Questions for Chap. 7

1. List the three types of wireless communication protocols that are available with ROBOTIS systems.
2. How many NIR communication channels are available with the combination of RC-100 and OIR-10 modules?
3. Describe how to set up a chosen RC-100 Channel inside a TASK program.
4. The MANAGER tool can be used to set up ZigBee communication IDs. (T-F)

5. The MANAGER tool can be used to set up BlueTooth communication IDs. (T-F)
6. Describe how to set up a chosen RC-100 Channel on an RC-100 Remote Controller.
7. On ROBOTIS systems, NIR communications are strictly on a 1-to-1 basis. (T-F)
8. On ROBOTIS systems, BlueTooth communications are strictly on a 1-to-1 basis. (T-F)
9. Describe the current options for getting BlueTooth communications among ROBOTIS systems.
10. Describe the hardware hook-up needed to communicate between two CM-530s via BlueTooth communications.
11. Describe the hardware hook-up needed to communicate between a CM-530 and a PC via BlueTooth communications.
12. Describe the hardware hook-up needed to communicate between a CM-530 and an RC-100 via BlueTooth communications.
13. How many COM port(s) are open on the PC side for each BlueTooth module?
14. How many COM port(s) are open on the PC side for each ZigBee module?
15. Describe the hardware hook-up needed to communicate between two CM-5s via NIR communications.
16. Describe the hardware hook-up needed to communicate between two CM-5s via ZigBee communications.
17. Describe the hardware hook-up needed to communicate between a CM-5XX and an RC-100 via ZigBee communications.
18. Describe the hardware hook-up needed to communicate between a CM-5XX and a PC via ZigBee communications.
19. What are the three modes of communications for ROBOTIS ZigBee protocols?
20. Can the Zig2Serial sub-tool inside the MANAGER tool be used with BlueTooth communications between the PC and a robot?
21. Describe the standard packet configuration used by ROBOTIS for its wireless communications network.
22. Within a standard communication packet, how many bits does the actual "message" component contain?
23. Assuming that Parameter B=0000 0000 0000 1011 , what is the result for Parameter A, after this TASK command was executed?

    A=B & 0001 0000 0000 1100
24. Assuming that Parameter B=0000 0000 0000 1011, what is the result for Parameter A, after this TASK command was executed?

    A=B* 0000 0100 0000 0000
25. Assuming that Parameter B=0110 0100 0000 0000 , what is the result for Parameter A, after this TASK command was executed?

    A=B/ 0000 0100 0000 0000
26. Describe procedure(s) to achieve broadcast communications with ROBOTIS NIR hardware and software.

27. Describe procedure(s) to achieve broadcast communications with ROBOTIS ZigBee hardware and software.
28. Describe procedure(s) to achieve broadcast communications with ROBOTIS BlueTooth hardware and software.
29. Review the information found from the following web links and design an alternate approach to obtain ZigBee broadcast capability from the PC to a group of ZIG-110A devices:

   (a) http://support.robotis.com/en/product/auxdevice/communication/zigbee_manual.htm.
   (b) http://support.robotis.com/en/product/auxdevice/communication/zig2serial_manual.htm.
   (c) http://support.robotis.com/en/software/roboplus/roboplus_manager/testandconfigure/etc/roboplus_manager_zig2serial.htm.

7.7 Review Exercises for Chap. 7

1. Adapt the programs "Gerwalk_L_ClosedLoop.tsk" and "Gerwalk_F_ClosedLoop.tsk" to your own favorite multi-linked robots.
2. Add to the previous programs features of audio alarms using the ideas proposed in "CM5XX-MimicGripperL_ClosedLoop.tsk" and "CM5XX-MimicGripperF_ClosedLoop.tsk".
3. Starting from the Visual C++ files inside "WirelessRobot.zip", the goal of this assignment is twofold:

   (a) To create a new TSK file "WirelessCarbot.tsk" that can accept "U-D-L-R" (for motion directions) and "1–2" (for Low-High speed settings) commands from the PC keyboard via ZigBee (as already shown in the example "CMDatLogger.TSK" and "main.CPP" files included in "WirelessRobot. zip"). Additionally and independent of the previous "direction" and "speed" commands, when receiving a special command from the PC, it should respond back to the PC with the designated NIR sensor reading(s) corresponding to the "Left-Center-Right" sensors of the AX-S1 module.
   (b) To modify the "main.cpp" file so that it can do the following procedures:

      (i) It does not have to "echo" the various "U-D-L-R-1-2" commands onto the run-time PC display anymore, as the user can see those effects directly on the Carbot's behaviors.
      (ii) Initially, it should display an option menu telling the user various keyboard actions that are available to the user:

         • "W-A-S-D" keys for "U-L-D-R" directions.
         • "1–2" keys for carbot speed setting.
         • "O" for a request to do a 1-time scan (and PC-side display) of the **user-chosen** NIR sensor(s) on the Carbot.

- "C" for a request to do a continuous scan (and PC-side display) of the **user-chosen** NIR sensor(s) on the Carbot.
- When either "O" or "C" options are in effect, the PC-side program should also ask the user for "how many" and "which" sensors for the Carbot to scan and transmit the data back to the PC.
- "S" should stop the "continuous scan" mode.

4. Mobile Wireless Sensor Network Project. Using the Thai and Paulishen (2011) IEEE paper, design a wireless mobile sensor network with 1 PC acting as the control station, and 3 Carbots acting as mobile scouts that can send back to the control station data from its three NIR sensors located in the AX-S1 module. The 2011 IEEE paper describes in fairly good details a possible approach to be used that you can adapt or come up with your own scheme as you wish:

- Your system should be able to issue commands for the three Carbots to disperse and spread themselves out as far as possible from the base station, but without losing ZigBee communications with each other (PC and Carbots). Essentially, the Carbots will string themselves out so as to form a relay system whereas the furthest-out Carbot will send its data to the middle Carbot which would then relay those messages to the one nearest to the base station which would next relay those messages to the base station as the last step. A similar relay scheme should be applicable for the middle and nearest Carbots with less data hops to perform.
- The base station would then display on the PC display the information that it got from each specific Carbot.
- Not implemented in the 2011 IEEE paper was the capability to steer a particular Carbot from the PC (a laptop would be best) – as the human user would be able to see where the Carbots were going down a hallway for example (see enclosed video file "Video 7.9" to see a completed project).

References

Huang AS, Rudolph L (2007) Bluetooth essentials for programmers. Cambridge University Press, Cambridge

Hughes C, Hughes T, Watkins T, Kramer B (2013) Build your own team of robots with LEGO MINDSTORMS NXT and Bluetooth. McGraw-Hill, New York

Milne B et al (2013) Design and development of teleoperation for forest machines: an overview. In: Habib MK, Davim JP (eds) Engineering creative design in robotics and mechatronics. IGI Global, Hershey, pp 186–207

Minh VT (2013) Development and simulation of an adaptive control system for the teleoperation of medical robots. In: Habib MK, Davim JP (eds) Engineering creative design in robotics and mechatronics. IGI Global, Hershey, pp 173–185

Thai CN, Paulishen M (2011) Using ROBOTIS BIOLOID systems for educational robotics. http://docs.wixstatic.com/ugd/714442_db5509e2dd144d41a206ca44d0dea5ef.pdf. Accessed 11 June 2017

Vertut J (2012) Teleoperation and robotics: applications and technology. Springer, Heidelberg

Chapter 8
Advanced Sensors

In Chap. 3, advanced sensors such as the AX-S20, GS-12, FPS and HaViMo2 were briefly described. In this chapter, more application details would be provided whereas the CallBack function was a critically needed tool.

This chapter's main topics are listed below:

* Static balance of humanoid robot using the AX-S20 (two-Leg and one-Leg versions).
* Dynamic balance of humanoid robot using the GS-12 for Walking Enhancement and Fall Detection.
* FPS application to a humanoid robot's feet (one Leg balance).
* HaViMo2 application to CarBot.

8.1 Humanoid Static Balance with AX-S20

In this section, a Humanoid A robot from the BIOLOID PREMIUM kit was used as the test platform (see Fig. 8.1). It actually had the AX-S20 mounted in its head and the GS-12 mounted inside the robot, just above the hip's actuators.

As the AX-S20 had a magnetometer, it had to be installed as far away as possible from the actuators which were generating strong and fluctuating magnetic fields which would interfere with the workings of this sensor. Thus it was mounted in its head (fortunately there was enough room in it). On the other hand, as the GS-12 (designed by ROBOTIS Vice-President In-Yong Ha) was based on MEMS sensors, there was no need to worry about magnetic interferences from the actuators, so it was mounted as much as possible near the center of gravity of the robot (i.e. on top of the hip's actuators).

© Springer International Publishing Switzerland 2017
C.N. Thai, *Exploring Robotics with ROBOTIS Systems*,
DOI 10.1007/978-3-319-59831-4_8

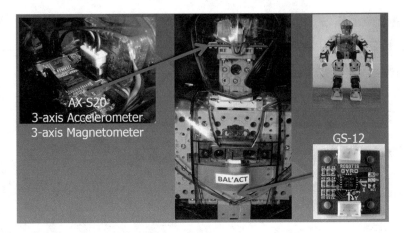

Fig. 8.1 PREMIUM Humanoid A with AX-S20 and GS-12

8.1.1   Two-Leg Static Balance with AX-S20

Figure 8.2 showed how the AX-S20 would appear inside the MANAGER tool as a Dynamixel-compliant sensor. Please note the various parameters (i.e. addresses) accessible to the user from inside a TASK program (using "Custom" commands with a Word size):

- Azimuth angle with respect to Z-axis, in degrees [0–359] (address 26).
- Pitch (with respect to Y-axis) and Roll (with respect to X-axis) angles in degrees, [−89 to 89], at addresses 28 and 30 respectively. Please see Fig. 8.3 for the coordinates system used by the AX-S20.
- "Real-time" X-Y-Z acceleration values could also be read via addresses 38, 40 and 46 respectively, [−2,048 to 2,048], i.e. at 0.01225 m/s^2 per count. The AX-S20 has about a 20 Hz refresh rate (i.e. rather slow, as it was designed around 2009).

The first AX-S20 application derived from a ROBOTIS demo program (c. 2009) which described how to read its Pitch and Roll angles and use those real-time data to adjust the Joint Offsets of selected actuators of both legs so that the Humanoid robot could maintain its "static" balance even though the platform, where it was standing on, was changing its inclination angles with respect to the Pitch and Yaw axes (see Video 8.1 and Fig. 8.4).

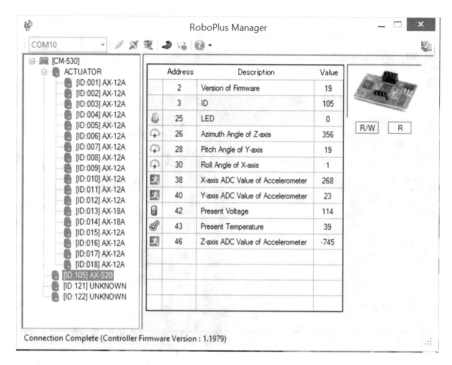

Fig. 8.2 AX-S20 panel inside MANAGER tool

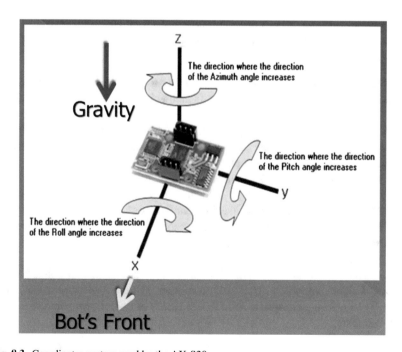

Fig. 8.3 Coordinates system used by the AX-S20

Fig. 8.4 Humanoid type A
two-leg balancing using
AX-S20

This application used the following files: "BalAct_2Legs_AXS20_Balance.tsk/
tskx" and "BalAct_2Legs.mtn/mtnx". The basic steps in the balance control algorithm used were as follows:

1. Get the robot into the initial pose as smoothly as possible by "playing" Motion
   Page 1 and then Motion Page 129 (statements 6–9 in the "BalAct_2Legs_
   AXS20_Balance.tsk/tskx" file).
2. Collect ten consecutive values of the X and Y components of gravity (addresses
   38 and 40) and save their average values respectively into the parameters
   "FBBalCenter" and "RLBalCenter" (statements 10–20).
3. Next play Motion Page 128 (balance position at statement 27) which had the
   same Goal Position values for all the actuators as Motion Page 129, except that
   "128" would keep on calling itself indefinitely, essentially enabling the Joint
   Offset functions to work in maintaining the robot balance position, when the user
   started to modify the inclination angles of the supporting platform, as shown in
   Video 8.1.
4. Effectively, "FBBalCenter" and "RLBalCenter" became the set points for this
   control algorithm which would seek to minimize future deviations of these values by triggering appropriate Joint Offset Values of the following actuators with
   ID numbers:

(a) (11, 13, 15) for the right leg and (12, 14, 16) for the left leg. Please note that these actuators influenced most directly the forward-backward motions (see Fig. 8.5).

(b) (17) and (18) which respectively influenced the right-left motions of the "ankles".

(c) In other words, this robot would seek to balance itself by adjusting its "ankles" and "knees" and by "crouching" up or down differentially between its right and left legs. Of course other solutions would also be possible by bringing actuators 9 and 10 in the mix for side-shifting of the hip.

5. This control algorithm was actually carried out "in parallel" of the main program by a CALLBACK function (statements 47–117) (http://support.robotis.com/en/ software/roboplus/roboplus_task/programming/command/roboplus_task_cmd_ callback.htm). ROBOTIS designed CALLBACK to be activated every 8 ms which happened to be the hardware refresh cycle time for all Dynamixels. Because of this rather short time period, the programmer could not use logical constructs such as loops and was limited to a maximum of two hardware calls, and furthermore the size of the CALLBACK function could not exceed 512 bytes.

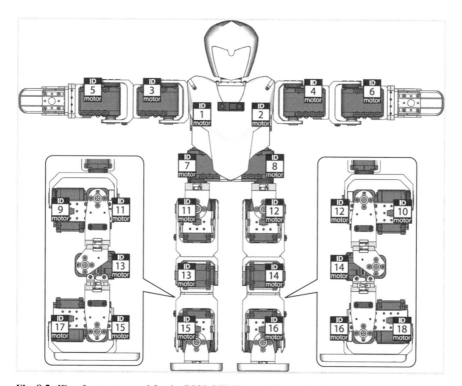

Fig. 8.5 IDs of actuators used for the BIOLOID Humanoid type A robot

The main sections of the CALLBACK function were as follows:

(a) Re-init "FBBalCenter" and "RLBalCenter" (statements 54–55).
(b) Read in new data for "FBBalData" and "RLBalData" (statements 59–60).
(c) Compute the current errors, "FBBalError" and "RLBalError" (statements 63–64) and ignore small errors by dividing them with 16, i.e. low byte discarded, to get the "Scaled" error parameters (statements 67–68).
(d) Next, these "Scaled Balance Error" parameters got "summed up" (i.e. integrated over time) with an upper limit (=2,048) and a lower limit (=−2,048) for both the FB and RL "Scaled Balance Error" parameters (statements 70–88).
(e) Statements 91 through 104 represented a complex "Gain Application" algorithm that sought to convert the "Balance Error Sums" parameters (i.e. angle values in degrees) into "appropriate" AX-12 Joint Offset values [−255 to 255] that depended on which part of the leg did that particular servo belong to. For example, adjustments to the "side-to-side ankle" servos (IDs 17 and 18) depended only on the "RL" errors and were numerically small as they represented the pivot point of an "inverted pendulum" (statements 92, 98, 102, 115 and 116). While adjustments to the "hip", "knee" and "forward-backward ankle" servos (IDs from 11 to 16) depended on <u>both "FB" and "RL" error terms</u> (statements 107–113), and they were numerically larger as they were further away from the pivot point of the "inverted pendulum". The reader could also note the "mirror image" effect used in the mathematical expressions used for the "right" leg (statements 107–109 for servos 11, 13 and 15) vs. the "left" leg (statements 111–113 for servos 12, 14 and 16). For more advanced analysis of bipedal walking motions, the reader would need to consult such works as Chevallereau et al. (2009) or Kajita et al. (2014).

8.1.2 One-Leg Static Balance

This work was done by Mr. Matthew Paulishen during his UGA student years. It used the two files "BalAct_1Leg_AXS20.mtn/mtnx" and "BalAct_1Leg_AXS20_Balance.tsk/tskx" (see Fig. 8.6 and Video 8.2).

This application did all the data acquisition and joint-offset computing inside the main TSK program and no CALLBACK function was used:

1. After zeroing out all previous joint-offsets (statement 13), this TASK program waited for the user to push the "Up" button for the first time to set the robot into Motion Page 13 which made the robot stand on its left leg while curling up its right leg. The user could now set the 1-legged robot onto the test platform (statements 18–22).
2. When ready to begin the actual balancing act, the user would push the "Up" button for the second time (statements 25–27) to trigger the function "ReCenterAXS20" (statements 128–154) which would acquire ten consecutive values of the FB and RL acceleration parameters (addresses 38 and 40 respectively) and saved the

Fig. 8.6 Humanoid Type
A one-leg balancing using
AX-S20

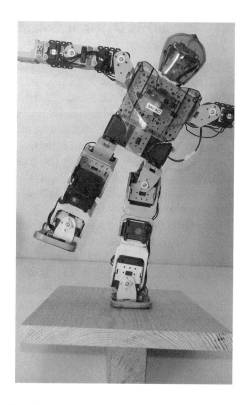

average values into "FBAccelCenter" and "LRAccelCenter" parameters (statements 148–149).

3. The robot was then ready to play Motion Page 12 (which called itself indefinitely) and got into an infinite loop to balance itself (statement 34–96) as the user changed the angles of the platform triggering changes in the values of the FB and RL acceleration parameters.

4. Similarly to the approach used in Sect. 8.1.1, the FB and RL acceleration error terms were then converted into appropriate joint-offsets for the legs' servos. Please note that while the "left knee" (ID = 14) was locked in place, the "left hip" (IDs = 10, 12) and "left ankle" (IDs = 16, 18) were "key players" in this balancing act, while the "right leg" lent its support via servos 9 and 11 for larger FB and RL error values, i.e. for larger angles of the standing platform (statements 77–90).

As previously mentioned in Chap. 3, the AX-S20 is no longer available from ROBOTIS, and the author is not aware if ROBOTIS is working on a similar sensor for the CM-5XX series. For the OpenCM-9.04-B controller, a more recent inertia-measuring device such as the SEN-11486 (MPU-9150) (https://www.sparkfun.com/products/11486) was incorporated into the OpenCM IDE V.1.0.2, but so far the author is not aware of an adaptation of this IMU device to the OpenCM-9.04-C or the CM-5XX series.

8.2 Humanoid Dynamic Balance with GS-12

A "2-legs balance" application, similar to the one made for the AX-S20 (see Sect. 8.1.1), could be created for the GS-12 also. Please review the enclosed files "BalAct_2Legs.mtn/mtnx" and "BalAct_2Legs_Gyro_Balance.tsk/tskx". Video 8.3 explained the algorithm used and showed the performance obtained on a Humanoid Type A and a CM-530 controller.

A more extensive application of the GS-12 sensor to the operations of a Humanoid A robot was also illustrated in the example TSK and MTN files provided by ROBOTIS (see files "BIO_PRM_Humanoid_A.tsk/tskx" and "BIO_PRM_ Humanoid_A.mtn/mtnx"). Fig. 8.7 showed Ports 3 and 4 being used for the GS-12 in our demonstration robot Bal'Act.

The GS-12 measured angular rates with respect to the (X, Y) axes (see Fig. 8.8) and the range of possible values for its outputs were from "45" (i.e., −300 /s) to "455" (i.e. +300 /s), with "250" corresponding to "0 deg./s" (a condition very hard to obtain in practice as the GS-12 was very sensitive to vibrations and signal noises).

In the TASK file "BIO_PRM_Humanoid_A.tsk/tskx", the function "InitializationGyro" (statements 840–873) collected ten consecutive readings of Ports 3 and 4 and computed their average values "FBBalCenter" and "RLBalCenter" respectively. Next, it used these average values to ascertain whether the GS-12 actually existed or not (statements 861–864). If the GS-12 existed, the parameters

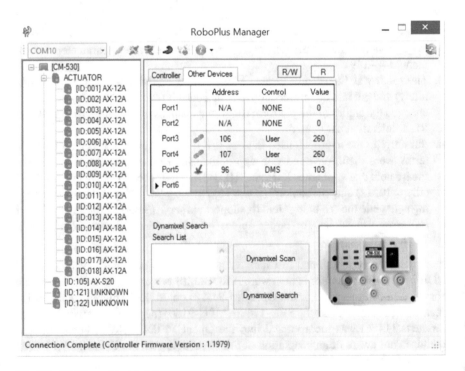

Fig. 8.7 GS-12 panel inside MANAGER tool

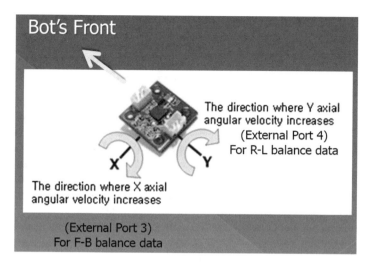

Fig. 8.8 Coordinates system used by the GS-12 for Bal'Act robot

"ExistGyro" and "GyroUse" would be set to TRUE, also the "Slip" parameter would be set to "0" (statements 860, 871 and 872). The "Slip" parameter was used to detect whether a "fall" had occurred to the robot (discussed later in Sect. 8.2.2).

8.2.1 Walk Enhancement with GS-12

If the GS-12 sensor was properly installed and working (parameters "ExistGyro" and "GyroUse" set to TRUE), the CALLBACK function inside the example code "BIO_PRM_Humanoid_A.tsk/tskx" was designed to stabilize the FB and RL motions when the robot was performing its programmed actions.

This CALLBACK function illustrated a similar data processing approach to the one used for the AX-S20, but with some important differences:

1. Read in current data from Ports 3 and 4 and compute the error signals "FBBalError" and "RLBalError" (statements 879–883).
2. Scale these error values appropriately and apply these scaled values to the joint offsets of the servos corresponding to the "hip" (IDs = 9, 10), "knee" (IDs = 13, 14) and "ankle" (IDs = 15 to 18) parts of the robot (statements 890–907).
3. The reader could still see the "mirror" characteristic in setting the joint-offsets of the "left" and "right" matching servos (for example, statements 899 and 901 for the "knee" servos).
4. Importantly, the reader should note that FB error signals were applied only to servos affecting the FB rotational direction, i.e. servos 13, 14, 15 and 16 (statements 899–902). While the RL error signals were applied only to those servos affecting the RL rotational direction, i.e. servos 9, 10, 17 and 18 (statements 904–907). Please contrast this approach to the one used for the AX-S20 which incorporated both FB and RL error terms in computing the servos joint-offset values (Sect. 8.1.1).

8.2.2 Fall Detection with GS-12

The CALLBACK function inside the "BIO_PRM_Humanoid_A.tsk/tskx" program
was also designed to detect a "fall" of the robot via the parameter "FBBalError"
(statements 885–888):

1. If the robot fell forward, its FBBalData's value would be close to 45 (see
   Fig. 8.8), i.e. smaller than the value for FBBalCenter (which should be around
   250) (statement 882). Thus FBBalError's value would be less than "−200", and
   consequently parameter "Slip" would be set to "1" (statements 887–888).
2. Conversely, if the robot fell backward, its FBBalData's value would be close to
   455, thus FBBalError's value would be more than "200", and consequently
   parameter "Slip" would be set to "−1" (statements 885–886).
3. Once the controller finished its CALLBACK cycle and got back to the main
   code, an IF structure (statements 307–319) would detect this condition and acted
   on it by first stopping all walking motions (statements 309–310), and then played
   either Motion Page 27 or 28 depending on whether the robot fell forward (Slip
   == 1) or backward (Slip == −1), respectively. Once the robot finished with either
   Motion Page 27 or 28, parameter Slip was reset to 0 and the controller continued
   on with this TSK program. Please note that there was no way for the robot to
   check if it actually stood back up successfully into its normal standing position,
   after playing Motion Pages 27 or 28, without the use of a sensor equivalent to the
   AX-S20.

8.3 Humanoid Balance with FPS

As previously mentioned in Chap. 3, the foot pressure sensor FPS from HUV
Robotics is no longer available commercially, but it illustrates an important class of
sensor needed for bipedal motion, so it still has some instructional value. This work
was also done by Mr. Matthew Paulishen.

The HUV-FPS sensor is Dynamixel-compliant and it actually had four sensing
pads, thus acquiring data from them presented a small challenge as the CALLBACK
function would only allow two device calls in its code section. Sect. 8.3.1 showed
how to adapt this constraint to the FPS sensor.

8.3.1 FPS Data Acquisition

As an illustration of the approach used, the reader is referred to the two files
"BalAct_1Leg_FPS.mtn/mtnx" and "BalAct_RLeg_FPS_DA.tsk/tskx" files.
Figure 8.9 showed the Dynamixel IDs and addresses scheme used in these demon-
stration codes.

In the "BalAct_1Leg_FPS.mtn/mtnx" file, Motion Pages 1 and 2 were used for a right instrumented foot during the "initialization" and "balance" phases respectively. While Motion Pages 5 and 6 were used in case of a left instrumented foot. Motion Pages 2 and 6 were set to get into a continuous play mode.

The "BalAct_RLeg_FPS_DA.tsk/tskx" program started out by making the robot go through Motion Pages 1 then 2, i.e. to put the robot into a similar posture as in Fig. 8.6, but on its right leg (see Video 8.4). Next, if the user pushed the "Up" button, the bot would collect data from the four pressure pads (addresses 26, 28, 30 and 32) and average them out into four parameters:

1. IFFSR (address 26) for the Inside-Front FSR.
2. OFFSR (address 28) for the Outside-Front FSR.
3. IRFSR (address 30) for the Inside-Rear FSR.
4. ORFSR (address 32) for the Outside-Rear FSR.
5. These initial values were then displayed in the Output Window (statements 57–60).

Then, if the user also pushed the "Right" button, the CALLBACK function would be activated. As there was a limit of two device calls in each 8 ms cycle of the CALLBACK function (statements 80–81), the data collecting and processing scheme used was to rotate through the relevant addresses (26 >> 28 >> 30 >> 32), via the setting of parameter "CurAdd", one at a time, during each consecutive cycle (statements 84–103). Thus during each CALLBACK cycle, two values of the "current" FSR data were collected (statements 80–81) and its average value saved into the appropriate parameters IFFSR, OFFSR, IRFSR or ORFSR as controlled by the value of "CurAdd" parameter via the IF-ELSE-IF structure.

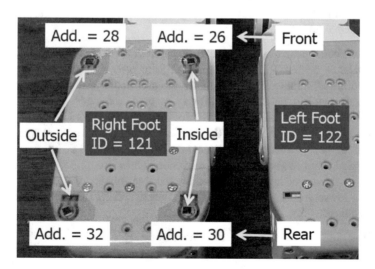

Fig. 8.9 Dynamixel IDs and addresses scheme for HUV FPS sensor

8.3.2   Humanoid One-Leg Balance with FPS

The "BalAct_1Leg_FPS_Balance.tsk/tskx" program was designed to be used with the "BalAct_1Leg_FPS.mtn/mtnx" file to achieve a one-Leg balance solution for a Humanoid A robot, using the HUV FPS sensors instead of the AX-S20 as previously done in Sect. 8.1.2.

This TSK program started out by initializing parameters UseBalance and HUMANOID and by zeroing out the joint-offsets of all 18 servos (statements 19–21).

Next, the robot needed to detect how many feet were instrumented and to help the user in deciding which one to serve as the balance foot. It accomplished this task by reading from address 0 of each of the two possible Dynamixel IDs (121 and 122, via statements 24 and 25). Then the robot went through an IF-ELSE-IF structure to figure out four possible outcomes (statements 26–74):

1. If no FPS was detected, the robot would play Melody 16 twenty times and quit the TSK program entirely (statements 26–36).
2. If the right FPS was detected, the robot would set parameter FOOT to 1 and play Melody 0 (statements 38–45).
3. If the left FPS was detected, the robot would set parameter FOOT to 5 and play Melody 1 (statements 46–53).
4. If both FPS were detected, the robot would play Melody 0 and then waited on the user to enter his or her choice of either the Left or Right leg to balance on, via the Left or Right buttons respectively (statements 61–73).

The robot next settled into its initial Motion Page (1 or 5 depending on the balance foot – statements 77–81) and waited for the user to push button "Up" (statement 83). And once "Up" was pushed, the robot activated its Balance Motion Page (2 or 6) and called the function "NOIBAL" (statements 109–279).

The function "NOIBAL" was essentially an endless loop wherein the robot could execute several possible tasks:

1. The user could push the "Down" button to make the robot stop balancing (i.e. set UseBalance to FALSE), and to reset all joint offsets by calling the function ZeroJoints (statements 268–276). Then the user could push the "Up" button to set UseBalance to TRUE and call for a new set of reference values for the FSR by calling function ReferenceFootSensorValues (statements 113–119).
2. When UseBalance was set to TRUE, the robot enter a large code segment (statements 120–277), whereas it computed the average value of 5 FSR readings and compared them to the respective reference values (i.e. "nom" parameters) in order to figure out how much to adjust the joint-offsets of appropriate servos among the servos 11, 12 and 15 for FB balance, and the servos 9, 10 and 17 for LR balance.

Video 8.5 showed the actual performance of this solution.

8.4 HaViMo2 Applications

The HaViMo2 color video camera (Fig. 8.10) was developed by Dr. Hamid Mobalegh for RoboCup applications (http://www.havisys.com/?page_id=8). It was Dynamixel compliant (3-pin TTL) and had a locked ID=100, which happened to be the default ID for a typical AX-S1 sensor also, thus the user would need to make proper ID adjustment on the AX-S1 if he or she planned to use these two sensors on the same robot.

This camera was compatible with the CM-5/CM-510/CM-700/CM-530 controllers via RoboPlus Task or directly via ROBOTIS's Embedded C tools (see Chap. 9). It was known that this camera had some latency issue with the CM-530 firmware (V. 1.1969) and users would need to install an alternate CM-530 firmware (http://www.havimo.com/?p=130).

It also worked with the Open-CM IDE on the CM-9.00 and CM-9.04 controllers (see Chap. 9).

8.4.1 HaViMo2 Features and Usage

The HaViMo2 camera was based on the CMOS image sensor HV7131RP (MagnaChip Semiconductor, c. 2005) and had on-chip image processing capabilities (http://www.havimo.com/?page_id=32).

The following RoboSavvy.com web site had practical documentation and software downloads for this camera (http://robosavvy.com/store/product_info.php/manufac-

Fig. 8.10 Dynamixel-compliant color camera HaViMo2

turers_id/21/products_id/639). Its video frame resolution was at 160x120 pixels, with color depth at 12 bits YCrCb and it had a maximum frame rate of 19 fps (interlaced). Its capabilities were better realized on a faster MCU such as the ARM controller used in the CM-530 and Open-CM-9.00/9.04. This camera required a PC-side application called HaViMoGUI (downloadable from RoboSavvy) to get itself calibrated to the light source used (a very important step) and to set its color look-up table LUT (up to eight distinct colors, i.e. background +7 user-created). Once the LUT was set, up to 15 contiguous objects (regions) could be identified during run-time. For each of these regions, the camera could report on its Color, row-column coordinates of its Center of Mass, Number of Pixels and its Bounding Box. The user is recommended to consult the enclosed file "HaViMo2UserManual.pdf" for more details.

8.4.2 HaViMo2 Application to a CM-5 CarBot

Figure 8.11 showed the CM-5 robot platform used to illustrate how to interface and program a HaViMo2 camera using the RoboPlus software suite.

As the HaViMo2 camera required rather lengthy procedures for its calibration and programming tasks, two video tutorials were made to inform the reader about those procedures:

1. In Video 8.6, the goal was to show how to access the camera via an USB2Dynamixel module (a more direct approach preferred by the author over other "through-connections" via the controllers CM-5XX). The reader is referred to page 3 of the document "HaViMo2UserManual.pdf" for more details on other possible hardware setups. This video also showed how to use the HaViMoGUI application to calibrate the camera and define the LUT to characterize the colors of four painted wooden dowels. Additionally, it illustrated the use of the program "Carbot_Visual_Track.tsk/tskx" to track those same wooden dowels.
2. In Video 8.7, more fine-tuning of the controlling parameters inside the program "Carbot_Visual_Track.tsk/tskx" was performed, and the program "Carbot_Find&Approach.tsk/tskx" was demonstrated. Essentially, the Carbot was programmed to find a user-designated colored dowel and drove up to get closer to it.

The reader readily noticed that the color tracking work was performed rather slowly, this was because the CM-5 was an Atmel AVR MCU running only at 16 MHz and we were using the TASK tool which had some computing overhead. When the HaViMo2 was used on a pan-tilt apparatus using a CM-530 and a TASK program, the performance was improved perceptively (https://www.youtube.com/watch?v=p og2gzpjo7g&list=UUGIds85x7Q_nBOReZ818LJQ) because of a faster clock rate and a more efficient MCU were used. When switching to an embedded C version on a CM-510, the performance got further improved showing the power of direct C control of the Atmel AVR MCU (https://www.youtube.com/watch?v=pMbSqkshN Zo&list=UUGIds85x7Q_nBOReZ818LJQ&index=35). Finally, on an OpenCM-9.04B, the performance was much more improved, combining faster MCU clock

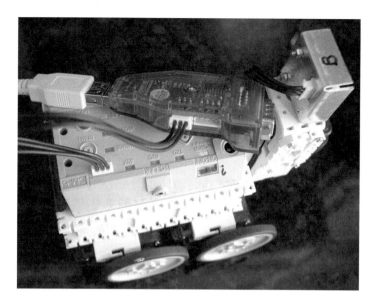

Fig. 8.11 HaViMo2 camera with CM-5 and USB2Dynamixel

rate and efficient architecture (https://www.youtube.com/watch?v=kCH8F4lXXZ
M&list=UUGIds85x7Q_nBOReZ818LJQ).

As the tutorial videos might not have enough resolution to allow the readers to
see individual lines of codes, the author would like to revisit some of the more
important image capture and processing concepts shown in the TSK programs
"Carbot_Visual_Track.tsk/tskx" and "Carbot_Find&Approach.tsk/tskx" (both created by Mr. Matthew Paulishen), in a more text-based manner:

1. **"Carbot_Visual_Track.tsk/tskx"**. The most important statement in this TSK
   program was Line 21 as shown in Fig. 8.12. It was used to capture a new video
   frame and to start the process of finding the regions as defined by the "colors"
   chosen by the user in the calibration process with the HaViMoGUI tool (see
   Video 8.6).

Parameter CamID was hardware-set to 100 and Address **0** was used to trigger
the frame capture process, although the function recommended in the
"HaViMo2UserManual.pdf" was "CAP_REGION" (i.e. Address **0x0E**) as shown in
Table 1 of Page 6 of that PDF document. Next was a "do-nothing" loop for 0.256 s
(lines 22–23) to allow the first image capture process to go through before the TSK
program went into its Endless Loop to find the wanted color object and track it with
the Carbot's AX-12s.

Line 28 called the function "Get_Bounding_Box" (defined at lines 81–113)
which looked at the data structure returned by the HaViMo2 to see if any valid
region, defined by the user's "Color" parameter as set on Line 14 of this TSK program, was actually found. Figure 8.13 displayed the FOR loop used to go through
the 15 possible regions provided by the HaViMo2 firmware for each frame captured
and processed (see line 84). Thus, for example, "Index" equaled "1" for the first
region and "Addr" correspondingly equaled "16". Next the TSK program looked at

Fig. 8.12 Line 21 – CUSTOM Command to capture and process a new video frame with HaViMo2

the "Byte" found at Address "Addr" (line 87), and if this "Byte" was non-zero, this was a valid region. Essentially, this "Byte" corresponded to the parameter "Index" as shown in Fig. 3 of Mobagleh (2010).

The information in this Fig. 3 (Mobagleh - 2010) was also reproduced below for the reader's convenience:

- Index (1 byte) – zero if the region is invalid and nonzero otherwise.
- Color (1 byte) – color code of the detected region (0 = Unknown, or 1 through 7).
- Pixels (2 bytes) – number of detected pixels inside the region.
- SumX (4 bytes) – sum of the X coordinates of the detected pixels.
- SumY (4 bytes) – sum of the Y coordinates of the detected pixels.
- MaxX (1 byte) – bounding box right margin.
- MinX (1 byte) – bounding box left margin.
- MaxY (1 byte) – bounding box bottom margin.
- MinY (1 byte) – bounding box top margin.

Now that the TSK/TSKX program had read in the value for parameter Index (line 87) and found that Index was non-zero, it would next increment Addr by 1 to get to the Color member of this 16-byte data structure. If this Color member matched with the Color parameter (set back in line 14), it incremented Addr again by 1 to get to the Size of the current detected region (2 bytes = 1 word, see lines 92–93). In this particular example, we were interested in tracking only ONE object/region, thus the use of the IF structure to find the biggest one among the 15 regions (see lines 94–98). At the completion of the FOR LOOP, parameter Max would contain the Size of the largest valid region with the matching Color, and parameter MaxAddr would contain the current pointer address (i.e. still pointing at the Pixels data member of the 16 byte data structure for this largest valid region).

Next, if Max was non-zero (i.e. valid region), the TSK program started from the MaxAddr value and jumped ahead by 10 bytes to get to the MaxX data member and saved that value in parameter Maxx (lines 104–105). Then it incremented Addr by 1 byte to get to MinX and therefore Minx (lines 106–107). Parameters Maxx and Minx would be used later in the main function to compute the appropriate steering commands for the CarBot in order to keep the user-defined object in front of the HaViMo2 camera as much as possible (lines 29–73).

```
81   FUNCTION  Get_Bounding_Box
82   {
83       Max = 0
84       LOOP FOR ( Index = 1 ~ 15 )
85       {
86           Addr = Index * 16
87           IF ( ID[CamID]: ADDR[Addr(b)]   != 0 )
88           {
89               Addr = Addr + 1
90               IF ( ID[CamID]: ADDR[Addr(b)]   == Color )
91               {
92                   Addr = Addr + 1
93                   Size = ID[CamID]: ADDR[Addr(w)]
94                   IF ( Size > Max )
95                   {
96                       Max = Size
97                       MaxAddr = Addr
98                   }
99               }
100          }
101      }
```

Fig. 8.13 Function "Get_Bounding_Box" to extract data out of the 15 regions

2. **"Carbot_Find&Approach.tsk/tskx"**. This TSK program used the same function "Get_Bounding_Box" to find the largest region of interest, but this implementation used the RC-100's buttons 1-2-3-4 to set the color wanted by the user
   on the fly (lines 31–61). It also added new codes to check on the current Size of
   the target (i.e. region of interest) and from there it could decide whether to command the carbot to get closer or to back away from the target to maintain a user-
   given TargetSize (lines 103–126).

8.5   Review Questions for Chap. 8

1. Why was the AX-S20 sensor mounted in the head of the Humanoid robot?
2. Why was the GS-12 sensor mounted in the abdomen of the Humanoid robot?

3. How many device calls are allowed in the CALLBACK function?
4. Is there a size limitation on the CALLBACK function?
5. What was the resolution in degree of the Azimuth, Pitch and Roll angles provided by the AX-S20?
6. How many bits were contained in the X-Y-Z acceleration data provided by the AX-S20?
7. What is the refresh rate for data provided by the AX-S20 sensor?
8. What is the refresh rate for data provided by the GS-12 sensor?
9. What was the procedure used to activate the Joint Offsets of the actuators used in the balancing of the Humanoid robot whether one uses the AX-S20 or GS-12?
10. The AX-S20 is a Dynamixel-compliant sensor. (T-F)
11. The GS-12 is a Dynamixel-compliant sensor. (T-F)
12. What is the value range for the digital output from the GS-12 sensor?
13. The HUV Robotics FPS sensor is a Dynamixel-compliant sensor. (T-F)
14. The HaViMo2 camera is a Dynamixel-compliant sensor. (T-F)
15. What is the pixel resolution of the HaViMo2 camera?
16. What is the maximum video frame rate for the HaViMo2 camera?
17. What is the color space used by the HaViMo2 camera? What is its color depth in terms of binary bits?
18. How many distinct colors can be saved in the color look-up table for the HaViMo2 camera?
19. How many contiguous objects/regions can the HaViMo2 search for during each cycle of operation?

8.6   Review Exercises for Chap. 8

1. Starting from the program "Carbot_Find&Approach.tsk/tskx", the reader could develop a slalom negotiating Carbot project using the HaViMo2 camera to weave around colored dowels, as shown in this video clip from a University of Georgia student (see Video 8.8).
2. Other National Taiwan University students mounted the HaViMo2 camera onto a PREMIUM GERWALK to perform the same Find and Approach feature for a given color patch. This work was more challenging as the GERWALK was always "jiggling" the camera, thus it could only capture and process images during a short time when the camera was level. The students also used the DMS sensor to command the robot to stop when close enough to the target (see video at https://www.youtube.com/watch?v=HWxMwvFriMc).

References

Chevallereau C et al (2009) Bipedal robots. Wiley, Hoboken

Kajita et al (2014) Introduction to humanoid robotics. Springer, Heidelberg

Mobagleh H (2010) HaViMo2 image processing module. (http://robosavvy.com/RoboSavvyPages/Support/Hamid/HaViMo2.pdf). Accessed 29 Dec 2014

Chapter 9
Embedded C Options

Chapter 8 showed that in order to get the maximum performance out of ROBOTIS systems, ones must consider the "Embedded C" routes which happened to be quite numerous. Some of the resources that I know of are listed below (and I know that I have missed many):

- Vanadium Labs had been supporting the ArbotiX RoboController line, Atmel AVR-based (http://www.vanadiumlabs.com/arbotix.html). It is supported by the ROS organization (http://wiki.ros.org/arbotix).
- The Humanoid Lab provides C libraries for the Bioloid Premium (http://apollo.upc.es/humanoide/trac/wiki/bioloidCframeworks) via Windows and Linux.
- BioloidCControl is an alternative firmware for BIOLOID PREMIUM Humanoids A/B/C (https://code.google.com/p/bioloidccontrol/).
- "Software Souls" also offer another approach (http://softwaresouls.com/software-souls/series/programming-robotis-bioloid-hardware/).
- Another interesting product is the USB2AX which is equivalent to the USB2Dynamixel module (http://www.xevelabs.com/doku.php?id=product:usb2ax:usb2ax) for Windows, Linux and MacOS systems.
- As far as Arduino books go, there are many in existence to suit the reader's skills and goals, but I would recommend the classic "Arduino Cookbook" by Margolis (2011).

However this chapter's main topics would stay with the official ROBOTIS Embedded C routes. Historically speaking, Embedded C was first available for the CM-5 via its BIOLOID Expert Kit (c. 2006, but it is no longer available) and currently Embedded C functionality is offered only for the CM-510/700/530 and the OpenCM-9.00/9.04 series.

Chronologically and for the international market, the Embedded C utilities for the CM-510 came out first in early 2010, next were the ones for the CM-530 in Spring 2012. While the OpenCM-9.00/9.04 systems came out in 2012–2013 and the last hardware component OpenCM-485-EXP came out in Summer 2014. However, for a better flow of the topics in this book, I would start with the OpenCM IDE for

© Springer International Publishing Switzerland 2017
C.N. Thai, *Exploring Robotics with ROBOTIS Systems*,
DOI 10.1007/978-3-319-59831-4_9

the OpenCM-9.00/9.04 series and end this chapter with Embedded C for the CM-510/530 systems, as I expect that most readers of this book would start with the RoboPlus system and the CM-5XX series for their ROBOTIS journey. I also realized that I would have to assume that the reader had some prior knowledge of C/C++ programming or had access to a good C/C++ programming resource, as that topic would be outside the scope of this book. Thus my overall goal for this chapter was to contrast similarities and differences in usage between RoboPlus and Embedded C tools regarding:

- General use differences to note between RoboPlus TASK and Embedded C.
- OpenCM IDE for the OpenCM-9.00/9.04 family.
- Mobile Manipulator applications with OpenCM IDE and PLAY700 smartphone app.
- Embedded C options for the CM-510 and CM-530 systems.
- Motion Programming and Embedded C.

9.1 Embedded C vs. RoboPlus' TASK

The TASK tool was designed for beginners in robotics and also in computer programming skills, thus its interface shielded the user from important details that could not be ignored when switching to Embedded C interfaces:

1. **Data Types**. A TASK programmer only needed to declare a parameter's name and started to use it anywhere in the TSK code, and this parameter could only handle integer values anyhow. Embedded C would allow all the standard data types – integer, floating-point and more complex data structures. The TASK programmer could already get a taste of things to come with the CUSTOM command for Controller and Dynamixel (see Figs. 8.12 and 8.13) whereas one had to choose between a BYTE (8 bits) or a WORD (16 bits) data type to properly read or write to a given parameter or device (such as for the HaViMo2 camera in Chap. 8).

2. **Assignments and Function (Method) Calls**. Typically, about 90% of all the statements used in a TASK program would be of the Assignment type (A = B, i.e. the value of parameter B is assigned to parameter A). For example, to set a Goal Position on an AX-12 with ID = 3 to a value of 800, the TASK user would type in the following statement:

ID[3]: Goal Position = 800

Switching to Embedded C, the C/C++ user would use a Function (Method) Call instead:

Dxl.writeWord(3, 30, 800);

where argument "3" corresponded to the ID, while argument "30" corresponded to the address of the Goal Position parameter as defined in the Control Table of the AX-12 (see web link at http://support.robotis.com/en/product/actuator/dynamixel/ax_series/dxl_ax_actuator.htm), and argument "800" was the desired value for the Goal Position. In other words, the majority of the statements used

Fig. 9.1 OpenCM-9.00V. 1 (*left*) and OpenCM-9.04/B (*right*)

in Embedded C would be Method Calls which required the proper setting of many formal parameters to properly use that Method. Therefore it would require the user to be familiar with the Control Table of each type of ROBOTIS actuators and sensors that were used in the robot being considered. This Control Table would also inform the user of the proper Data Type to use (BYTE or WORD). An alternative Method Call, with a syntax much closer to the one used in the previous TASK statement, could also be used:

Dxl.goalPosition(3, 800);

3. **Devices**. In the previous sub-section, the reader might have noticed the "Dxl." notation used in the Method Call Dxl.goalPosition(3, 800). Actually, there were 2 other statements that must have been asserted before a C/C++ programmer could use the Method goalPosition() properly:

   (a) **Dynamixel Dxl(1);** to define the object "Dxl" as a "Dynamixel" device that was associated with Serial Bus "1" (i.e. argument "1" within the parentheses) which corresponded to the 3-pin TTL bus on each ROBOTIS controller (from CM-5 to OpenCM-9.04 – see Fig. 9.1).
   (b) **Dxl.begin(3);** to initialize the communication with the object "Dxl" (i.e. with the Dynamixel 3-pin TTL bus) at a baud rate of 1 Mbps (i.e. argument "3" within the parentheses).

For devices using the 5-pin bus (see Fig. 9.2), a similar process needed to be used, but with the OLLO device:

   (a) **OLLO myOLLO;** to define the object "myOLLO" as an "OLLO" device corresponding to the GPIO 5-pin bus on ROBOTIS controllers such as CM-510/530 and OpenCM-9.04/A/B/C.

RS485 4-pin
Serial 3

TTL 3-pin
Serial 3

OpenCM-9.04/C

Fig. 9.2 OpenCM-9.04/C on *top* on a 485 EXP shield

  (b) **myOLLO.begin(2);** to initialize the particular "myOLLO" device that was hooked up to the GPIO port "2" (as an example).

  (c) **myOLLO.read(2);** to read the device's current response at port "2" as a digital value.

Please note that a TASK programmer would never have to worry about this kind of details, as the TASK tool took care of associating and initializing Dynamixel and OLLO devices in the background. Video 9.1 was a side-by-side comparison of a TSK program (on a CM-530 – "TSKvsIDE.tsk") and a C program (on an OpenCM-9.04/B + 485-EXP combo – "TSKvsIDE.ino") designed to do the same operations on an AX-12 actuator and a DMS sensor. This video file also showed how to adapt this "TSKvsIDE.ino" sketch to the situation when the AX-12 was connected to the TTL bus on the 485-EXP board (see next paragraph on communication ports for more details).

4. **Communication Ports**. Just as "Dynamixel" devices were connected via Serial Bus 1, per-se "communication" devices such as the LN-101 (wired) and ZIG-110 or BT-110/210/410 (wireless) were programmed to connect via Serial Bus 2 (i.e. through the 4-pin connector (see Figs. 9.1 and 9.2). The OpenCM 485-EXP expansion shield is assigned to Serial Bus 3 (see Fig. 9.2) and it can be programmed with the OpenCM IDE but NOT FULLY with the TASK tool at present (Spring 2017 and also see Sect. 5.2.2).

5. **Dynamixel Communication Protocols 1 and 2**. Previously, all ROBOTIS Dynamixels used the same communication protocol (http://support.robotis.com/en/product/actuator/dynamixel/communication/dxl_packet.htm), but with the introduction of the PRO line in 2012 and then the X series starting in 2013, a

second communication protocol was needed with different instruction/status packet design and faster baud rates (http://support.robotis.com/en/product/actuator/dynamixel_pro/communication/instruction_status_packet.htm). At present (Spring 2017), ROBOTIS supports mixed-protocol programming only on the OpenCM IDE (i.e. with the OpenCM-9.04/A/B/C) and it would require the use of the Method setPacketType (DXL_PACKET_TYPE) whereas the parameter DXL_PACKET_TYPE is set (as expected) to "1" for Protocol 1 and "2" for Protocol 2. Most importantly, the Method setPacketType() had to be used to set the appropriate DXL_PACKET_TYPE **before** any communication to the respective type of Dynamixel (protocol 1 or 2) – see example code below:

```
// Dynamixel with ID_1 uses Protocol 1
Dxl.setPacketType(1);
Dxl.goalPosition(ID_1, 512);  // go to position 512
delay(1000);  // delay 1 second
// Dynamixel with ID_3 uses Protocol 2
Dxl.setPacketType(2);
Dxl.goalPosition(ID_3, 1023);
delay(500);  // delay 0.5 second
```

9.2   Embedded C for the OpenCM-9.00/9.04

The current version for the OpenCM IDE is version 1.04, available since early 2014 (http://support.robotis.com/en/software/robotis_opencm.htm) and a User's Manual for the OpenCM-9.04 system and other hardware information are available at http://support.robotis.com/en/product/controller/opencm9.04.htm. Although the OpenCM-9.00 controller is no longer available commercially, enclosed with this book is a ZIP file that has technical information for this system (OpenCM-900-Manuals.zip). Other "community" resources exist such as "trossenrobotics.com" and "robosavvy.com", just to name a few. Also as the OpenCM IDE was based on the Arduino architecture, there are numerous resources in print and on the web for Arduino that the user could consult to go beyond the ROBOTIS manuals.

9.2.1   Basic CarBot Applications

As the ROBOTIS manuals and web sites were already providing good instructions for installing the OpenCM IDE and for its basic and advanced uses, this section would only illustrate two example projects that hopefully will be useful to the beginning Embedded C programmer:

1. Video 9.2 described the remote control of a CarBot equipped with BT-210 by the Virtual RC-100 from inside the ROBOPLUS MANAGER (V.1) tool on the PC side. The corresponding Arduino Sketch file was named "CM9_Carbot_RC.ino" and included with this book.
2. Video 9.3 described the use of a 485-EXP expansion shield with an OpenCM-9.04/B and also the mixing of Dynamixel protocols. The corresponding Arduino Sketch file was named "CM9_Mixed_Protocols.ino" and included with this book.

In Sect. 8.4.2, an early model of OpenCM-9.04/B was shown to have worked well with the HaViMo2 camera using the ROBOTIS IDE V.1 (c. 2013) (https://www.youtube.com/watch?v=kCH8F4lXXZM&list=UUGIds85x7Q_nBORe-Z818LJQ). Unfortunately, the author had been not able to reproduce the same results using the current model of the OpenCM-9.04/B and the current ROBOTIS IDE V.1.0.4. The attempted sketch "CM9_CarbotHaViMo2.ino" was enclosed with this book for the reader's reference. This sketch compiled successfully but the HaViMo2 never responded to the first ping to this Dynamixel, invoked in a call to the method hvm2.ready().

9.2.2 Remote Control of Mobile Manipulator

In this section, the Mobile Manipulator system described in Fig. 6.30 would be controlled by an OpenCM-9.04/C + 485-EXP combo (see Fig. 9.3). Its USB port was used for downloading executable codes from the PC and also for printing various debugging information. The 4-pin communication port (Serial 2) was used with a BT-210 for run-time interactions with the user via the Virtual RC-100 facility found inside the TASK tools V.1 and V.2.

As Sect. 6.2.2.4 already provided a detailed approach to achieve Remote Control of this Mobile Manipulator platform, this section would only concentrate on the Embedded C aspects. The resulting sketch "CM9_MobileManipulator_RC.ino" was included in the Springer Extra Materials.

Figure 9.4 showed various parameter definitions that were needed:

• Statements 22 to 24 pertained to the RC-100 Remote Controller characteristics and RC-100's messages were stored in Parameter RcvData. Statement 23 associated the name "Controller" to the RC-100 device which would need to be initialized later in the setup() function.
• Statements 26–28 defined the Constants used for communication ports Serial 1, Serial 2 and Serial 3. Serial 1 was not used in this application sketch. Serial 2 was used for the RC-100 Controller and set to 57,600 Kbps. Serial 3 corresponded to the 485-EXP board and was set to 1 Mbps.
• Statements 30–41 pertained to the IDs of the actuators used. The Manipulator Arm used 8 AX-12As (ID'ed from 1 to 8). The wheeled platform used 4 AX-12Ws (ID'ed from 11 to 14).

Fig. 9.3 Mobile
Manipulator controlled by
"OpenCM-9.04/C + 485-
EXP"

- Statements 42–45 listed the "addresses" (i.e. various actuator functions) from the
  Control Table of the AX-12s (http://support.robotis.com/en/product/actuator/
  dynamixel/ax_series/dxl_ax_actuator.htm   and   http://support.robotis.com/en/
  product/actuator/dynamixel/ax_series/ax-12w.htm).
- Statement 47 defined the Number of Actuators used for the arm (i.e. 8) and
  Statement 48 defined a 1-D array named "CurrentPos" to be used to record the
  current positions of these 8 actuators at the beginning of each iteration of the
  main control loop (see Fig. 9.7).
- Statement 49 defined the step used for each positional change of the AX-12As of
  the arm, while "NewPosA" and "NewPosB" (Statement 50) were temporary
  variables used in computing new Goal Positions for the actuators as the result of
  the user's inputs via the RC-100.
- Statement 52 associated the name "Dxl" to various actuator devices connected
  on the 485-EXP board via Port "Serial 3".

  Figure 9.5 described the customary setup() function for a typical Arduino sketch:

- Statement 56 initialized the object "Dxl" to a baud rate of 1 Mbps.
- Statement 57 initialized the object "Controller" to a BlueTooth mode with a baud
  rate of 57.6 Kbps.

```
CM9_MobileManipulator_RC.ino ⊠
22   #include <RC100.h>
23   RC100 Controller;
24   int RcvData =0;
25
26   #define DXL_BUS_SERIAL1 1   //Dynamixel on Serial1(USART1)   <-OpenCM9.04
27   #define DXL_BUS_SERIAL2 2   //Dynamixel on Serial2(USART2)   <-LN101,BT210
28   #define DXL_BUS_SERIAL3 3   //Dynamixel on Serial3(USART3)   <-OpenCM 485EXP
29
30   #define ID_NUM_1 1
31   #define ID_NUM_2 2
32   #define ID_NUM_3 3
33   #define ID_NUM_4 4
34   #define ID_NUM_5 5
35   #define ID_NUM_6 6
36   #define ID_NUM_7 7
37   #define ID_NUM_8 8
38   #define ID_NUM_11 11
39   #define ID_NUM_12 12
40   #define ID_NUM_13 13
41   #define ID_NUM_14 14
42   #define GOAL_POSITION 30
43   #define GOAL_SPEED 32
44   #define PRESENT_POSITION 36
45   #define MAX_TORQUE 14
46
47   #define NUM_ARM_ACTUATOR 8
48   word CurrentPos[NUM_ARM_ACTUATOR];
49   word Step = 10;
50   word NewPosA = 0, NewPosB = 0;
51
52   Dynamixel Dxl(DXL_BUS_SERIAL3);
```

Fig. 9.4 Parameter definitions for "CM9_MobileManipulator_RC.ino"

- Statements 59–66 set the arm's actuators to Joint Mode.
- Statements 68–71 set the rover's actuators to Wheel Mode.
- Statement 73 called the function initpage() (see Fig. 9.6) which was equivalent to activating Motion Page 2 in Statement 12 of Fig. 6.32, but at a "slow" Moving Speed of "128".
- The For loop (Statements 74–78) put the arm's actuators to Maximum Torque and Speed of "1023" for use during the user's input phase (i.e. function loop() as shown in Fig. 9.7).

Figure 9.7 showed the beginning of the main function loop() which contained the control algorithm used:

- If there was a user input via the RC-100 (i.e. "Controller"), the "IF" statement 84 would be activated and Statement 85 would be activated next and saved the user's input in Parameter "RcvData".
- Statement 87–89 saved the Present Positions of the arm's actuators into the array CurrentPos[]. The important point to notice was that the IDs of the actuators

```
54  void setup() {
55     // Dynamixel 2.0 Protocol -> 0: 9600, 1: 576C
56     Dxl.begin(3);
57     Controller.begin(1); // RC100 initialized as
58
59     Dxl.jointMode(ID_NUM_1);  // Set Joint Mode f
60     Dxl.jointMode(ID_NUM_2);
61     Dxl.jointMode(ID_NUM_3);
62     Dxl.jointMode(ID_NUM_4);
63     Dxl.jointMode(ID_NUM_5);
64     Dxl.jointMode(ID_NUM_6);
65     Dxl.jointMode(ID_NUM_7);
66     Dxl.jointMode(ID_NUM_8);
67
68     Dxl.wheelMode(ID_NUM_11);  // Set Wheel Mode
69     Dxl.wheelMode(ID_NUM_12);
70     Dxl.wheelMode(ID_NUM_13);
71     Dxl.wheelMode(ID_NUM_14);
72
73     initpage();
74     for (int i = 1; i <= NUM_ARM_ACTUATOR; i++)
75     {
76        Dxl.writeWord( i, MAX_TORQUE, 1023 ); // se
77        Dxl.writeWord( i, GOAL_SPEED, 1023 ); // se
78     }
79  }   // end of setup()
```

Fig. 9.5 Function setup() for "CM9_MobileManipulator_RC.ino"

```
203  void initpage(){
204     for (int i = 1; i <= NUM_ARM_ACTUATOR; i++)
205     {
206        Dxl.writeWord( i, GOAL_SPEED, 128 ); // s
207     }
208     Dxl.writeWord(1, GOAL_POSITION, 512); //Com
209     Dxl.writeWord(2, GOAL_POSITION, 500); //Com
210     Dxl.writeWord(3, GOAL_POSITION, 524); //Com
211     Dxl.writeWord(4, GOAL_POSITION, 225); //Com
212     Dxl.writeWord(5, GOAL_POSITION, 799); //Com
213     Dxl.writeWord(6, GOAL_POSITION, 795); //Com
214     Dxl.writeWord(7, GOAL_POSITION, 512); //Com
215     Dxl.writeWord(8, GOAL_POSITION, 512); //Com
216  }   // end of Direct Init Page
```

Fig. 9.6 Function initpage() for "CM9_MobileManipulator_RC.ino"

```
82  void loop() {
83
84      if(Controller.available()){
85          RcvData = Controller.readData();
86          // fetch current positions of arm's actuators
87          for(int i = 1; i <= NUM_ARM_ACTUATOR; i++){
88          CurrentPos[i-1] = Dxl.readWord (i, PRESENT_POSITION);
89          }
90
91      // For Rover Motions and only 1 U-D-L-R button at any one time
92          if(RcvData == RC100_BTN_U){
93              Dxl.goalSpeed(ID_NUM_12, 500);          //Dynamixel 12 -> Left Front motor
94              Dxl.goalSpeed(ID_NUM_11, 500+1023);     //Dynamixel 11 -> Right Front motor
95              Dxl.goalSpeed(ID_NUM_14, 500);          //Dynamixel 14 -> Left Rear motor
96              Dxl.goalSpeed(ID_NUM_13, 500+1023);     //Dynamixel 13 -> Right Rear motor
97          }
98          else if(RcvData == RC100_BTN_D){
99              Dxl.goalSpeed(ID_NUM_12, 500+1023);
100             Dxl.goalSpeed(ID_NUM_11, 500);
101             Dxl.goalSpeed(ID_NUM_14, 500+1023);
102             Dxl.goalSpeed(ID_NUM_13, 500);
103         }
```

Fig. 9.7 Start of Function loop() for "CM9_MobileManipulator_RC.ino"

were labeled from i = 1 to 8, but the index in C arrays would vary from 0 to 7, thus the use of "i-1" as the index of the array CurrentPos[].

- Next an IF-ELSE-IF structure was used to trigger the appropriate actions for this Mobile Manipulator robot corresponding to each specific user's input. For example if the Button "UP" was pushed, then Actuators 11 through 14 were set to make the platform go forward, while if the Button "DOWN" was pushed then the platform would be made to go backward, etc... The reader would be referred to the actual INO file to see the codes used for "LEFT" and "RIGHT" maneuvers.

Using the same approach described in Sect. 6.2.2.4, the REACTOR arm was considered as having three joints plus hand grasp (i.e. gripper):

- The Shoulder Joint was represented by Actuators 1, 2 and 3.
- The Elbow Joint was represented by Actuators 4 and 5.
- The Wrist Joint was represented by Actuators 6 and 7.
- The Gripper was represented by Actuator 8.

The Shoulder Joint's actions were designed with a combination of the "U-D-L-R" buttons with the "1" button:

- "L + 1" and "R + 1" would slew the arm left and right respectively and only Actuator 1 was involved.
- "U + 1" and "D + 1" would move Actuators 2 and 3 "up" or "down" respectively.

Figure 9.8 showed the implementation of this approach for the Shoulder Joint, for example:

```
116  // For controlling arm's joints
117      else if (RcvData == (RC100_BTN_L + RC100_BTN_1)) {  // ID = 1
118         NewPosA = CurrentPos[0] + Step;
119         if ((NewPosA >= 0) && (NewPosA <= 1023)) {
120            Dxl.writeWord( 1, GOAL_POSITION, NewPosA ); // set new GoalPos(1)
121         }
122      }
123      else if(RcvData == (RC100_BTN_R + RC100_BTN_1)) {  // ID = 1
124         NewPosA = CurrentPos[0] - Step;
125         if ((NewPosA >= 0) && (NewPosA <= 1023)) {
126            Dxl.writeWord( 1, GOAL_POSITION, NewPosA ); // set new GoalPos(1)
127         }
128      }
129      else if(RcvData == (RC100_BTN_U + RC100_BTN_1)) {  // ID = 2 and 3
130         NewPosA = CurrentPos[1] - Step;
131         NewPosB = CurrentPos[2] + Step;
132         if ((NewPosA >= 0) && (NewPosA <= 1023) && (NewPosB <= 1023) && (NewPosB >= 0)) {
133            Dxl.writeWord( 2, GOAL_POSITION, NewPosA ); // set new GoalPos(2)
134            Dxl.writeWord( 3, GOAL_POSITION, NewPosB ); // set new GoalPos(3)
135         }
136      }
137      else if (RcvData == (RC100_BTN_D + RC100_BTN_1)) {  // ID = 2 and 3
138         NewPosA = CurrentPos[1] + Step;
139         NewPosB = CurrentPos[2] - Step;
140         if ((NewPosA >= 0) && (NewPosA <= 1023) && (NewPosB <= 1023) && (NewPosB >= 0)) {
141            Dxl.writeWord( 2, GOAL_POSITION, NewPosA ); // set new GoalPos(2)
142            Dxl.writeWord( 3, GOAL_POSITION, NewPosB ); // set new GoalPos(3)
143         }
144      }
```

Fig. 9.8 Control of shoulder joint motions for "CM9_MobileManipulator_RC.ino"

- If Buttons "L + 1" were pushed then Statements 118–121 would be executed whereas the "NewPosA" parameter would be computed from the Current Position of Actuator 1 (i.e. CurrentPos[0]) and an added "Step" amount. Next, "NewPosA" would be checked against the "legal" range of 0–1,023 for an actuator's position value, and if it satisfied this criterion, it would be used as Goal Position to Actuator 1 in Statement 120.
- Whereas if Buttons "U + 1" were pushed then Statements 130–136 would be executed. As we would be dealing with two actuators (ID = 2 and 3), "NewPosA" and "NewPosB" would now be used in a similar manner as shown in the previous paragraph for these two actuators. A reminder from Sect. 6.2.2.4 was that these two actuators were mounted as a mirror image of each other, thus the appropriate "subtraction" or "addition" operations would have to be used depending on the actuator's configuration (Statements 130 and 131).

Figure 9.9 showed the implementation of this approach for the Elbow Joint while Fig. 9.10 showed a similar implementation for the Wrist Joint.

Figure 9.11 showed how the Gripper (Actuator 8) was closing down by "Step" via each push of Button 5, but it would be fully open in one move (Button 6).

Figure 9.11 also showed that the default control behavior was to stop the roving platform when no "legal" user's inputs were found (Statements 195–198), with the arm holding its last commanded position.

The key difference between the programs "MobileReactor_RC_Demo.tsk/tskx" (Chap. 6) and "CM9_MobileManipulator_RC.ino" was that in "MobileReactor_RC_Demo.tsk/tskx" **all** arm's servos were updated during each RC loop by calling

```
145      else if(RcvData == (RC100_BTN_U + RC100_BTN_2)) {  // ID = 4 and 5
146        NewPosA = CurrentPos[3] + Step;
147        NewPosB = CurrentPos[4] - Step;
148        if ((NewPosA >= 0) && (NewPosA <= 1023) && (NewPosB <= 1023) && (NewPosB >= 0)) {
149          Dxl.writeWord( 4, GOAL_POSITION, NewPosA ); // set new GoalPos(4)
150          Dxl.writeWord( 5, GOAL_POSITION, NewPosB ); // set new GoalPos(5)
151        }
152      }
153      else if (RcvData == (RC100_BTN_D + RC100_BTN_2)) {  // ID = 4 and 5
154        NewPosA = CurrentPos[3] - Step;
155        NewPosB = CurrentPos[4] + Step;
156        if ((NewPosA >= 0) && (NewPosA <= 1023) && (NewPosB <= 1023) && (NewPosB >= 0)) {
157          Dxl.writeWord( 4, GOAL_POSITION, NewPosA ); // set new GoalPos(4)
158          Dxl.writeWord( 5, GOAL_POSITION, NewPosB ); // set new GoalPos(5)
159        }
160      }
```

Fig. 9.9 Control of elbow joint motions for "CM9_MobileManipulator_RC.ino"

```
161      else if (RcvData == (RC100_BTN_U + RC100_BTN_3)) {  // ID = 6
162        NewPosA = CurrentPos[5] + Step;
163        if ((NewPosA >= 0) && (NewPosA <= 1023)) {
164          Dxl.writeWord( 6, GOAL_POSITION, NewPosA ); // set new GoalPos(6)
165        }
166      }
167      else if(RcvData == (RC100_BTN_D + RC100_BTN_3)) {  // ID = 6
168        NewPosA = CurrentPos[5] - Step;
169        if ((NewPosA >= 0) && (NewPosA <= 1023)) {
170          Dxl.writeWord( 6, GOAL_POSITION, NewPosA ); // set new GoalPos(6)
171        }
172      }
173      else if (RcvData == (RC100_BTN_R + RC100_BTN_3)) {  // ID = 7
174        NewPosA = CurrentPos[6] + Step;
175        if ((NewPosA >= 0) && (NewPosA <= 1023)) {
176          Dxl.writeWord( 7, GOAL_POSITION, NewPosA ); // set new GoalPos(7)
177        }
178      }
179      else if(RcvData == (RC100_BTN_L + RC100_BTN_3)) {  // ID = 7
180        NewPosA = CurrentPos[6] - Step;
181        if ((NewPosA >= 0) && (NewPosA <= 1023)) {
182          Dxl.writeWord( 7, GOAL_POSITION, NewPosA ); // set new GoalPos(7)
183        }
184      }
```

Fig. 9.10 Control of wrist joint motions for "CM9_MobileManipulator_RC.ino"

```
185      else if(RcvData == (RC100_BTN_5)) {  // ID = 8
186        NewPosA = CurrentPos[7] - Step;
187        if (NewPosA >= 150) {
188          Dxl.writeWord( 8, GOAL_POSITION, NewPosA ); // closing gripper(8) by Step
189        }
190      }
191      else if(RcvData == (RC100_BTN_6)) {  // ID = 8
192        Dxl.writeWord( 8, GOAL_POSITION, 512 ); // open gripper fully
193      }
194      else{
195        Dxl.goalSpeed(ID_NUM_12, 0);
196        Dxl.goalSpeed(ID_NUM_11, 0);
197        Dxl.goalSpeed(ID_NUM_14, 0);
198        Dxl.goalSpeed(ID_NUM_13, 0);
199      }
200    }
201  } // end of loop()
```

Fig. 9.11 Control of gripper moves for "CM9_MobileManipulator_RC.ino"

the "Execute" function, while in "CM9_MobileManipulator_RC.ino", only the servos that needed updating were updated during each RC loop.

Video 9.4 showed the performance of this OpenCM-IDE's sketch "CM9_MobileManipulator_RC.ino".

9.2.3   Using Phone Camera with Mobile Manipulator

This project used the front camera of an Android phone to search for a red dowel and an IR sensor to let the arm approach the dowel to a given distance and then to pick it up and place it in another location (see Fig. 9.12).

Because of the extent of the arm, a cable extension was added to the standard 5-pin ribbon cable used with the IRSS-10 module which was attached to the gripper's end (so that it could reach the OpenCM-904/C controller). This cable extension modified the voltage presented to the A/D converter of the OpenCM-904/C and resulted in unreliable readings during run time of the created code for this project (when the arm was moving around).

This project used the SMART DEVICE functions of the OpenCM IDE tool and the Android PLAY700 App which would be described in more details in Chap. 11 and

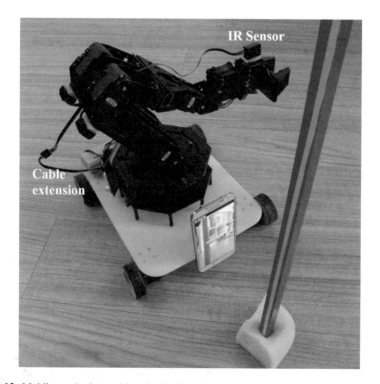

Fig. 9.12 Mobile manipulator with android phone

```
18   #include <OLLO.h>
19   OLLO myOLLO;
20
21   #define DXL_BUS_SERIAL1 1   //Dynamixel on Serial1(USART1)   <-Op
22   #define DXL_BUS_SERIAL2 2   //Dynamixel on Serial2(USART2)   <-LN
23   #define DXL_BUS_SERIAL3 3   //Dynamixel on Serial3(USART3)   <-Op
24
25   Dynamixel Smart(DXL_BUS_SERIAL2);  // phone connects to 904 via
26   Dynamixel Dxl(DXL_BUS_SERIAL3);  // Actuator control done via 4
27
28   #define ID_NUM_1 1
29   #define ID_NUM_2 2
30   #define ID_NUM_3 3
31   #define ID_NUM_4 4
32   #define ID_NUM_5 5
33   #define ID_NUM_6 6
34   #define ID_NUM_7 7
35   #define ID_NUM_8 8
36   #define ID_NUM_11 11
37   #define ID_NUM_12 12
38   #define ID_NUM_13 13
39   #define ID_NUM_14 14
40   #define GOAL_POSITION 30
41   #define GOAL_SPEED 32
42   #define PRESENT_POSITION 36
43   #define MAX_TORQUE 14
44   #define NUM_ARM_ACTUATOR 8
45
46   word CurrentPos[NUM_ARM_ACTUATOR], GoalPos[NUM_ARM_ACTUATOR];
47   word PosStep = 10, Speed = 350, SpeedL = 200;
48   word NewPosA = 0, NewPosB = 0;
49   byte LineColor = 3; // red = 3, black = 2, blue = 5, green = 4
50   byte LinePosition = 0; // valid line positions 1,2,3,4,5
```

Fig. 9.13   Parameters Definition in "CM9_MobileManipulator_PhoneCam.ino"

in particular Sect. 11.8, thus the reader may have to skip to Chap. 11 first before reading on the rest of this section.

Figure 9.13 described various parameter definitions used in the resulting sketch "CM9_MobileManipulator_PhoneCam.ino". Besides the actuators related parameters already explained in the previous Sect. 9.2.2, the "new" parameters were:

- Statements 18 and 19 were needed to use the IRSS-10 sensor which was attached to Port 1 of the OpenCM-9.04/C controller used (i.e. "OLLO" type of sensor).
- The Android phone (and its components) was considered a SMART "Dynamixel" associated with SERIAL 2 (Statement 25) – please see Chap. 11 for more details. Physically this connection was made via a BT-210 module.
- "LineColor" and "LinePosition" were parameters to be passed back and forth between the 9.04 controller and the Android phone (Statements 49 and 50). The PLAY700 phone app could only search for "vertical stripes" of four different

```
52  void setup() {
53    // Dynamixel 1.0 and 2.0 Protocol -> 0: 9600, 1: 57600, 2: 115200, 3: 1Mbps
54    Smart.begin(1);  // Phone initialized to 57600 bps via SERIAL 2
55    myOLLO.begin(1,IR_SENSOR);  // IR Sensor on Port 1
56    Dxl.begin(3);  // 485-EXP board initialized to 1 Mbps
57
58    Dxl.jointMode(ID_NUM_1);  // Set Joint Mode for ID=1 to 8 for Reactor arm
59    Dxl.jointMode(ID_NUM_2);
60    Dxl.jointMode(ID_NUM_3);
61    Dxl.jointMode(ID_NUM_4);
62    Dxl.jointMode(ID_NUM_5);
63    Dxl.jointMode(ID_NUM_6);
64    Dxl.jointMode(ID_NUM_7);
65    Dxl.jointMode(ID_NUM_8);
66
67    Dxl.wheelMode(ID_NUM_11);  // Set Wheel Mode for ID=11 to 14 for rover's wheels
68    Dxl.wheelMode(ID_NUM_12);
69    Dxl.wheelMode(ID_NUM_13);
70    Dxl.wheelMode(ID_NUM_14);
71
72    initpage();  // get arm in init pose
73    readyphone();  // get phone connection ready
74    readypage();  // get arm to ready pose
75
76  }  // end of setup()
```

Fig. 9.14 Function setup() in "CM9_MobileManipulator_PhoneCam.ino"

"colors": black, red, green and blue. "LineColor" was set to three because a red dowel was used. "LinePosition" also had a limited range of values, between one and five only (see Sect. 11.4 for more information) and it would be returned by the phone to the 9.04 controller during the sketch's execution. "LinePosition" represented five zones in the top fifth of the phone screen, with "3" being the center zone and "1" being the zone bordering the screen's left edge.

Figure 9.14 described the standard Arduino setup() function where:

- Statement 54 showed that communications between the phone and the 904C was set at 57.6 Bkps (i.e. BlueTooth).
- Statements 58–70 defined the operating modes for each of the twelve actuators used.
- The function initpage() was called to set the arm in its initial pose (Statement 72), i.e. aiming the gripper to the left of the manipulator's platform.
- The function readyphone() was used to synchronize the starting activities of the phone (start the PLAY700 app and connect via BlueTooth) and the 904-C controller.
- The function readypage() slew the arm towards the center front of the platform.

Figure 9.15 showed the details of the functions initpage() and readypage():

- The For loop (Statements 132–136) set the Maximum Torque for each of the eight arm's servos to its maximum value of 1,023, and their Goal Speed to 512 (i.e. at half power).
- Next, each servo from 1 to 8 was given a positional value [0–1,023] to go to, resulting in a pose captured in Fig. 9.12.

```
131  void initpage(){
132    for (int i = 1; i <= NUM_ARM_ACTUATOR; i++)
133    {
134       Dxl.writeWord( i, MAX_TORQUE, 1023 ); // se
135       Dxl.writeWord( i, GOAL_SPEED, 50 ); // set
136    }
137    Dxl.writeWord(1, GOAL_POSITION, 800); //Compa
138    Dxl.writeWord(2, GOAL_POSITION, 383); //Compa
139    Dxl.writeWord(3, GOAL_POSITION, 631); //Compa
140    Dxl.writeWord(4, GOAL_POSITION, 264); //Compa
141    Dxl.writeWord(5, GOAL_POSITION, 747); //Compa
142    Dxl.writeWord(6, GOAL_POSITION, 641); //Compa
143    Dxl.writeWord(7, GOAL_POSITION, 512); //Compa
144    Dxl.writeWord(8, GOAL_POSITION, 512); //Compa
145  }  // end of Direct Init Page
146
147  void readypage(){
148    for (int i = 1; i <= NUM_ARM_ACTUATOR; i++)
149    {
150       Dxl.writeWord( i, GOAL_SPEED, 50 ); // set
151    }
152    Dxl.writeWord(1, GOAL_POSITION, 512); //Compa
153 // Dxl.writeWord(2, GOAL_POSITION, 383); //Cor
154 // Dxl.writeWord(3, GOAL_POSITION, 631); //Cor
155 // Dxl.writeWord(4, GOAL_POSITION, 264); //Cor
156 // Dxl.writeWord(5, GOAL_POSITION, 747); //Cor
157 // Dxl.writeWord(6, GOAL_POSITION, 641); //Cor
158 // Dxl.writeWord(7, GOAL_POSITION, 512); //Cor
159 // Dxl.writeWord(8, GOAL_POSITION, 512); //Cor
160  }  // end of Ready Page
```

Fig. 9.15 Functions initpage() and readypage() in "CM9_MobileManipulator_PhoneCam.ino"

- The function readypage() was similar to the function initpage(), but only Servo ID = 1 needed to be set to a new position (i.e. Statement 152). The other Dxl. WriteWord () functions (Statements 153–159) were commented out and were only listed for the reader's reference.

Figure 9.16 explained the workings of Function readyphone() which was used to access text-display and text-to-speech services of the phone:

- Statement 194 showed that after getting to its "initpage()" pose, the arm would wait for the user to wave a hand in front of the IR Sensor so as to trigger a reading greater or equal to 50.
- Then a series of SMART functions would be called to trigger different kind of services on the phone (considered as a Dynamixel, ID = 100):
  - Statement 195 set the phone to its portrait mode.
  - Statement 196 cleaned all text on the phone's screen.

```
193  □void readyphone(){
194      while(myOLLO.read(1, IR_SENSOR) < 50) {} /
195      Smart.writeByte(100, 10010, 1);  // set ph
196      Smart.writeDword(100, 10160, 0);  // clea1
197      Smart.writeDword(100, 10180, 1);  // TTS i
198      Smart.writeDword(100, 10160, 73662733);  /
199      Smart.writeByte(100, 10020, 2); // using f
200      Smart.writeByte(100, 10030, 0); // set can
201      while (Smart.readByte(100, 10180) != 0){}
202  └}
```

Fig. 9.16 Function readyphone() in "CM9_MobileManipulator_PhoneCam.ino"

```
79  □void loop() {
80   // user waves in front of IR sensor to start operations
81     while(myOLLO.read(1, IR_SENSOR) < 50) {} // Waiting for user's signal
82   // Phone announces search and sets camera to Line Detection mode
83     Smart.writeDword(100, 10180, 2);  // TTS item 2
84     Smart.writeDword(100, 10160, 107217421);  // text display item 2 at po:
85     while (Smart.readByte(100, 10180) != 0){} // waiting for TTS Item 2 to
86     Smart.writeDword(100, 10040, 4);  // set Line Detection mode
87     Smart.writeDword(100, 10050, LineColor);  // looking for LineColor
88
89   // Rover rotates right until camera sees color dowel
90     turn_right();
91     while (Smart.readByte(100, 10110) == 0){} // waiting for color dowel t(
```

Fig. 9.17 Part 1 of main loop() in "CM9_MobileManipulator_PhoneCam.ino"

– Statement 197 started the Text-To-Speech process for Text Item 1 (i.e. voicing the word "ready").
– Statement 198 displayed the same Text Item 1 ("ready").
– Statement 199 told the phone to use its front camera and Statement 200 set no zoom on this camera.
– Statement 201 waited for the Text-To-Speech process to finish before moving on with the rest of the control algorithm (i.e. calling the readypage() function).

After setting the arm to its "readypage()" pose, the sketch entered the main loop() function which had 4 parts and Part 1 was described in Fig. 9.17:

• Once again, the controller would wait for the user to waive a hand in front of the IR Sensor when ready to start the dowel scan process (Statement 81).
• Statements 83–85 started "Text-To-Speech" for Text Item 2 ("front camera"), displayed it and waited for the "Text-To-Speech" process to finish.
• Statement 86 set the phone camera to its Line Detection mode and Statement 87 told it to look for LineColor (=3, i.e. red color) to show up in one of its five zones labeled 1 to 5 (see Fig. 11.25).

```
93    // Rover maneuvers to keep color dowel in view while appro
94    while(myOLLO.read(1, IR_SENSOR) < 700) {
95         LinePosition = Smart.readByte(100, 10110); // LinePo
96         if (LinePosition == 3){
97             forward();
98         }
99         else if (LinePosition <= 2){
100            turn_right();
101        }
102        else if (LinePosition >= 4){
103            turn_left();
104        }
105        else if (LinePosition == 0){ // when lost track of c
106            stop_rover();
107            Smart.writeDword(100, 10180, 3);  // TTS item 3
108            Smart.writeDword(100, 10160, 90440461);  // text
109            while (Smart.readByte(100, 10180) != 0){} // wa:
110        }  // end of if-else-if
111    }  // end of while IR_SENSOR < 700
112    stop_rover();
```

Fig. 9.18 Part 2 of main loop() in "CM9_MobileManipulator_PhoneCam.ino"

- The platform was then instructed to turn right (Statement 90) and to keep on turning right as long as the phone did not report seeing anything "red" in its camera's view (Statement 91). Eventually, the camera would see "red" and returned a non-zero zone number via the function Smart.readByte(100, 10110). Next the 904 controller would enter Part 2 of the dowel search algorithm.

Figure 9.18 showed the steps used in Part 2 of the algorithm which used the camera to keep the platform centered on the red dowel and to approach it up to a certain distance controlled by the IR Sensor (i.e. its reading should be less than 700):

- Statement 94 started a While loop based on the IR Sensor's reading (i.e. MyOLLO.read() on Port 1). If this reading was less than 700, the While loop would be entered.
- Next, it would save in "LinePosition" the "zone number" returned by the function Smart.readByte(100, 10110) corresponding to the zone where the phone's camera had found some "red" blob inside of it (Statement 95).
- The IF-ELSE-IF structure (Statements 96 to 110) was designed for the platform to turn right or left to keep the red dowel centered in the camera's view (i.e. if LinePosition was on the "right" or "left" sides of Zone 3), and to go forward and approach the dowel if it was found in Zone 3.
- Statements 105–109 were designed to catch a "serious" error condition when suddenly the camera would not see the red dowel anymore and the rover would stop in its track and then voice and display Text Item 3 (i.e. "dowel lost") via the phone.

```
232 ⊟void stop_rover(){
233     Dxl.goalSpeed(ID_NUM_12, 0);
234     Dxl.goalSpeed(ID_NUM_11, 0);
235     Dxl.goalSpeed(ID_NUM_14, 0);
236     Dxl.goalSpeed(ID_NUM_13, 0);
237  }
238
239 ⊟void stop_arm(){
240     Dxl.goalSpeed(BROADCAST_ID, 0);
241 // Set all servos to their current positions wherever they were stopped at
242 ⊟  for(int i = 1; i <= NUM_ARM_ACTUATOR; i++){
243        NewPosA = Dxl.readWord (i, PRESENT_POSITION);
244        Dxl.writeWord(i, GOAL_POSITION, NewPosA);
245 //     Dxl.writeWord(i, GOAL_POSITION, Dxl.readWord (i, PRESENT_POSITION));
246     }
247  }
```

Fig. 9.19 Functions stop_rover() and stop_arm() in "CM9_MobileManipulator_PhoneCam.ino"

```
113     extendpage(); // start to extend arm out towards dowel
114     while(myOLLO.read(1, IR_SENSOR) <= 1600) {} // Waiting 1
115     stop_arm(); // stop all servos and keep them at current
116
117 //  gripper-only action
118     Dxl.goalSpeed(ID_NUM_8, 256); // restart gripper
119     Dxl.writeWord(8, GOAL_POSITION, 120);  // close gripper
120     delay(2000); // waiting for gripper to close
```

Fig. 9.20 Part 3 of main loop() in "CM9_MobileManipulator_PhoneCam.ino"

- If the platform would approach the red dowel close enough to make the IR Sensor return a reading more than 700, then the While loop would be exited and Statement 112 would be executed and the rover would stop (see Fig. 9.19 for details of Function stop_rover()). Then the controller entered Part 3 of the control algorithm.

Figure 9.20 described Part 3 where the controller would only use the IR Sensor to decide when to grab the dowel while the arm was extending towards the dowel:

- With Statement 113, Function extendpage() (see Fig. 9.21) was triggered to stretch out the arm to get the gripper and thus the IR Sensor closer to the dowel. Please note that Function extendpage() unfolded the arm from its "ready" pose to its "extended" pose using only Servos with IDs = 2, 3, 4 and 5.
- Eventually, the IR Sensor would report a value greater than 1,600 (Statement 114) and this While loop would be exited and Statement 115 would stop all the arm's servos. This was done by the function stop_arm() which first set their GOAL SPEEDs to zero with a broadcast command, then to make all their PRESENT POSITIONs become their GOAL POSITIONs (see Fig. 9.19 for details of Function stop_arm()). This was done to prevent the arm's servos from trying to go to their original goal positions defined by Function extendpage() which was executed earlier (Statement 113 in Fig. 9.20), whenever their GOAL

```
162  void extendpage(){
163     for (int i = 1; i <= NUM_ARM_ACTUATOR; i++)
164     {
165        Dxl.writeWord( i, GOAL_SPEED, 25 ); // set
166     }
167  //  Dxl.writeWord(1, GOAL_POSITION, 512); //Com
168     Dxl.writeWord(2, GOAL_POSITION, 539); //Compa
169     Dxl.writeWord(3, GOAL_POSITION, 474); //Compa
170     Dxl.writeWord(4, GOAL_POSITION, 430); //Compa
171     Dxl.writeWord(5, GOAL_POSITION, 580); //Compa
172  //  Dxl.writeWord(6, GOAL_POSITION, 641); //Com
173  //  Dxl.writeWord(7, GOAL_POSITION, 512); //Com
174  //  Dxl.writeWord(8, GOAL_POSITION, 512); //Com
175  }  // end of Extend Page
176
177  void storepage(){
178     for (int i = 1; i <= NUM_ARM_ACTUATOR; i++)
179     {
180        Dxl.writeWord( i, GOAL_SPEED, 50 ); // set
181     }
182     Dxl.writeWord(1, GOAL_POSITION, 800); //Compa
183     Dxl.writeWord(2, GOAL_POSITION, 383); //Compa
184     Dxl.writeWord(3, GOAL_POSITION, 631); //Compa
185     Dxl.writeWord(4, GOAL_POSITION, 264); //Compa
186     Dxl.writeWord(5, GOAL_POSITION, 747); //Compa
187     Dxl.writeWord(6, GOAL_POSITION, 800); //Compa
188  //  Dxl.writeWord(7, GOAL_POSITION, 512); //Com
189  //  Dxl.writeWord(8, GOAL_POSITION, 512); //Com
190     delay(2000);  // waiting for arm to get to St
191  }  // end of Store Page
```

Fig. 9.21 Functions extendpage() and storepage() in "CM9_MobileManipulator_PhoneCam.ino"

SPEEDs were reset to non-zero values in the later Part 4 of the algorithm. This was an important and subtle issue to note.

- Once the arm was completely stopped, only the gripper (ID = 8) was re-enabled by Statement 118 in Fig. 9.20.
- Statement 119 closed the gripper and Statement 120 waited for 2,000 ms as the gripper was closing in on the dowel.

Part 4 of the algorithm was described in Fig. 9.22 whereas:

- All servos were re-enabled with a broadcast command to set all GOAL SPEEDs to 50 (Statement 122).
- Next, the arm was reset to the readypage() pose and then moved to the storepage() pose (Statements 123 and 124).

```
121    // get whole arm back on line
122    Dxl.goalSpeed(BROADCAST_ID, 50);
123    readypage();  // pull arm back to ready pose
124    storepage();  // go to storage position
125    Dxl.writeWord(8, GOAL_POSITION, 512);  // open gripper
126
127    while(1);  // DONE so wait here for user to push reset button to restart
128
129  }  // end of loop()
```

Fig. 9.22 Part 4 of the main loop() function in "CM9_MobileManipulator_PhoneCam.ino"

- Statement 125 finally opened the gripper.
- Statement 127 was only a "trick" to make the standard Arduino loop() function to essentially execute only once (like the regular C main() function).

Please also note that Function storepage() swung the arm (with the dowel in its gripper) back to the left side of the manipulator's platform and used Servos with IDs = 1 to 6 (see Fig. 9.21). The "2000 ms" delay in Statement 190 of Fig. 9.21 was used to wait for the arm to reach its "storage" location.

Video 9.5 showed the performance of this algorithm on a complete run of the algorithm which only happened occasionally during the author's actual tests. The reader might recall from the beginning of this section that the author had to use the IRSS-10 sensor beyond its design constraints because a much longer 5-pin cable had to be used in order to connect it from the gripper's station down to the controller GPIO Port 1, and along the entire arm. The trouble spot was with Statement 114 in Fig. 9.20. At this stage of the algorithm, the arm was extending forward and also tugging on the 5-pin cable. Thus the IR Sensor could not function properly at times and therefore returned a reading that was always less than 1,600, even though physically the IR Sensor was getting very close to the dowel, but it just could not "see" it. Therefore externally, the author could see that the arm kept on extending to its final position as defined inside the extendpage() function and nothing else would happen afterwards (as the controller was still stuck inside the While loop of Statement 114). Perhaps some readers could find a better sensor than the IRSS-10? ROBOTIS has not offered an ultrasonic sensor in its shop yet.

9.3 Embedded C for the CM-510 and CM-530

The current C/C++ programming SDK (V. 1.02) for the CM-510/CM-700 is accessible at http://support.robotis.com/en/software/embeded_c/cm510_cm700.htm while the corresponding one for the CM-530 can be found at http://support.robotis.com/en/software/embeded_c/cm530.htm.

These two web sites also have detailed information about installing the needed tool chains for these two controllers:

1. The CM-510/CM-700 tool chain recommended by ROBOTIS is WinAVR® and Atmel Studio® (http://support.robotis.com/en/software/embeded_c/cm510_cm700/embeded_c_start.htm). Although it is also possible to use the Eclipse IDE with WinAVR (see Sub-Sect. 9.3.1).
2. The CM-530 tool chain recommended by ROBOTIS is JRE, WinARM and Eclipse® (http://support.robotis.com/en/software/embeded_c/cm530/embeded_c_start_stm.htm).

9.3.1 Tutorials for CM-510

ROBOTIS provided tutorial information for the programming of the CM-510 at (http://support.robotis.com/en/software/embeded_c/cm510_cm700/programming.htm) using the Atmel Studio tool. ROBOTIS also provided many worked out examples at (http://support.robotis.com/en/software/embeded_c/cm510_cm700/example.htm).

If the reader is more interested in using Eclipse with the CM-510, the reader needs to review the series of four video tutorials created by Dr. Yanfu Kuo from National Taiwan University on YouTube:

1. https://www.youtube.com/watch?v=csgotzBhbmI – Tutorial 1 described the basic architecture of the Atmel AVR ATmega2561 and showed how it was implemented on the CM-510. It also showed the detailed steps for programming the CM-510 from creating an Eclipse project, coding and downloading to the controller using the ROBOTIS Terminal tool. Tutorial 1 showed how to control devices such as LEDs, buttons, serial communication, buzzer and microphone already built-in on the CM-510 controller.
2. https://www.youtube.com/watch?v=OIUkl6iBUfM – Tutorial 2 dealt with sensor interfacing issues using the OLLO NIR sensor as the example. It continued on showing how to control Dynamixel actuators such as the AX-12, and ended with a ZigBee communications programming example.
3. https://www.youtube.com/watch?v=6jZtmN3PXEY – Tutorial 3 showed how to program the HaViMo2 video camera to track colored objects using a pan-tilt platform constructed with 2 AX-12 actuators. The source code is enclosed in the ZIP file "HaViMo2.zip".
4. https://www.youtube.com/watch?v=GGjeCOuua9M – Tutorial 4 demonstrated how to interface the Gyro sensor on a CarBot platform and how to integrate the Gyro's angular rate data to obtain an estimate of how much of an angle (in degrees) that the CarBot had rotated after a given maneuver. The source code is enclosed in the ZIP file "GyroCompass.zip".

9.3.2 Tutorials for CM-530

ROBOTIS provided tutorial information for the programming of the CM-530 at (http://support.robotis.com/en/software/embeded_c/cm530/programming_stm. htm) using the Eclipse-WinARM tool chain. ROBOTIS also provided several programming examples at (http://support.robotis.com/en/software/embeded_c/cm530/ example_stm.htm).

Unfortunately, there no video resources, known to the author, showing how to use of the Eclipse-WinARM tool chain for the CM-530, but the CM-510 tutorial series provided by Dr. Yan-Fu Kuo should serve as a good starting point as they were also using Eclipse.

9.3.3 Future Support for Embedded C for CM-510/530?

At present, ROBOTIS had not been updating the CM-510/530 libraries for the new sensors such as the IR Sensor Array, Color and Magnetic sensors, nor for the 485-EXP Expansion module, but they did recently release the new Windows and Linux Dynamixel Communications Protocol 2.02 (http://support.robotis.com/en/software/ dynamixel_sdk/usb2dynamixel/usb2dxl_windows.htm#bc-1).

9.4 Motion Programming and Embedded C

With Embedded C SDKs, ROBOTIS had graciously shared their knowledge and expertise for their robotic systems (BIOLOID and OpenCM) with robotics enthusiasts everywhere. However, ROBOTIS still kept their Motion Programming technologies pretty much proprietary, and this move was very understandable by anyone who had a closer look at what Motion V.2 could do (more on this tool in Chap. 10). At present, there are only a few resources dealing with Motion Programming with the OpenCM IDE.

Included with the OpenCM IDE (V.1.0.2) were 2 ROBOTIS examples showing the basics of creating Motion Pages and how to synchronize-play these Motion Pages:

1. The first sketch "q_Motion_Page_Play" is accessible via the "File" pull-down menu >> Examples >>06.Dynamixel >>q_Motion_Page_Play.
2. The second sketch "s_Manipulator_4DOF" is accessible via the "File" pull-down menu >> Examples >>06.Dynamixel >>s_Manipulator_4DOF.

Video 9.6 demonstrated the operation of the "q_Motion_Page_Play" sketch, modified to work with the RC-100 on a set of 4 AX-12s mounted in a CarBot, but these actuators could also be constructed into a robotic arm such as the one shown in Fig. 9.23.

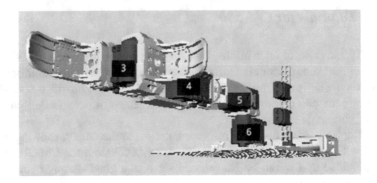

Fig. 9.23   4-DOF robotic arm from BIOLOID STEM EXPANDED kit

Mr. Matthew Paulishen had contributed many advances for our UGA robotics laboratory as shown in earlier Chaps. 7 and 8. In 2013, he adapted the PyPose tool from Vanadium Labs (http://vanadiumlabs.github.io/pypose/) into the OpenCM IDE as the library CM9_BC which could convert a standard MTN file into PyPose data sets. A presentation of his work can be found on YouTube at https://www.you-tube.com/watch?v=iPD5oOFYq3Y. The application of the CM9_BC tool to an OpenCM-9.00V.1 can be viewed at this link https://www.youtube.com/watch?v=4X-KbiOPE4s, while an application on the OpenCM-9.04/B can be watched at this web site https://www.youtube.com/watch?v=EdlTAVn2d3s.

More recently (April 2014), Dr. Dr. Hamid Mobalegh created a direct walking engine into an OLLO biped, with 4 XL-320s, as a sketch for the OpenCM-9.04/B (http://www.havisys.com/?p=148).

In conclusion, the Open-CM IDE seemed to be well accepted by a still small community but interested in developing Open Motion Programming capabilities for the OpenCM-9.04 systems.

9.5   Review Questions for Chap. 9

1. Variables in TASK programs can be of the floating-point types. (T-F)
2. An assignment statement in a TASK program would be equivalent to a function call in Embedded C. (T-F)
3. What ROBOTIS database would an Embedded C programmer consult in order to user proper data types for variables used?
4. What is the most important object in Embedded C for ROBOTIS hardware?
5. Which type of data bus is the device type Dynamixel associated to?
6. Which type of data bus is the device type OLLO associated to?
7. How many serial channels does the OpenCM9.04 systems provide?
8. Which serial channel is the shield 485 EXP board associated with?

9. Which serial channel are the GPIO ports associated with?
10. Which serial channel are the Dynamixel ports associated with?
11. Which serial channel is associated with communication devices such as ZigBee and BlueTooth?
12. Which Dynamixel Communication Protocol does an MX-28 actuator respond to?
13. Which Dynamixel Communication Protocol does an XL-320 actuator respond to?
14. Describe the tool chain needed to use Embedded C on the CM-510.
15. Describe the tool chain needed to use Embedded C on the CM-530.
16. Describe the tool chain needed to use Embedded C on the OpenCM-9.04/B.

9.6   Review Exercises for Chap. 9

ROBOTIS had created extensive Embedded C example programs for their controllers CM-510, CM-530 and OpenCM-9.04/A/B/C that the readers can refer to when they install these resources on their own PCs.

Reference

Margolis M (2011) Arduino cookbook. O'Reilly Media, Sebastopol

Chapter 10
ROBOTIS-MINI System

In the Spring of 2014, the DARWIN-MINI system was released internationally using a new controller (OpenCM-9.04-C), with a new communication device (BT-210) and the new RoboPlus 2.0 suite: TASK, R+ DESIGN and R+ MOTION (V.2). The RoboPlus Manager tool (currently at V.1.0.34.1) could only serve as a firmware updater to the 9.04-C, but the RoboPlus Dynamixel Wizard tool (currently at V.1.0.19.21) had been modified to work with OpenCM-9.04-C quite well as a Firmware Update/Recovery tool for the XL-320 actuators as well as for the 9.04-C controller (see Sect. 4.2.2). The OpenCM-9.04-C could also be used with the OpenCM IDE V.1.0.4, but it would need a firmware recovery (via Manager only and the micro USB port) to make it work again with the RoboPlus 2.0 suite. In November 2014, the 2nd edition of this system with an improved XL-320 (metal pinion gear) was renamed ROBOTIS-MINI.

The robot programming concepts behind ROBOTIS MOTION tools could be generalized to a research area named Programming by Demonstration (PbD), also known as Imitation Learning or Apprentice Learning. The interested reader is referred to Billard et al. (2008), Calinon (2009) and also these web links:

- http://www.scholarpedia.org/article/Robot_learning_by_demonstration
- http://programming-by-demonstration.org/

This chapter's main topics are listed below:

- PC wireless options for the MINI.
- New motion concepts in MOTION V.2.
- PC motion control with TASK and MOTION V.2.
- LED integration with TASK and MOTION V.2.
- Choreography for 2 MINIs.

© Springer International Publishing Switzerland 2017
C.N. Thai, *Exploring Robotics with ROBOTIS Systems*,
DOI 10.1007/978-3-319-59831-4_10

10.1   PC to MINI Wireless Options

On a mobile device, such as an Android tablet, BlueTooth comes standard thus mobile users can only use BlueTooth with the ROBOTIS-MINI. However on the PC, users could choose between ZigBee (ZIG-110A) and BlueTooth (BT-110A or BT-210).

The ZIG-110A by default worked at 57.6 Kbps but it could only go up to 115.2 Kbps as it was an older ROBOTIS product (c. 2009), but it could emulate 3 modes of communications: 1 to 1 (1:1), 1 to many (1:N) or broadcast (N:N). In the author's experiences, upon powering up ZigBee usually achieved connections much quicker and more reliably than BlueTooth. On the PC side, the user had to use a combination of three modules (USB2Dynamixel + Zig2Serial + ZIG-100) to make the connection, but it used only one Serial COM port on the PC side. The ZigBee N:N option, once set up properly (see Sects. 7.1 and 7.2), could be particularly convenient if ones needed multiple robots to communicate with each other without involving the PC and this work could be done via the current TASK tool.

The BT-110A (BT specification 2.0) by default was also set to 57.6 Kbps but could reach out to 230.4 Kbps, while the BT-210 (BT specification 2.1) could get up to 400 Kbps. On the PC side, most new PCs would come with a built-in BlueTooth server (specification 4.0 at present) or a small USB device could be purchased to fulfill this role on an older PC. Upon connection to the PC, the PC OS would use two Serial COM ports per BT device (see Fig. 10.1) and the user would have to take care to pick only the OUTGOING COM ports to connect between the various ROBOPLUS software tools (TASK, MANAGER, DYNAMIXEL WIZARD) and multiple MINIs. However, since V.2.2.3, the R+MOTION tool filtered out the

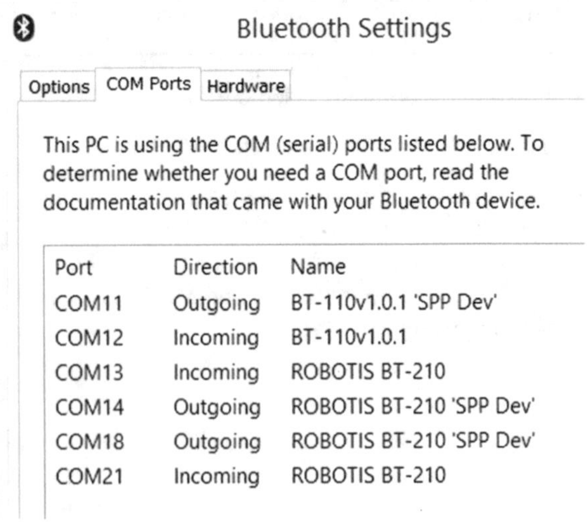

Fig. 10.1 BT settings on PC with 1 BT-110A and 2 BT-210 connected

Bluetooth Settings

Options COM Ports Hardware

This PC is using the COM (serial) ports listed below. To determine whether you need a COM port, read the documentation that came with your Bluetooth device.

Port	Direction	Name
COM11	Outgoing	BT-110v1.0.1 'SPP Dev'
COM12	Incoming	BT-110v1.0.1
COM13	Incoming	ROBOTIS BT-210
COM14	Outgoing	ROBOTIS BT-210 'SPP Dev'
COM18	Outgoing	ROBOTIS BT-210 'SPP Dev'
COM21	Incoming	ROBOTIS BT-210

INCOMING COM port, and thus presented only the OUTGOING COM port to the user, which was a good step forward.

Figure 10.1 also implied that BT communications between multiple robots would have to be mediated via the BT server on the PC, which would require more programming resources and skills beyond the TASK tool. As a matter of fact, ROBOTIS recently released a technical note regarding the pairing of BT-210s using the ROBOTIS IDE and an OpenCM-9.04/B controller (http://support.robotis.com/en/techsupport_eng.htm#product/auxdevice/communication/bt-210_setting.htm).

Summing up, it really depended upon the user's needs, monetary funds and current programming expertise to choose the proper wireless protocol to use with the ROBOTIS-MINI system. Working with only 1 MINI, the BT-210 would be the most economical way to go for PC and Android platforms.

Around Summer 2015, ROBOTIS plans to release the new BT series BT-410 Master and Slave modules to allow 1:1 and 1:N communications. The BT-410 series would be based on BlueTooth 4.0 Low Energy.

10.2 New Motion Concepts in MOTION V.2

Between Version 1 and Version 2 of the MOTION tool, ROBOTIS had made quite a few fundamental changes such as: file suffix change from MTN to MTNX, removing size limitation on the physical file, changing internal data structures for easier motion design and editing, synchronization between the 3-D simulated robot moves and the actual physical moves of the demonstration robot.

The English version of the user manual for MOTION V.2 is available at (http://support.robotis.com/en/software/roboplus2/r+motion2/rplus_motion2.htm) and has many detailed procedures that the reader should review as needed. The ROBOTIS Development Team also hosted a YouTube channel where the reader could watch more tutorial videos at https://www.youtube.com/playlist?list=PLEf1s0tzVSnQSIj5MP0Uuss86BdjnlGAY.

10.2.1 Unlimited File Size for MTNX

The old MTN motion file had a maximum file size that was linked to the working memory size of the respective CM-5XX controllers, i.e. 127 motion pages for CM-5 and 255 motion pages for the CM-510 and CM-530.

The new MTNX file no longer has an upper limit for its physical size thanks to a new Motion Data structure being implemented (see next Sub-Sect. 10.2.2). The MTNX file was also now referred to as a "Project" in ROBOTIS' technical documents.

10.2.2   Efficient Motion Data Structure

Motion data in Version 1 (see Fig. 10.2) could be described hierarchically as:

1. POSE – specify a set of user-defined goal position values for all actuators used by the demonstration robot and at an instant in time.
2. STEP – specify a set TIME interval for the robot to reach a given POSE. Up to a maximum of 7 STEPS could be defined per motion PAGE which could be considered as a small move by the robot. A PAUSE time interval could also be defined for "between" STEPS.
3. PAGE – each motion PAGE can be linked to a NEXT page to create more elaborate robot gestures. An EXIT page could also be defined to ensure a stable position for emergency stops. The PAGE NUMBERS defined could be triggered or "played" from a controlling TASK program, resulting in the actual performance of the robot's moves.

Figure 10.3 was a screen capture of the main Motion Editing interface for MOTION V.2 which used the Unity® graphics engine (http://unity3d.com/).

MOTION V.2 used a Global Time Line where the smallest Time Frame allowed was 8 ms which corresponded to the refresh cycle time for all ROBOTIS Dynamixels (see Fig. 10.3). This 8 ms timing also explained the synchronization process between the simulation graphics and the real timing of the robot's executed moves.

Each robot POSE still stood for a set of goal position values of the robot's actuators which could be manipulated singly or as a group using the POSITIONING tool, located in the lower right corner of Fig. 10.3, with full 3-D graphical feedback on the robot model. Once satisfied with a given POSE, the user could insert it into a

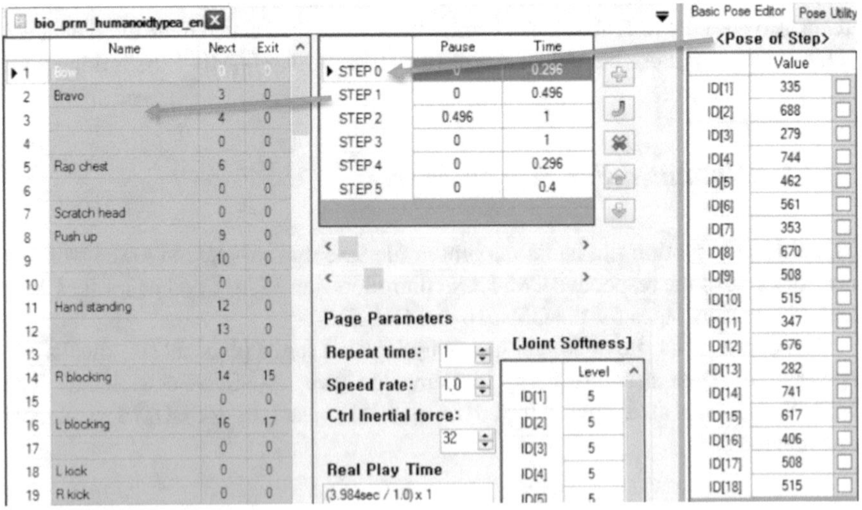

Fig. 10.2 Motion Data Structure used in Motion V.1

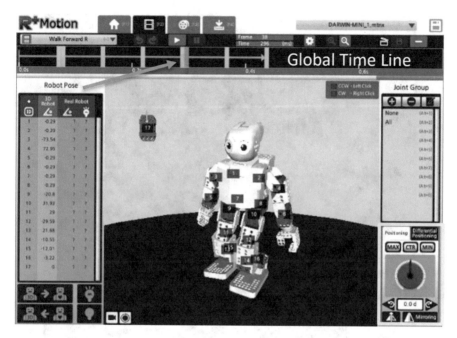

Fig. 10.3 Motion unit editing interface in motion V.2

wanted time frame on the Global Time Line to make it become a KEY FRAME. Several KEY FRAMES would form a MOTION UNIT (see Fig. 10.4).

Several MOTION UNITS could then be edited into a MOTION LIST which essentially performed the Flow Control task for the selected MOTION UNITS (see Fig. 10.5).

The next step for the user was to create a custom MOTION GROUP which had user-selected MOTION LISTS as "independent" members of this Motion Group (see Fig. 10.6).

The user could create several MOTION GROUPS to be saved in the same MTNX file (because its physical size on the PC is now unlimited), but the user could DOWNLOAD only ONE Motion Group to the MINI at any one time, because the working memory on the MINI was still finite (see Fig. 10.7).

The INDEX parameter (see Fig. 10.7) was the one that the TASK tool can access to activate selected robot moves from inside a companion TSK program (INDEX was therefore equivalent to the PAGE NUMBER when using V.1).

The video file "Video 10.1" showed how to use MOTION V.2 for various functions.

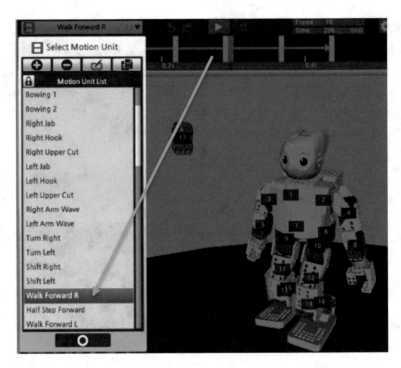

Fig. 10.4 Listing of user-created motion units

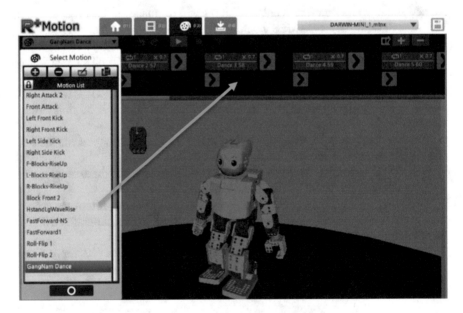

Fig. 10.5 Editing motion units into a motion list

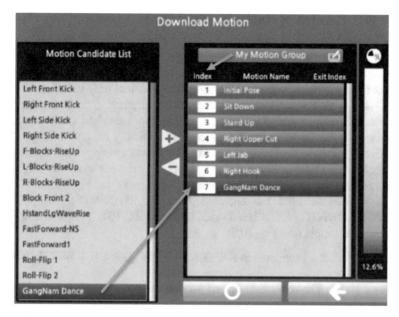

Fig. 10.6 Creating motion group with user-selected motion lists

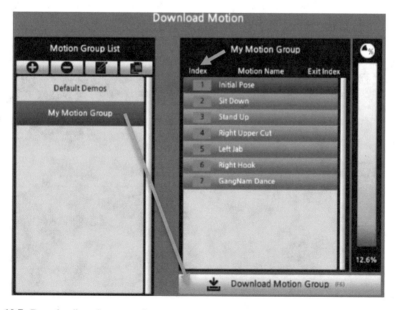

Fig. 10.7 Downloading chosen motion group

10.3   PC Control of Robot Moves

This little exercise was created to illustrate the basics on integrating PC to MINI communications, TASK and MOTION programs all together in one application.

Most folks likely had used the RoboPlus Manager tool only to update firmware or to have a quick check on actuators and sensors attached to the controller in use (as it was originally intended to). But the Manager tool also had a very handy sub-tool called Zig2Serial Management which was originally created to help manage the ZIG-100 (circa 2009). But as it turned out, this was a very general communications tool that can be used on the PC regardless whether ones use ZigBee or BlueTooth (just use the appropriate Windows COM port – see Fig. 10.8).

This application used the enclosed files "DARWIN-MINI-1.MTNX" and "DARWIN-MINI-RC.TSK". The DARWIN-MINI-RC.TSK programming structure was quite simple (see Fig. 10.9):

(a) Lines 6–7. Set all actuators to JOINT MODE (i.e. "2") and TORQUE to be ON (TRUE).
(b) Lines 10–11. The robot next played Motion Index "1" which was the READY Pose.
(c) Lines 14–23. Then the robot entered an endless loop where it waited for an input number coming from the PC and saved it in parameter "MotionGroupNo" (lines 16–19). The robot sent this "MotionGroupNo" value back to the PC for confirmation (line 20) and triggered this Motion Group's moves and waited until that was done (lines 21–22).

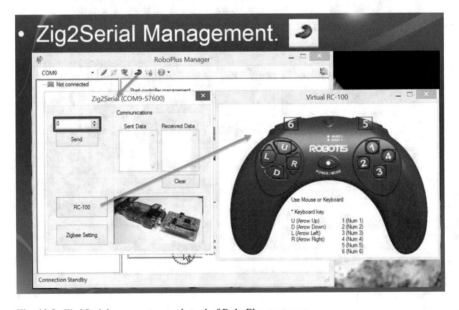

Fig. 10.8 Zig2Serial management sub-tool of RoboPlus manager

```
3    START PROGRAM
4    {
5        // Set all ID to joint mode
6        ● ID[254]: ADDR[11(b)]    = 2
7        ● ID[All]:  ◎ Torque ON/OFF  = TRUE
8
9        // Get into Init Pose
10       ● Motion Index Number  = 1
11       WAIT WHILE  ( ● Motion Status   ==  TRUE  )
12
13       // Endless loop for RC input from user
14       ENDLESS LOOP
15       {
16           MotionGroupNo  = 0
17           WAIT WHILE  ( ✎ Remocon Data Received   ==  FALSE  )
18
19           MotionGroupNo  = ✎ Remocon RXD
20           ✎ Remocon TXD  = MotionGroupNo
21           ● Motion Index Number  = MotionGroupNo
22           WAIT WHILE  ( ● Motion Status   ==  TRUE  )
23       }
24   }
```

Fig. 10.9 Program DARWIN-MINI-RC.TSK

From practice, the author had found that there was a very particular order that these files had to be downloaded to the MINI for them to work together properly:

1. Download the DARWIN-MINI-RC.TSK file first, via the TASK tool and the appropriate COM port. Close the TASK window to release this COM port.
2. Download the DARWIN-MINI-1.MTNX file next, via the MOTION V.2 tool and the same COM port. Close the MOTION window and make sure to turn the POWER OFF the MINI.
3. Turn power back on the MINI so that the TASK program get executed before starting the Zig2Serial sub-tool from inside the Manager tool. If the reader used BlueTooth, it might take 10–15 s for the Zig2Serial sub-window to come up (see Fig. 10.8) – ZigBee would connect much quicker than BT.
4. The user could now enter a number into the SEND field of the Zig2Serial sub-window and clicked away on the SEND button. The same number should appear under the SENT DATA list and then also in the RECEIVED DATA list (this indicated that 2-way communications had been established between PC and MINI). This "number" of course had to correspond to a valid INDEX number of the MOTION GROUP that had been downloaded (see Fig. 10.7).

The video file "Video 10.2" illustrated such as a session as described above.

10.4   Synchronizing LEDs to Motion

The XL-320 actuators on the MINI were equipped with programmable LEDs. In this next exercise, these LEDs would be programmed to turn on only for those actuators involved in a chosen robot's move to emphasize this particular move to the audience.

This application would use the "DM-LED-Synch.tsk" files and the Motion Group List named "LED Sync 1" in the "DARWIN-MINI-1.MTNX" file (see Fig. 10.10). This Motion Group List had two Motion Groups labeled "Initial Pose" (Index = 1) and "Test Moves" (Index = 2).

Figure 10.11 displayed the timing of the Key Frames used in the "Test Moves" Motion Group, and also the ON/OFF timings of the LEDs of the right arm (IDs = 1, 3, 5) and of the left arm (IDs = 2, 4, 6):

1. Time = [0, 390] ms – The robot turned its head to the right and all LEDs OFF.
2. Time = [390, 796] ms – The robot raised its right arm, thus right LEDs ON.
3. Time = [796, 1,195] ms – The robot raised its left arm, thus right LEDs OFF and left LEDs ON.
4. Time >1,195 ms – The robot brought both arms down, thus all LEDs OFF.

These 4 (actually only the first 3) time periods were monitored using the Hi-Resolution Timer of the OpenCM-9.04-C to trigger the needed ON/OFF actions for the LEDs involved. In the "DM-LED-Synch.tsk" file, the control logic implemented was quite simple:

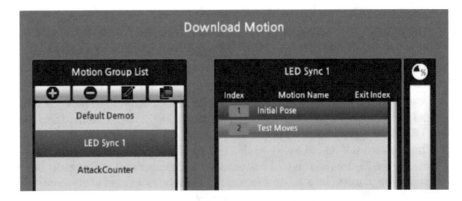

Fig. 10.10   Motion Group List "LED Sync 1"

Fig. 10.11   LEDs ON/OFF timings for right and left Arms

1. Play the Motion Group "2" (line 33), and start the Hi-Res Timer for 390 ms and essentially do nothing during this time period as the LEDs were set to OFF at the beginning of this program already (lines 35–36).
2. Lines 39–46 – Start the Hi-Res Timer for 406 ms (line 39) and during this time period, turn ON the LEDs of the right arm (IDs = 1, 3, 5). When the Hi-Res Timer "timed out", turn OFF all LEDs (line 46).
3. Lines 49–56 – Start the Hi-Res Timer for 399 ms and turn ON the left arm LEDs (IDs = 6, 4, 2) and then turn them OFF at time-out (line 56).

The video file "Video 10.3" illustrated the operation of this program using the Zig2Serial Management tool to trigger user-wanted events.

10.5 Fight Choreography for 2 MINIs via ZigBee

This last application illustrated a possible choreography framework for coordinating the interactions between two MINIs performing some Karate moves. One MINI served as the Lead Fighter while the other acted as the Counter Fighter.

This application used the Motion Group List named "AttackCounter" found in the "DARWIN-MINI-1-ZB.mtnx" file (see Fig. 10.12).

This Motion Group List had seven Motion Groups:

(a) Index 1 corresponded to the Ready Pose.
(b) Index 2 corresponded to Attack1 moves while Index 3 corresponded Counter1 moves.
(c) Indices 4 and 5 corresponded to Attack2 and Counter2 moves, while indices 6 and 7 corresponded Attack3 and Counter3 moves.

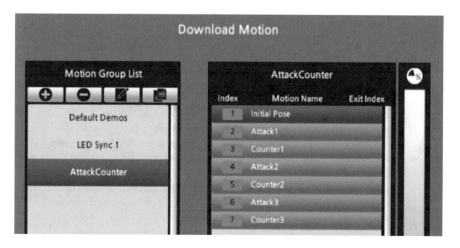

Fig. 10.12 Motion group list "AttackCounter" with seven Motion Groups

The TASK files "DM-LeadFighter-ZB.tsk" and "DM-CounterFighter-ZB.tsk" were designed to work with the previous 7 Motion Groups. This application was designed for a ZigBee Broadcast environment whereas the operator/judge would issue a "Ready" command (i.e a "1") or a "Fight" command (i.e. an "11") from the PC via the Zig2Serial Management tool. The interesting feature of the Lead-Fighter program was that it used the Random Number utility available on the OpenCM-9.04 series to trigger at random one of its 3 possible attack moves, while the Counter-Fighter program would trigger the appropriate non-randomized counter moves (of course the reader can modify the Counter-Fighter code to provide randomized counter moves or added more sensor-based counter moves).

The main logic in the "DM-LeadFighter-ZB.tsk" program was implemented via an endless loop that would "listen" for a ZigBee packet and processed it to figure out whether a "1" or an "11" was actually received:

(a) If a "1" was received, parameter "MotionGroupNo" was set to "1".
(b) If an "11" was received, the controller would throw the dice once and got a value for "RandomNo" between 0 and 255 (line 27). Next, a UNIFORM statistical distribution was assumed, thus there would be a 33% chance for the parameter "MotionGroupNo" to get its final value of 2, 4 or 6 (lines 28–43 and see Fig. 10.13).

```
21          IF ( DataIn  ==  1  )
22          {
23              MotionGroupNo   = DataIn
24          }
25          ELSE IF ( DataIn  ==  11  )
26          {
27              RandomNo  = 2 Random Number
28              IF ( RandomNo  <=  85  )
29              {
30                  MotionGroupNo  = 2
31              }
32              ELSE IF ( RandomNo  >  85  &&  RandomNo  <=  170  )
33              {
34                  MotionGroupNo  = 4
35              }
36              ELSE IF ( RandomNo  >  170  &&  RandomNo  <=  255  )
37              {
38                  MotionGroupNo  = 6
39              }
40              ELSE
41              {
42                  MotionGroupNo  = 1
43              }
```

Fig. 10.13 Logic for the LeadFighter program to decide on a randomized Attack move to perform

(c) The next step (line 45) was crucial because as we were using a broadcast environment, therefore the PC and the Counter-Fighter would receive all communication packets. Thus we had to "special-code" the information meant for the Counter-Fighter, by shifting it left by 4 bits (i.e. multiply it with 16) and saved this special information as parameter "MotionNoCounterFighter".

(d) Then the "LeadFighter" controller was instructed to broadcast the parameter "MotionNoCounterFighter" (line 47) which was really meant for the CounterFighter, and lastly triggered its own Attack move as reflected in the actual value of "MotionGroupNo" (line 50 and see Fig. 10.14).

The main logic in the "DM-CounterFighter-ZB.tsk" program was also implemented via an endless loop that would "listen" for a ZigBee packet and processed it (i.e. divide it by 16 – statement 20) to figure out whether "DataIn" was "0" (i.e. coming from the PC) or a "2", "4" or "6" (i.e. coming from the Lead-Fighter). Next this program used that information to set parameter "MotionGroupNo" with the correct Counter Motion Group's Index value and finally triggered the appropriate Motion Group (see Fig. 10.15).

The video file "Video 10.4" showed the execution of the above programs with 2 MINIs and within a broadcast ZigBee environment as described above.

10.6   Fight Choreography for 2 MINIs via BlueTooth

The author also had designed an alternate procedure using R+Motion V.2 under a BlueTooth environment to perform this choreography application.

First, ROBOTIS wrote the R+Motion V.2 in such a way that the Windows PC user can run multiple instances of this tool on their computer (this was not possible with the RoboPlus Motion V.1 and Manager tools). Thus the author's approach was to spawn out two instances of the R+Motion V.2 application: one instance would

```
45        MotionNoCounterFighter  = MotionGroupNo  *  16
46        // Send to Counter Fighter
47        Remocon TXD  = MotionNoCounterFighter
48
49        // Start Fight
50        Motion Index Number  = MotionGroupNo
51        WAIT WHILE ( Motion Status  ==  TRUE )
52
53        // Back to Ready Pose
54        Motion Index Number  = 1
55        WAIT WHILE ( Motion Status  ==  TRUE )
56    }
```

Fig. 10.14 Logic for the LeadFighter program to send information to CounterFighter

```
20        DataIn  = DataIn  /  16
21
22        IF ( DataIn  ==  0 )
23        {
24             MotionGroupNo  = 1
25        }
26        ELSE
27        {
28             IF ( DataIn  ==  2 )
29             {
30                  MotionGroupNo  = 3
31             }
32             ELSE IF ( DataIn  ==  4 )
33             {
34                  MotionGroupNo  = 5
35             }
36             ELSE IF ( DataIn  ==  6 )
37             {
38                  MotionGroupNo  = 7
39             }
40
41        // Start Counter Fight
42        ▶ Motion Index Number  = MotionGroupNo
43        WAIT WHILE ( ▶ Motion Status  ==  TRUE )
44
45        // Back to Ready Pose
46        ▶ Motion Index Number  = 1
47        WAIT WHILE ( ▶ Motion Status  ==  TRUE )
```

Fig. 10.15 Logic for the Counter-Fighter program to decide on an appropriate counter move to perform

control the Lead-Fighter via a given BT outgoing COM port, while the second instance would control the Counter-Fighter via a separate BT outgoing COM port. As the commands to each fighter now came from a single PC, it was more a matter of how fast the user could switch from one window application to the next and click on the appropriate Motion Unit to activate   the Attack and Counter moves onto the respective robots (see file "DARWIN-MINI-1-BT.mtnx"). The video file "Video 10.5" illustrated such an episode where the reader could see that "human" manual synchronization of robot moves did not work very well, but that the ROBOTIS-MINI system had lots of potentials.

Thus it remains a challenge for all users and ROBOTIS to come up with a ZigBee or BlueTooth environment that could be used for multiple MINIs. The author is looking forward to the release of the BT-410 Master-Slave series.

10.7   Review Questions for Chap. 10

1. What are the wireless communication options available with the ROBOTIS-MINI system?
2. What is the highest baud rate achievable with ROBOTIS ZigBee devices?
3. What is the highest baud rate achievable with ROBOTIS BlueTooth devices?
4. List advantages of ZigBee over BlueTooth.
5. List advantages of BlueTooth over ZigBee.
6. List advantages of the MTNX file format over the MTN file format.
7. List the data components found in an MTN file.
8. Describe the data architecture used in an MTN file.
9. List the data components found in an MTNX file.
10. Describe the data architecture used in an MTNX file.
11. Describe how the Zig2Serial Management sub-tool can be used to execute motion pages or groups stored on the ROBOTIS-MINI.
12. List the TASK commands controlling the LEDs found on the XL-320 actuator.

10.8   Review Exercises for Chap. 10

1. Create a custom MTNX file to allow the ROBOTIS-MINI go up and down a set of stairs (see video file "Video 10.6 and Fig. 10.16").
2. Create a custom MTNX file to allow the ROBOTIS-MINI to dance and synchronize to your favorite music, see example video at https://www.youtube.com/watch?v=4VsNyzABXsQ.
3. Combine the example MOTION UNITS into MOTION GROUPS of your own, similarly to the approach used to choreograph the MINI fighters.
4. Integrate sensors such as NIR and DMS sensors to help the MINI avoid obstacles. Later in 2015, ultrasonic and proximity sensors should also be available for the MINI.
5. Practice programming the OpenCM-9.04-C using the ROBOTIS IDE and practice recovering the firmware to get back to programming with TASK and MOTION again.

Fig. 10.16 ROBOTIS-MINI going up and down stair steps

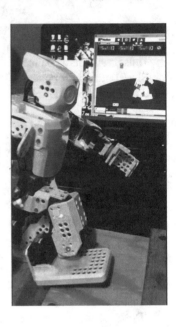

References

Billard A et al (2008) Robot programming by demonstration. In: Siciliano B, Khatib O (eds) Springer handbook of robotics. Springer, Heidelberg, pp 1371–1394
Calinon S (2009) Robot programming by demonstration: a probabilistic approach. EPFL Press, Lausanne

Chapter 11
Using Multimedia with Firmware 2.0 Systems

During 2016, ROBOTIS released two new tools for users who would be interested in using multimedia tools such as graphics, video and audio beyond the simple musical scales and melodies offered by the microphone/speaker system built-in the current CM and OpenCM controllers:

- In April 2016, "R+SCRATCH" was released as an MS Windows helper application to allow the user to interface ROBOTIS robots with the off-line SCRATCH 2 IDE tool from MIT, but only for the CM-50 and OpenCM-7.00 at that time. The R+SCRATCH tool allowed users to apply all the multimedia and event programming tools of SCRATCH 2 to work with ROBOTIS hardware.
- In December 2016, "R+m.PLAY700" was released as a mobile app (Android and iOS versions) to accompany the newly minted OLLO PLAY700 kit (i.e. CM-50), but it turned out that this mobile app was applicable to all Firmware 2.0 systems (CM-50/150/200 and OpenCM-7.00/9.04), which was a boon to ROBOTIS users (*NOTE: CM-150 required a special procedure to be described in Sect. 11.3*). This mobile app allowed a TASK program to access various tools on the typical mobile device such as cameras, graphical/audio/video systems and sensors such as the touchscreen, gesture and tilt sensor (see Sect. 5.6.3). Furthermore, the OpenCM-IDE tool for embedded C with the OpenCM-9.04C (see Sect. 9.2) was also found to be compatible with the R+m.PLAY700 app.

In this chapter, the goals were to illustrate the added multimedia dimensions to a robotics project to allow users to tell a more complex (and hopefully more engaging) "story" with robots. We'll first go through a SCRATCH 2 project to showcase the use of multimedia and event programming tools with ROBOTIS hardware, and also to point out important timing issues to consider in the overall robot behavior at run time. The second goal was to explain the structure and operational details of a SMART project that could employ the multimedia services of an Android device such as a SAMSUNG Galaxy S4® or ASUS ZenPad S8® (but some observations when using an iOS device like an iPhone® 4S or 7 would be also included for other

© Springer International Publishing Switzerland 2017
C.N. Thai, *Exploring Robotics with ROBOTIS Systems*,
DOI 10.1007/978-3-319-59831-4_11

users' needs) – **using both TASK and OpenCM IDE tools**. The third goal was to contrast selected "multimedia" SMART projects with their "SCRATCH 2" equivalents.

11.1 "Moody DogBot" SCRATCH® 2 Project

Figure 11.1 described the CM-50 DogBot used in this SCRATCH 2 project. For a more thorough training in using SCRATCH 2, the reader would be referred to more complete works such as Ford (2014) and Warner (2015), as only selected SCRATCH 2 features, most applicable to robotics, would be presented in this section.

Figure 11.2 showed that the communication lines between the SCRATCH 2 IDE software (running on the Windows PC) and the actual robot was mediated by the application R+SCRATCH (which was also running as a Windows PC executable). This extra communication layer would produce run-time timing issues documented in Sect. 5.6.2 which would not be repeated here (please note that the limiting "speed" of wireless BlueTooth communications should also be taken into account).

One important notice to pass along to the reader was that when R+SCRATCH connected to the robot, it would set the robot to the MANAGE mode whereas the existing TASK program would be disabled from running, although it would stay "intact" inside the memory space of the controller.

Although SCRATCH 2 was designed for beginner programmers aged 8 to 16, it had quite a sophisticated structure for its "project":

- Each SCRATCH project could have several actors called "SPRITES" and the Moody DogBot project would use a single Sprite called "Sprite 1" (see Fig. 11.3 – bottom left panel). The large area on the top left panel, where a larger picture of Sprite 1 was found, was called the STAGE.

Fig. 11.1 CM-50 DogBot used in the "Moody DogBot" SCRATCH 2 project

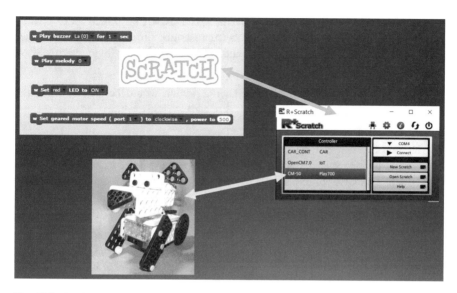

Fig. 11.2 Communication lines between SCRATCH 2 IDE and DogBot

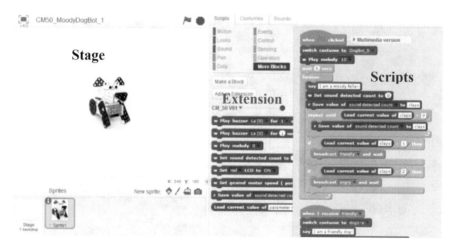

Fig. 11.3 Components of the "Moody DogBot" SCRATCH 2 project

- Attached to each Sprite were its SCRIPTS (the actual codes), its COSTUMES (graphics components triggered by its SCRIPTS), and its SOUNDS (audio components also controlled by its SCRIPTS) – see top of mid-panels of Fig. 11.3.
- To allow SCRATCH 2 to interact with the robot via R+SCRATCH, the user must also mount appropriate "EXTENSION" packages (i.e. a *.JSON file – see bottom of mid-panel in Fig. 11.3). Please see Video 11.1 for a primer on how to set up your Windows PC to create a SCRATCH 2 project that would interact with this example DogBot.

- The right panel contained the actual project's SCRIPTS which were created using different types of colored blocks (representing different data and behavior structures) that the user could stack and unstack to build the logic required for a project.

Figure 11.4 displayed the MAIN SCRIPT which would be started when the "GREEN FLAG" block was clicked:

1. Next, Sprite 1 switched to its neutral "DogBot_S" costume (see Fig. 11.5)
2. The next block was a WRITE (W) command to the CM-50 to play Melody 10 using its built-in speaker, and as this sound clip would last 5 s, hence the "WAIT 5 SECS" block was executed next (essentially making the PC wait for the CM-50 to finish its job of playing Melody 10).

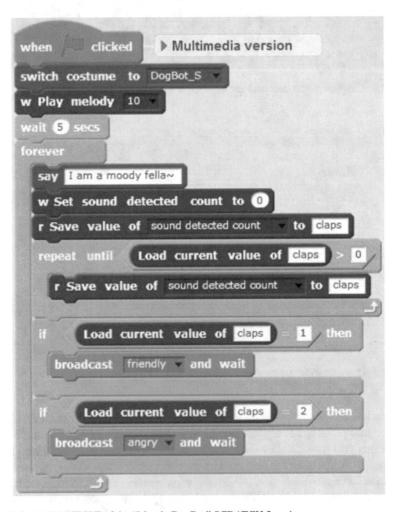

Fig. 11.4 MAIN SCRIPT of the "Moody DogBot" SCRATCH 2 project

Fig. 11.5 COSTUMES used in the "Moody DogBot" SCRATCH 2 project

3. The FOREVER LOOP formed the main sensing and control algorithm for this "Moody DogBot" to switch between its "Friendly" or "Angry" moods based on how many claps did the user make at run-time:

   (a) The first step in this loop was to SAY (i.e. display) the text "I am a moody fella~" next to Sprite 1 in the STAGE area.
   (b) Next was a "W" (i.e. WRITE) command sent to the CM-50 to reset its local parameter "Sound Detected Count" to zero.
   (c) Then was an "R" (i.e. READ) command for the robot to send back to the PC the current value of its local parameter "Sound Detected Count" and save it onto the PC (i.e. inside SCRATCH 2) as the parameter named "Claps".
   (d) The "REPEAT UNTIL" loop block was used next to keep on checking whether the user had clapped his or her hands at run time. Once the user had clapped his hands any number of times, this "REPEAT UNTIL" loop terminated, and one of the next two "IF" blocks would be executed depending on the actual number of claps determined and recorded by the CM-50.
   (e) The first "IF" block would be triggered if the user happened to clap ONCE, then the MAIN SCRIPT would BROADCAST the "FRIENDLY" message

and then "WAITed" at this block for the "FRIENDLY" SUB SCRIPT (i.e. parallel thread) to finish what it was supposed to do (see Fig. 11.6). When the "FRIENDLY" SUB SCRIPT was executed, Sprite 1 would switch to its "Dog1-A" costume and SAY "I am a friendly dog!". Then the next "R" command along with the "REPEAT UNTIL" loop on parameter "Sensor" implemented an Input Check process on the built-in Center IR Sensor of the CM-50 controller (See Fig. 11.1). For example, when the user approached his hand close enough to the Center IR Sensor to cause its reading to go above 35, then the PC would play the "Small-dog-barking.mp3" audio clip. Next the PC would send two "W" commands to the CM-50 making it turn on the two motors located on PORT 1 and PORT 2 (at speed 500) to effectively move the bot forward for 2 s. Then the PC would send two other "W" commands to turn off these two motors, and waited for 5 s before switching back to its neutral "DogBot_S" costume and finally shifting back the logic control to the MAIN SCRIPT which would then continue its FOREVER LOOP, waiting for the user to clap his or her hands all over again.

(f) The second "IF" block on the MAIN SCRIPT (Fig. 11.4) would be triggered if the user happened to clap TWICE, then the MAIN SCRIPT would BROADCAST the "ANGRY" message and then "WAITed" at this block for the "ANGRY" SUB SCRIPT (i.e. thread) to finish what it was supposed to do (see Fig. 11.7). When the "ANGRY" SUB SCRIPT was executed, Sprite 1 would switch to its "Dog2-C" costume and "SAY" "Angry! Angry! Stay

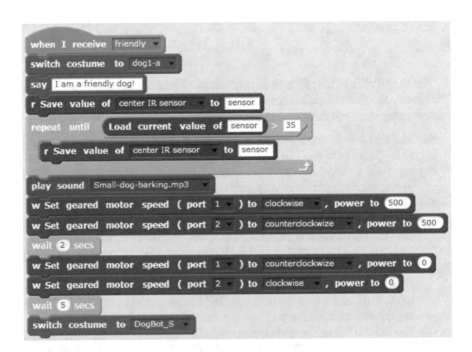

Fig. 11.6 "FRIENDLY" SUB SCRIPT triggered when user clapped once

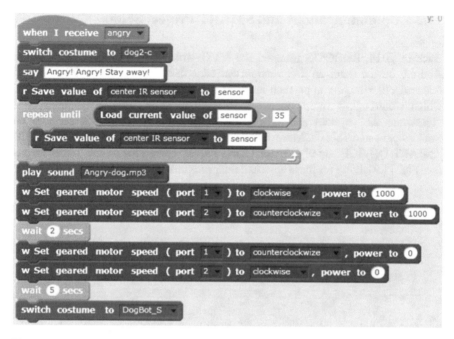

Fig. 11.7 "ANGRY" SUB SCRIPT triggered when user clapped twice

away!". Then the next "R" command along with the "REPEAT UNTIL" loop on parameter "Sensor" implemented an Input Check process on the built-in Center IR Sensor of the CM-50 controller. If the user approached his hand close enough to the Center IR Sensor to cause its reading to go above 35, then the PC would play the "Angry-dog.mp3" audio clip. Next, the PC would send two "W" commands to the CM-50 making it turn on the two motors located on PORT 1 and PORT 2 (at speed 1,000) to effectively jump the bot at the user for 2 s. Then the PC would send two other "W" commands to turn off these two motors, and waited for 5 s before switching back to its neutral "DogBot_S" costume and finally shifting flow control back to the MAIN SCRIPT which would then continue its FOREVER LOOP, waiting for the user to clap his or her hands again.

Video 11.2 showed a demonstration of this SCRATCH 2 project, and its source file "CM50_MoodyDogBot_1.sb2" was included in the Extra Materials for this book. Also as a chapter exercise, the file "CM50_MoodyDogBot_2.sb2" was also included, whereas a simple "BROADCAST" command was used, instead of the previous "BROADCAST and WAIT" command.

In the next sections, we would take on a different multimedia framework created by ROBOTIS to achieve similar goals as for SCRATCH 2 on the PC, but now with Android and iOS devices.

11.2   Communications and SMART Project Setup

Back in 2014, ROBOTIS released the MINI which could be operated from an Android device (later an iOS version was added), and since then ROBOTIS had released other mobile apps such as R+m.SMART, R+m.SMART2 and R+m.IoT which would allow interconnections between a TASK program running on a "Firmware 2.0" system and various audio, vision and messaging services on Smartphones and Tablets. More specifically, they were commands available in the "SMART DEVICE" and "SMART CONSTANT" categories on the TASK menus (see Fig. 11.8), however these apps and their programming tutorials were accessible

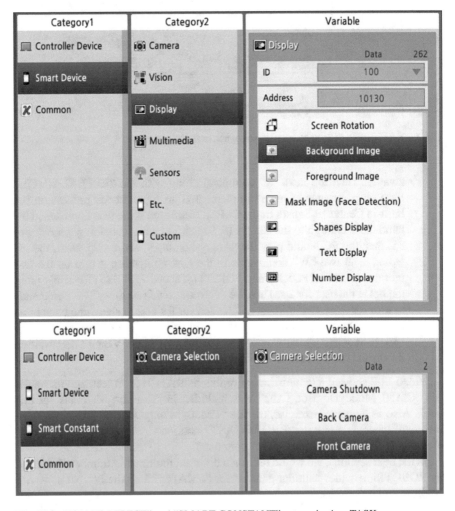

Fig. 11.8 "SMART DEVICE" and "SMART CONSTANT" categories in a TASK program

only to the Korean speaking community (http://support.robotis.com/ko/software/ mobile_app/r+smart/smanrt_manual.htm,   http://support.robotis.com/ko/software/ roboplus2/r+task2/programming3_paramerter/r+task2_parameter2.htm and http:// support.robotis.com/ko/software/roboplus/roboplus_task/programming/parameter/ remoteterminal/smart_parameter.htm), however if the reader used Chrome for web browser, these pages could be translated to "understandable" English.

Then "the clouds parted" in December 2016 when the PLAY700 system based on the CM-50 controller was released exclusively for the USA market, and with it the R+m.PLAY700 app was also made available for both Android and iOS devices, and most importantly with an English interface! Although the R+m.PLAY700 app showed only four example projects made for the PLAY700 system, the author quickly realized that this mobile app could actually be used with any "Firmware 2.0" controllers such as the CM-150 (requiring special procedure to be described in Sect. 11.3) and OpenCM-9.04 which were already available to the English speaking community.

Currently, the following R+m apps are available for Android devices: DESIGN, PLAY700, TASK, TASK2, MOTION2, MINI, SMART, SMART2, SMART3 and IoT. For iOS devices, the list is shorter: DESIGN, PLAY700, TASK2, MOTION2 and MINI.

11.2.1   BlueTooth Devices Usage

All Android and iOS devices use BlueTooth technologies to communicate with the ROBOTIS Firmware 2.0 systems:

- iOS devices have to use the BT-410 module (http://www.robotis.us/bt-410/), but the BT-410 will not work with MS Windows PCs.
- Android devices can use either the BT-410 or the BT-210 modules (http://www. robotis.us/bt-210/). The BT-210 works with MS Windows PCs and therefore is also compatible with ROBOTIS software tools such as MANAGER, TASK, MOTION and OpenCM IDE.

As the author was more productive within a desktop PC environment than with the much smaller touch screen of a Smartphone, he used the following setup:

1. A desktop PC (Windows 10) was used to develop and debug appropriate TASK codes for a given project. BlueTooth services were used to pair and connect to a BT-210 module attached to the robot to download TASK codes as needed via the TASK V.2 tool.
2. Since TASK V.2 did not stay connected to this BT-210 module once it finished downloading a TSKX file, the same BT-210 could be used without any re-pairing by an Android device to interact with the robot at run time. In other words, a PC/ Android environment is recommended by the author for an efficient "SMART project" process.

3. The author also used two iPhones, 4S and 7, to verify the "iOS" run-time behavior of the same TSKX code on the robot and the chosen iOS device. In this case, the BT-210 module, previously used in Step 2, had to be disconnected from the robot and a BT-410 module was then connected to the robot in order to test again this same TSKX code, but now with the iOS device. Interestingly, sometimes there would be differences in the behavior of the mobile devices used (Android or iOS) for the same TSKX code (will be shown in later examples in this chapter).
4. The needed multimedia files needed for the SMART project were also prepared on the desktop PC and then downloaded to the mobile Android or iOS device as appropriate (see Sect. 11.2.2).

11.2.2   File Management on Mobile Devices

Unfortunately, the file management process was quite different on an Android device than on an iOS device. For example, the author would hook up his Samsung S4 (Android 4.4.2) to his desktop PC via a micro USB port and just waited for Windows OS to mount it as a Multimedia Transfer Protocol (MTP) device and Windows Explorer would display this phone as a portable storage drive. Then the author would have full access to all folder and files residing on the Android device's SD cards from the Windows PC. If the ASUS ZenPad S8 (Android 6) was used, files transfer had to be done via HTTP protocols and WiFi through its File Manager app (the USB cable used to work when this ZenPad was still on Android 5). Thus overall, MS Windows and Android OSes worked together well.

However, hooking up an iPhone to a Windows PC via USB would allow access only to the Camera Roll, and to be able to work with all the folders and files on the iPhone, the user must acquire and install non-OS file manager software tools, of which some are "free" and some come at a cost. The author happened to be using "iMazing" in this book and this software needed to be installed only on the Windows PC side.

A ROBOTIS SMART project has many components represented by the folder structure as shown in Fig. 11.9 for an Android device and Fig. 11.10 showed how this project structure would be arranged on an iPhone 4S.

Figure 11.9 showed that ROBOTIS R+m apps would all be installed under a main folder named RoboPlus right of the root directory of the main SD card of the Android device. As ones drilled into a specific app, for example the PLAY700 folder, ones would encounter three sub-folders "Custom", "System" and "Temp". The "System" sub-folder would contain the four example projects provided by PLAY700. The "Custom" sub-folder would contain the user-initiated projects, for example the "MoodyDogBot" project where the project's components were stored (whether actually needed or not):

- **Audio** – where audio files to be played by the TASK code were stored.
- **Captured** – where photos captured by the PLAY700 app via the mobile device's front and back cameras were kept.

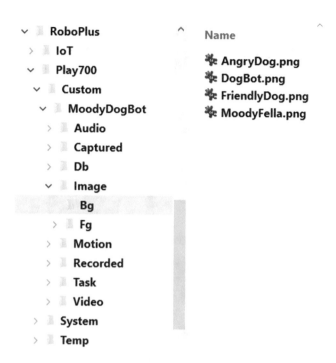

Fig. 11.9 Folders and Files structure for ROBOTIS apps on an Android device

- **Db** – where the SMART databases were stored, including "Character" and "Audio Input" lists used for Speech Recognition (these items could only be modified by users via the editing of the Project's Tools – see Sects. 11.2.4 and 11.3).
- **Image** – where the photos specifically used by the "MoodyDogBot" project were first "registered" (see Sect. 11.2.3) and then "stored" as background images (i.e. "**Bg**" sub-folder) or as foreground images (i.e. "**Fg**" sub-folder).
- **Motion** – where the MTNX files were stored. Please note that any MTNX code could be edited at the Windows PC level then downloaded onto the robot directly, so a copy of it did not need to be stored on the Android device, unless the user planned to modify it directly on the Android device at a later time.
- **Recorded** - where videos captured by the PLAY700 app via the mobile device's front and back cameras were kept.
- **Task** – where the TSKX files were stored. Please note that any TSKX code could be edited at the Windows PC level then downloaded onto the robot directly, so a copy of it did not need to be stored on the Android device, unless the user planned to modify it directly on the Android device at a later time
- **Video** – where video files to be played by the TASK code were stored.

The "Temp" folder was only used by the PLAY700 app for its own purposes.

For an iOS device, Fig. 11.10 showed that ROBOTIS R+m apps would all be installed under a main folder named "Apps" off the root directory of the main SD card of the device. As ones drilled into a specific app, for example the PLAY700 folder, ones would encounter a "Documents" sub-folder which further branched out

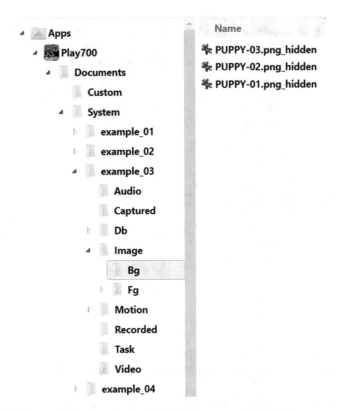

Fig. 11.10   Folders and Files structure for ROBOTIS apps on an iOS device

into two sub-folders "Custom" and "System". The "System" sub-folder would contain the four example projects provided by PLAY700 and Fig. 11.10 showed the components of the "example_03" project where the standard project's components were laid out: Audio, Captured, Db, Image, Motion, Recorded, Task and Video. These components were described earlier, except that the readers should notice that the actual files had "hidden" added to their normal file extension name, but the readers should not worry as the projects seemed to be displaying them fine inside the PLAY700 app). On an Android device, there was an app-level Setting that would make all these files visible to the Windows Explorer tool on the PC. These R+m app's settings are described in the next section.

11.2.3   Settings for a Typical R+m App

Figure 11.11 showed a screenshot of the main menu window for PLAY700 (V.0.9.4.0) on an Android device where the reader can see a "gear" icon on the top right corner. Ones would see a similar interface on an iOS device for the PLAY700

Fig. 11.11 Main Menu of
PLAY700 app, opening on
Custom Project
"MoodyDogBot"

app (currently at V. 1.0.4). Once this "gear" icon was tapped once, the "Settings" screen would be shown (see Fig. 11.12).

The first item in Fig. 11.12 was "Connect to Robot" which was of course where the user could scan for new BT devices and pair them or to choose a specific BT device among those that were previously paired.

The second item "Reset example" was to reset all ROBOTIS example projects to their originally installed states (if the user happened to have modified some or all of their components and wanted to go back to the original projects). This setting does not affect the Custom projects.

The third item "Accessibility Set-up" only pertained to Android devices where the user could set-reset many OS-level settings such as "Auto Rotate Screen", "Screen Timeout", or for "Services" – i.e. PLAY700 could use TalkBack or work with the other R+m apps like MINI or IoT for example. This "Accessibility" setting was not available for iOS devices.

The fourth item "Range of Gesture Error Setting" would be used to set the accuracy of the gesture function (a number between 0 and 30 – available to both Android and iOS).

The fifth item "Display Example Image on Gallery" also pertained to Android devices only and did what its label implied.

The sixth item "Scanning Media" would also pertain to Android devices only and removed the original "hidden" status of the folders and files inside the PLAY700 main folder so that Windows Explorer on a PC could "see" them.

11.2.4   Previews of Tools That Can Be Used in a Project

Going back to Fig. 11.11, if the user tapped on the "Edit" button of any project, a screen such as the one shown in Fig. 11.13 would display all the components/tools that could be used for this project. The current Android version (0.9.4.0) had all

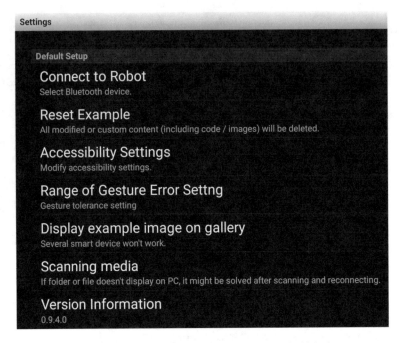

Fig. 11.12 App-level "Settings" for PLAY700 (V.0.9.4.0) on an Android device

these tools operational, but on the current iOS version (1.0.4), six tools were not yet functional:

- "Instrument" in the Multimedia group.
- "Illumination" in the "Sensor" group.
- "Received SMS", "Status Bar", "Vibration", "Application" in the "Other" group.

For example:

- The "R+m.TASK" item would allow the user to edit the TSKX files corresponding to this project.
- The "Face Detection" item would kick in the camera whereas the user could choose the front or back camera so that it could find a human face and then put a pair of sun glasses across the eyes, all in "real time".
- The "Color Detection" item would also kick in the camera and showed a live image with a small centered region of interest where it would try to detect for four "colors": Black, Red, Green and Blue. It was not a sophisticated tool as it would classify "Yellow" as being "Red"!
- The "Display" section of course pertained to "Background" and "Foreground" images, and other items like "Text" and "Number". For example the user could add (and register) Text Items if the user chose the "Text" item in this sub-menu "Display" (more details in Sect. 11.3).
- The "Multimedia" section could be used to record or play audio and video clips, and also for text-to-speech and voice recognition (using Google's web-based Speech API).

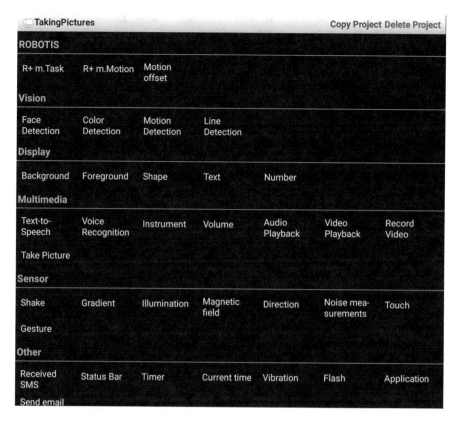

Fig. 11.13 Components/tools of a SMART project

- The "Sensor" section pertained to sensors embedded in the mobile device such as its Touch, Gesture, Gyro and Tilt sensors.
- The "Other" section pertained mostly Messages and other "Applications" that could be linked to this project.

The author would recommend that the reader spend some time experimenting with these options, as they would well prepare the reader for programming a custom project (i.e. a custom TSKX program) at a later time.

11.3 "Taking Pictures" SMART Project

This project illustrated the basics of accessing the back and front cameras of the mobile device and taking two snap shots out of each camera, one with no zoom, and one with 50% or 75% zoom in. This project also showed how to print texts and numbers and how to do "text-to-speech" on the mobile device and also how to sequence these mini-tasks efficiently. This project was applied to the following controllers: CM-50, CM-150, OpenCM-9.04 and OpenCM-7.00.

Figure 11.14 showed Part 1 of the TASK file "CM50_TakingPictures.tskx". A quick reminder here was that all the ROBOTIS controllers used in this section were designed such that they would immediately run the last working TSKX code that was downloaded to them earlier. Thus Statements 8 and 9 were needed to wait for the user (on the mobile device side) to finish connecting to the robot via the BT-210 or BT-410 modules (essentially waiting for a hand clap from the user to synchronize the robot code and the mobile device's services). Statement 10 set the mobile device screen into portrait mode. Statement 11 essentially cleared the screen, and the "zero" in the RHS meant "Do Not Show" if the user looked at it as a "Smart Constant" from inside the TASK tool.

Next, the author wanted to do "Text Display" and "Text-to-Speech" of the same "Text Item 1" (i.e. "Ready") on the mobile device. Recognizing that while "Text Display" would be practically "instantaneous" in execution, "Text-to-Speech" would take a longer time period to execute in order to have human users understand the "spoken" item (i.e. the eyes are much faster than the ears for most folks). Furthermore, right now "Text Item 1" was just one word, but what would happen if "Text Item 1" would be a longer sentence of several words? If hardware timing loops were used to match a specific text length like "Ready", then there would be much tweaking of various time delays involved to make the sequencing of these tasks efficient and independent of the word length of the Text Item. Thus the following coding approach was used to make Statements 12 through 19 work as a coordinated unit:

- First, the "Text to Speech" procedure was activated by Statement 12 (i.e. by "writing" Text Item 1 into the "Text to Speech" function), but as this process would take some time to finish, the check for the end of this process was pushed back to Statement 19 which was a "Wait While" command, essentially waiting for the "Text to Speech" function to return a "zero" when it finished its assigned task (essentially, here the "Text to Speech" function was "read from", instead being "written to" as before).

7	// User needs to make sure that robot is connected to mobile device. User claps hands when ready.
8	Result of Sound Counter = 0
9	WAIT WHILE (Result of Sound Counter == 0)
10	SMART: Screen Rotation = Portrait Mode (1)
11	SMART: Text Display = 0
12	SMART: Text to Speech (TTS) = Text Item 1
13	SMART: Text Display = [Position:(3,3)],[Item:1],[Size:100],[Color:Green]
14	IF (SMART: Camera Selection != Back Camera (1))
15	{
16	SMART: Camera Selection = Back Camera (1)
17	}
18	SMART: Camera Zoom = 0
19	WAIT WHILE (SMART: Text to Speech (TTS) != 0)

Fig. 11.14 Part 1 of TASK file "CM50_TakingPictures.tskx"

- In order to make the robot and mobile device "work" together more efficiently, the "Text Display", "Camera Selection" and "Camera Zoom" tasks were then "sandwiched" in-between Statements 12 and 19 (i.e. minimizing "wait" states when the robot and the mobile device were "doing nothing").

Statement 13 displayed Text Item 1 which was "Ready" at position (3, 3), its size set at 100 and its color set to Green (see Fig. 11.15 for details of the Right-Hand-Side of Statement 13). This project used five Text Items which were registered using the Project "Editor" (see Sect. 11.2.4 and Fig. 11.11). Figure 11.16 displayed these five Text Items.

A few words of clarification for the sake of the readers were needed here:

(a) The author created the five Text Items under the sub-menu of "Display >> Text" from Fig. 11.13.
(b) Once Step (a) was done, the author was much surprised to see the same five items under the sub-menu of "Multimedia >> Text-to-Speech" from Fig. 11.13 (see Fig. 11.17 also). This sub-menu also allowed the author to add new "Text-to-Speech" items by tapping on "Add" (top right of Fig. 11.17). In other words, the "Text" list and the "Text-to-Speech" list contained the same items. Importantly, as previously mentioned in Sect. 11.2.2, please note that these "Text" and "Text-to-Speech" lists were saved inside the "Db" folder, i.e. not directly accessible to the users (see Sect. 11.2.2).
(c) Compounding confusions, the author discovered that the same five items were referred to as "Text Item" inside the TASK tool (see Statement 12 of Fig. 11.14)!

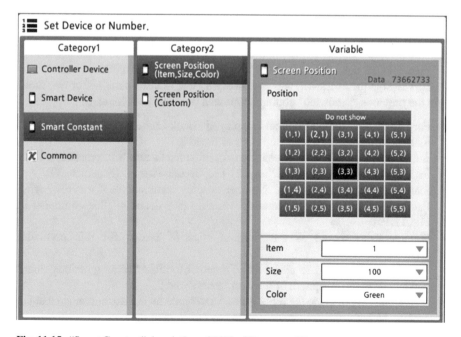

Fig. 11.15 "Smart Constant" description of RHS of Statement 13

Fig. 11.16 The five text
items used in "CM50_
TakingPictures.tskx"

Fig. 11.17 The five text-to-speech/text items used in "CM50_TakingPictures.tskx"

Statement 12 used "Text-to-Speech" services to "announce" Text/Text-to-Speech Item 1 out on the speaker of the mobile device with the default Google female voice.

(d) Moreover the same "Text-to-Speech" items will be found in the "Voice Recognition" tool which will be demonstrated in a "Voice Control" project later in Sect. 11.5.

Figure 11.18 showed Part 2 of the TASK file "CM50_TakingPictures.tskx".

Part 2 (Statements 21–42) was essentially a FOR LOOP operating for two times and the previous "sandwich" coding appproach was also used here:

- For both times, it announced and displayed "Back Camera" (Statements 23–24; Statement 23 serving as the "top slice of bread").
- But on Loop 1 (I = 1) it announced "Item 4", i.e. "First Picture" (Statement 29), while for Loop 2 (I = 2) it announced "Item 5", i.e. "Second Picture" (Statement 33).
- Statement 25 was a "Custom" Number Display command as it was displaying Parameter I which was a numerical variable (see details of this command in Fig. 11.19).
- Statement 26 served as the "bottom slice of bread" for this particular "sandwich".
- Statements 29, 33 and 35 also worked together to make "waiting periods" independent of the word-length of the Text Items 4 and 5.
- Statements 36–38 made the robot play a "La#" note for 0.5 s to announce that the mobile device's back camera was about to take a snap shot.

```
20    //  Start capturing images with Back Camera
21    LOOP FOR  (  I  =  1  ~  2  )
22    {
23        [ SMART: 🔊 Text to Speech (TTS) ]  =  Text Item 2
24        [ SMART: 🖼 Text Display ]  =  [Position:(3,3)],[Item:2],[Size:100],[Color:Yellow]
25        [ SMART: 🔢 Number Display ]  =  [Position:(3,4)],[Item:I],[Size:100],[Color:Red]
26        WAIT WHILE  ( [ SMART: 🔊 Text to Speech (TTS) ] != 0  )
27        IF  (  I  ==  1  )
28        {
29            [ SMART: 🔊 Text to Speech (TTS) ]  =  Text Item 4
30        }
31        ELSE IF  (  I  ==  2  )
32        {
33            [ SMART: 🔊 Text to Speech (TTS) ]  =  Text Item 5
34        }
35        WAIT WHILE  ( [ SMART: 🔊 Text to Speech (TTS) ] != 0  )
36        🔔 Buzzer Timer  =  0.5sec
37        🎵 Buzzer Index  =  La# (1)
38        WAIT WHILE  ( 🔔 Buzzer Timer  >  0.000sec  )
39        [ SMART: 📷 Photo Capture ]  =  Capture (1)
40        WAIT WHILE  ( [ SMART: 📷 Photo Capture ] != 0  )
41        [ SMART: 🔘 Camera Zoom ]  =  30
42    }
```

Fig. 11.18 Part 2: Back camera took 2 pictures (1 and 2)

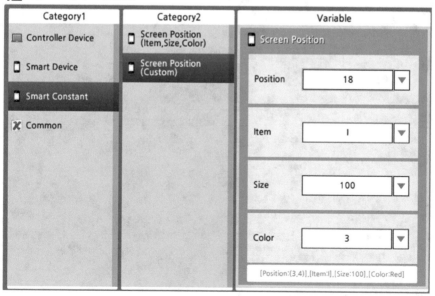

Fig. 11.19 "Smart Constant" description of RHS of Statement 25 (please note that Position 18 was equivalent to Position (3,4) (i.e. below Position (3,3), see Fig. 11.15 also)

- Statement 39 showed that the "Photo Capture" procedure could be started by "writing" a "1" into it. While Statement 40 showed that "Photo Capture" could also be "read" and this returned value compared to "zero" to check whether the "Photo Capture" process was finished or not. Picture 1 was taken here and saved into the phone's Gallery.
- Statement 41 set the new Zoom level to be used for Picture 2, and this "back camera" FOR LOOP ran for the second time to capture and save Picture 2.

Figure 11.20 showed Part 3 of the TASK file "CM50_TakingPictures.tskx".

Part 3 (Statements 44–67) was essentially another FOR LOOP, also operating two times in a similar way as in Part 2, with the Text Item 3, "Front Camera" appropriately substituted for "Back Camera", and using a "Do#" instead of a "La#" like in Part 2.

Video 11.3 showed how this project was actually working out on two Android devices, Samsung Galaxy S4 and ASUS ZenPad S8, and two iPhones, 4S and 7.

The code "CM150_TakingPictures.tskx" was written for the controller CM-150 (DREAM system) and it was identical to the one made for the CM-50 except that

43	// Switching to Front Camera
44	☐ SMART: ⚫ Camera Selection = Front Camera (2)
45	☐ SMART: ⚫ Camera Zoom = 0
46	LOOP FOR (I = 1 ~ 2)
47	{
48	☐ SMART: 🔊 Text to Speech (TTS) = Text Item 3
49	☐ SMART: 🖳 Text Display = [Position:(3,3)],[Item:3],[Size:100],[Color:Blue]
50	☐ SMART: 🔢 Number Display = [Position:(3,4)],[Item:I],[Size:100],[Color:Green]
51	WAIT WHILE (☐ SMART: 🔊 Text to Speech (TTS) != 0)
52	IF (I == 1)
53	{
54	☐ SMART: 🔊 Text to Speech (TTS) = Text Item 4
55	}
56	ELSE IF (I == 2)
57	{
58	☐ SMART: 🔊 Text to Speech (TTS) = Text Item 5
59	}
60	WAIT WHILE (☐ SMART: 🔊 Text to Speech (TTS) != 0)
61	⚫ Buzzer Timer = 0.5sec
62	🎵 Buzzer Index = Do# (4)
63	WAIT WHILE (⚫ Buzzer Timer > 0.000sec)
64	☐ SMART: ⚫ Photo Capture = Capture (1)
65	WAIT WHILE (☐ SMART: ⚫ Photo Capture != 0)
66	☐ SMART: ⚫ Camera Zoom = 40
67	}

Fig. 11.20 Part 3: front camera took 2 pictures (3 and 4)

the user needed to remember to set the Programming field to the "CM-150 (2.0)" value (see Fig. 11.21 – Left Top area). Video 11.4 showed the demonstration results for the CM-150 controller.

Starting with TASK V.2.1.4 (March 2017) and for unknown reasons, ROBOTIS removed access to the "Smart Device" sub-menu for the CM-150's programming interface, however the actual "SMART" commands were still functional within the CM-150 firmware (*at least for now, i.e. March 2017*). Thus in order to use SMART commands with a CM-150, the user would first have to temporarily change the "Programming Controller" to one of the other Firmware 2.0 controllers (i.e. CM-50/200 or OpenCM-7.00/9.04, see Fig. 11.21), then the user would be able to access the SMART commands as normal. However, the user would also need to remember to switch back to the CM-150 as the "Programming Controller" to use the CM-150's specific functions and before compiling/downloading the resulting TASK code to a CM-150 robot – so a bit of annoyance to put up with until ROBOTIS changes their policy towards the CM-150 in the future (hopefully!). The interesting point to note was that all those SMART commands would be retained in the user's TASK code from then on!

The code "CM904_TakingPictures.tskx" was written for the controller OpenCM-904 (MINI system) and as it was not equipped with a buzzer, an IR sensor on Port 1 was used to trigger the picture taking process, and instead of the La# and Do# used for the CM-50 and CM-150, the author turned ON and OFF an LED module on Port 2 instead (see Fig. 11.22 and Video 11.5 for a demonstration).

The code "CM700_TakingPictures.tskx" was written for the controller OpenCM-700 (IoT system) and it has a buzzer and a special function to check whether connections had been made between the robot and the mobile device (see Statement 8 in Fig. 11.23). Please also see Video 11.6 for a video demonstration.

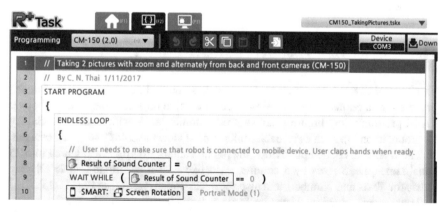

Fig. 11.21 Setting the CM-150 in the "Programming Controller" field for "CM150_TakingPictures. tskx"

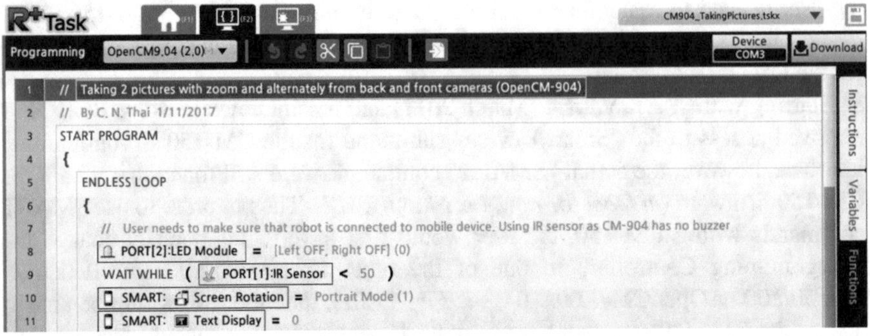

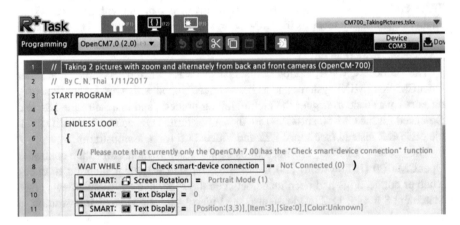

Fig. 11.22 Using IR sensor and LED modules on the CM-904 for "CM904_TakingPictures.tskx"

Fig. 11.23 Using "Check Smart Device Connection" function on the OpenCM-700 for "CM700_TakingPictures.tskx"

11.4 "Color Line Follower" SMART Project

The ROBOTIS SMART structure also allowed some rudimentary "real-time" image processing functions on the mobile device that would be accessible by the robot via TASK programming. Figure 11.24 showed a simple carbot with a Samsung Galaxy S4 mounted on top such as its back camera could view the color lines track beneath.

The R+m.PLAY700 app divided the phone camera's entire sensing area into 25 equal zones, i.e. 5 rows by 5 columns, just like how it partitioned its 25 Display Areas for Texts and Numbers and also for its 25 Touch Areas (see Fig. 11.25).

The current Color Line Detection tool used only the top row of the sensing zones, i.e. zones 1 through 5, mainly because the BlueTooth communication channel would be overwhelmed when data from all 25 zones were streamed back to the CM-50 robot to process. However, from the point of view of Machine Vision and Maneuver Control, the robot should use "relevant" and "fresh" data as much as possible, thus

Fig. 11.24 "Color Line Follower" carbot using Top of Display Screen scanning zones 1 through 5

Touch				
1	2	3	4	5
6	7	8	9	10
11	12	13	14	15
16	17	18	19	20
21	22	23	24	25

Fig. 11.25 The 25 zones used for Touch/Display/Camera-Sensing functions

Fig. 11.24 also showed that the camera was mounted such that the bottom half of the camera view was actually obstructed by the robot's frame. This "mechanical" intervention essentially brought the source of data "relevant" to the computation of appropriate robot maneuvers **closer** to the source of those maneuvers (i.e. the motors). In a way, we wanted the robot to "live in the moment" in a Zen like manner! This "mechanical" solution also helped simplify the control algorithm which turned out to be not much different from the one used to track an arbitrary curved black line with the IRSA module which had 7 IR channels (see Sect. 5.5).

The Color Line Detection process had three general steps:

1. Set SMART DEVICE "Camera Sensor" to "Line Detection Mode".
2. Set SMART DEVICE "Vision – Tracking Color" to a chosen color among 4 predefined "Colors" (Black, Red, Green and Blue).
3. Read SMART DEVICE "Vision – Line Detection Area" to obtain a numerical value which would be between 1 and 5 when the "assigned" color was found, otherwise it would return a zero value.

The complete code "CM50_ColorLineFollower_T.tskx" was included in the Extra Materials of the book and Fig. 11.26 showed Part 1 of its main Endless Loop.

Figure 11.26 showed that after waiting for the user to clap (Statements 10–11), the robot controller would send commands to the mobile device to perform the following tasks:

- Set display screen to portrait mode (Statement 12) and clear the screen (Statement 13).
- Start announcing Text Item 1 ("Ready") (Statement 14) and display same item in the middle of the screen (Statement 15).
- Make sure that the back camera was chosen (Statements 16–19) and set this camera to normal view (i.e. zero zoom) with Statement 20.

7	ENDLESS LOOP
8	{
9	// User needs to make sure that robot is connected to mobile device. User claps hands when ready.
10	🔔 Result of Sound Counter = 0
11	WAIT WHILE (🔔 Result of Sound Counter == 0)
12	⬜ SMART: 📱 Screen Rotation = Portrait Mode (1)
13	⬜ SMART: 🖼 Text Display = 0
14	⬜ SMART: 🔊 Text to Speech (TTS) = Text Item 1
15	⬜ SMART: 🖼 Text Display = [Position:(3,3)],[Item:1],[Size:100],[Color:Green]
16	IF (⬜ SMART: 📷 Camera Selection != Back Camera (1))
17	{
18	⬜ SMART: 📷 Camera Selection = Back Camera (1)
19	}
20	⬜ SMART: ◉ Camera Zoom = 0
21	WAIT WHILE (⬜ SMART: 🔊 Text to Speech (TTS) != 0)

Fig. 11.26 Part 1 of main endless loop in "CM50_ColorLineFollower_T.tskx"

- Statement 21 just waited for the Text-to-Speech process to be done.

Figure 11.27 showed Part 2 of the Endless Loop whereas the goal was to set the Camera Sensor into its Line Detection mode (Statement 23). Next, Statements 25 to 36 prompted the user to provide the proper number of hand claps by announcing and displaying the four Text Items 2 through 5 which were:

- "1 clap for black"
- "2 claps for red"
- "3 claps for blue"
- "4 claps for green"

Figure 11.28 showed Part 3 of the Endless Loop where the user's number of claps was collected (Statements 38–40) and then processed to provide the proper "color" value to Parameter "LineColor" (Statements 41 through 56).

Figure 11.29 showed Part 4 of the Endless Loop which was the important "Sensing and Control" loop (Statements 61–85) where:

- First, the display screen was cleared (Statement 58) and the current value of Parameter LineColor was "written" into the SMART DEVICE Tracking Color (Line-Tracer) function (Statement 59), effectively starting the process for the mobile phone to do its job of scanning Zones 1 through 5 to see if there was any group of pixels that could be labeled with the "color" value provided by Parameter LineColor.
- Statements 60 and 84 read in a numerical value returned from the function Line Detection Area and saved it into Parameter LinePosition. These actions could be labeled as the "Sensing" aspects of the "Sensing and Control" loop. Parameter LinePosition would have a numerical value between 1 and 5 if one of the 5

22	// Setting Camera Sensor to Line Detection mode
23	☐ SMART: 📷 Camera Sensor = Line Detection Mode (4)
24	// User claps to choose Color of Line to track: 1 for Black; 2 for Red; 3 for Blue; 4 for Green
25	☐ SMART: 🔊 Text to Speech (TTS) = Text Item 2
26	☐ SMART: 🖥 Text Display = [Position:(3,2)],[Item:2],[Size:100],[Color:Yellow]
27	WAIT WHILE (☐ SMART: 🔊 Text to Speech (TTS) != 0)
28	☐ SMART: 🔊 Text to Speech (TTS) = Text Item 3
29	☐ SMART: 🖥 Text Display = [Position:(3,3)],[Item:3],[Size:100],[Color:Yellow]
30	WAIT WHILE (☐ SMART: 🔊 Text to Speech (TTS) != 0)
31	☐ SMART: 🔊 Text to Speech (TTS) = Text Item 4
32	☐ SMART: 🖥 Text Display = [Position:(3,4)],[Item:4],[Size:100],[Color:Yellow]
33	WAIT WHILE (☐ SMART: 🔊 Text to Speech (TTS) != 0)
34	☐ SMART: 🔊 Text to Speech (TTS) = Text Item 5
35	☐ SMART: 🖥 Text Display = [Position:(3,5)],[Item:5],[Size:100],[Color:Yellow]
36	WAIT WHILE (☐ SMART: 🔊 Text to Speech (TTS) != 0)

Fig. 11.27 Part 2 of main endless loop in "CM50_ColorLineFollower_T.tskx"

Fig. 11.28 Part 3 of main endless loop in "CM50_ColorLineFollower_T. tskx"

```
38    Result of Sound Counter = 0
39    WAIT WHILE ( Result of Sound Counter == 0 )
40    Claps = Result of Sound Counter
41    IF ( Claps == 1 )
42    {
43        LineColor = Black Line (2)
44    }
45    ELSE IF ( Claps == 2 )
46    {
47        LineColor = Red Line (3)
48    }
49    ELSE IF ( Claps == 3 )
50    {
51        LineColor = Blue Line (5)
52    }
53    ELSE IF ( Claps == 4 )
54    {
55        LineColor = Green Line (4)
56    }
```

Fig. 11.29 Part 4 of main endless loop in "CM50_ColorLineFollower_T. tskx"

```
58    SMART: Text Display = 0
59    SMART: Tracking Color (Line-Tracer) = LineColor
60    LinePosition = SMART: Line Detection Area
61    LOOP WHILE ( LinePosition != 0 )
62    {
63        CALL DisplayColorPosition
64        IF ( LinePosition == 3 )
65        {
66            CALL Forward
67        }
68        ELSE IF ( LinePosition == 1 )
69        {
70            CALL RotateLeft
71        }
72        ELSE IF ( LinePosition == 2 )
73        {
74            CALL TurnLeft
75        }
76        ELSE IF ( LinePosition == 4 )
77        {
78            CALL TurnRight
79        }
80        ELSE IF ( LinePosition == 5 )
81        {
82            CALL RotateRight
83        }
84        LinePosition = SMART: Line Detection Area
85    }
```

scanned zones was found to contain the "color" being sought after (i.e. a valid result). If a zero was returned into Parameter LinePosition, this meant that the "chosen" color could not be "seen" by the camera (at least in none of the 5 scanned zones) and the LOOP WHILE (Statements 61 through 85) would not be taken or it would be terminated if it was looping for a while (no pun intended).

- Inside this LOOP WHILE construct, the function DisplayColorPosition was used to continuously display the parameter LineColor and LinePosition for run-time debugging purposes.
- The IF-ELSE-IF structures inside this LOOP WHILE construct represented the "Control" aspects, i.e. the proper actions/maneuvers for the robot to perform, depending on what was the actual value contained in Parameter LinePosition, at any moment during run-time.

Figure 11.30 showed Part 5 of the Endless Loop which handled the case when no "pre-chosen" color was found, whereas it would immediately stop the robot, next announced and displayed the Text Item 6 which was "no color found" and went back to the start of the Main ENDLESS LOOP at Statement 8 of Fig. 11.26.

Please see Video 11.7 for a demonstration of the TASK program "CM50_ColorLineFollower_T.tskx" (in this particular demo, the author only used the "Rotate" maneuvers for the cases when LinePosition was equal to 1, 2, 4 or 5).

During the writing of this book, Mr. Kimoon Kim from ROBOTIS had graciously let me test out a special version of the R+m.PLAY700 which scanned the bottom 5 zones of the display screen, i.e. Zones 21 to 25 in Fig.11.25. This scanning option allowed the author to shift the phone forward on the robot frame and thus allowed the use of the phone's assistive LED light during the execution of the TASK program "CM50_ColorLineFollower_B.tskx" (see Fig. 11.31).

Please see Video 11.8 for a demonstration of the TASK program "CM50_ColorLineFollower_B.tskx" (in this particular demo, the author also only used the "Rotate" maneuvers for the cases when LinePosition was equal to 1, 2, 4 or 5).

11.5 "Voice Control" SMART Project

From what the author gathered, the Voice Recognition tool in the R+m.PLAY700 App was designed with the web-based Google Speech® API (https://cloud.google.com/speech/), meaning that your mobile device needed to be connected to the web via WiFi in order to use this feature.

```
86   CALL  Stop
87   ☐ SMART: 🔊 Text to Speech (TTS) = Text Item 6
88   ☐ SMART: 📺 Text Display = [Position:(3,1)],[Item:6],[Size:125],[Color:White]
89   WAIT WHILE ( ☐ SMART: 🔊 Text to Speech (TTS) != 0 )
```

Fig. 11.30 Part 5 of main endless loop in "CM50_ColorLineFollower_T.tskx"

Fig. 11.31 "Color Line Follower" carbot using *bottom* of display screen scanning zones 21–25 and including Assistive LED light

As previously mentioned in Sect. 11.3, all Text Items registered in a SMART project were also potential Voice Recognition Items (see Fig. 11.32), and the user could tap on the greyed out Microphone icon (top right in Fig. 11.32) to test out success or failure of his or her particular speech patterns. The user could also train a particularly "tough" item by tapping on the Item Number to bring on a pulled down menu where the user could choose the "Train" option.

Figure 11.33 showed the Voice Recognition Items used in this Voice Control SMART Project.

Items 1 through 6 were trained using all the words in each item, but Item 7 was trained only on "INVALID" and Item 8 was trained only on "VALID", **thus the DISPLAY or TEXT-TO-SPEECH Items and the VOICE RECOGNITION Items did not have to match word for word at all.**

The TASK file "CM50_VoiceControlcarbot.tskx" has the complete solution for this Voice Control SMART Project, and Fig. 11.34 showed Part 1 of this program.

Figure 11.34 showed that after initializing parameters Speed, TurnSpeed, VoiceCommand and InvalidCommand to their proper values (Statements 5–8), the robot would wait on the user to whistle in the phone's microphone loudly enough for the Noise level to be above 50 (Statement 10). Then the function Menu was called to display the operating menu as shown in Fig. 11.35.

Figure 11.36 showed Part 2 of the algorithm used in the Voice Control Carbot project where the External ENDLESS LOOP would be entered at Statement 13:

- The "Whistle" label was an important "jump to" location that would be explained later in Part 3.
- Statement 16 showed that the robot/phone was waiting on the user to whistle in one more time and once this event occurred, the phone would display Text Item 1 (i.e. "ready") and the robot would wait for 2 s for the user to get ready to verbalize his or her command.

Figure 11.37 showed Part 3 of the algorithm used in the Voice Control Carbot project where the Inner Loop was entered (Statement 24):

Fig. 11.32 Further "Training" of voice recognition items available

Fig. 11.33 Voice recognition items used in "Voice Control" SMART project

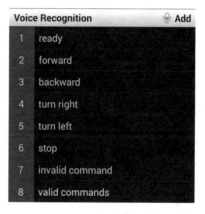

```
 3   START PROGRAM
 4   {
 5       Speed  =  180
 6       TurnSpeed  =  160
 7       VoiceCommand  =  Text Item 6
 8       InvalidCommand  =  FALSE (0)
 9       //  User needs to make sure that robot is connected to mobile device. User whistles loudly when ready.
10       WAIT WHILE  (  [ SMART: Noise (dB) ] <  50  )
11       CALL  Menu
```

Fig. 11.34 Part 1 of program "CM50_VoiceControlCarbot.tskx"

Fig. 11.35 Operating
Menu for Voice Control
Carbot Program "CM50_
VoiceControlCarbot.tskx"

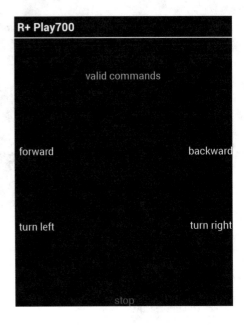

13	ENDLESS LOOP
14	{
15	Whistle :
16	WAIT WHILE (☐ SMART: 🔊 Noise (dB) **<** 50)
17	☐ SMART: 🖩 Text Display **=** 0
18	☐ SMART: 🔊 Text to Speech (TTS) **=** Text Item 1
19	☐ SMART: 🖩 Text Display **=** [Position:(3,3)],[Item:1],[Size:100],[Color:Green]
20	WAIT WHILE (☐ SMART: 🔊 Text to Speech (TTS) **!=** 0)
21	🕙 High-resolution Timer **=** 2.000sec
22	WAIT WHILE (🕙 High-resolution Timer **>** 0.000sec)

Fig. 11.36 Part 2 of program "CM50_VoiceControlCarbot.tskx"

1. To start the Voice Recognition task with Statement 27 followed by a WAIT WHILE for this process to finish (Statement 29). When Statement 27 was executed the mobile device would display "Speak to Me" (see Fig. 11.38) and waited for the user to verbalize the command (within a given time-out period). When the user was finished with the command and while Statement 29 was being executed, it would display "Recognizing" (see Fig. 11.39).
2. Next, the Result of Speech Recognition was saved in Parameter VoiceCommand (it was a number corresponding to the index of the "recognized" Text Item as shown in Fig. 11.33.

```
24    ENDLESS LOOP
25    {
26        //  Start Speech Recognition Process
27        [ ] SMART: [ ] Speech Recognition  =  Start (1)
28        //  Wait for SR to complete
29        WAIT WHILE  (  [ ] SMART: [ ] Speech Recognition  != 0  )
30        VoiceCommand  =  [ ] SMART: [ ] Result of Speech Recognition
31        IF  (  VoiceCommand != -1  )
32        {
33            CALL  ProcessCommand
34            [ ] SMART: [ ] Result of Speech Recognition  =  0
35        }
36        IF  (  InvalidCommand  ==  TRUE (1)  )
37        {
38            InvalidCommand  =  FALSE (0)
39            JUMP  Whistle
40        }
41    }
42    }
43  }
```

Fig. 11.37 Part 3 of program "CM50_VoiceControlCarbot.tskx"

Fig. 11.38 "Voice Recognition" process started, ready for user's verbal input

Fig. 11.39 User's verbal command being recognized

3. The first IF construct (Statements 31–35) checked to see if the value saved in VoiceCommand was valid (i.e. different from "-1"). If it was valid, the function ProcessCommand was called (Statement 33) and when this function finished, the parameter Result of Speech Recognition was cleared to zero (Statement 34).

4. The function ProcessCommand (see Fig. 11.40) started by checking if Parameter VoiceCommand was non-zero, i.e. an actual match was made with one of the Voice Recognition Items shown in Fig. 11.33.

5. Next, the IF-ELSE-IF construct (Statements 105–129) was used to trigger the appropriate robot maneuvers (Forward, Backward, Turn Right, Turn Left or Stop) depending on the value found in Parameter VoiceCommand at runtime.

6. If no match was found for the user's verbal command (for whatever reasons), Parameter VoiceCommand would have a zero value and this "Invalid Command" case was taken care of by the construct comprising Statements 131–140 (see Fig. 11.41). The robot would be stopped (Statement 133) and Parameter InvalidCommand set to TRUE (Statement 134). Next a Display and Text-to-Speech procedure would be performed with Text Item 7 (i.e. "Invalid Command" – Statements 135–138). Finally the Menu of Fig. 11.35 was re-displayed (Statement 139) and program flow control returned to the main program at Statement 34 which went on to Statement 36 (see Fig. 11.37).

```
101   FUNCTION  ProcessCommand

102   {

103       IF  (  VoiceCommand  != 0  )

104       {

105           IF  (  VoiceCommand  == Text Item 2  )

106           {

107               CALL  Forward

108               [] SMART: [] Text Display  =  [Position:(3,3)],[Item:2],[Size:100],[Color:White]

109           }

110           ELSE IF  (  VoiceCommand  == Text Item 3  )

111           {

112               CALL  Backward

113               [] SMART: [] Text Display  =  [Position:(3,3)],[Item:3],[Size:100],[Color:White]

114           }

115           ELSE IF  (  VoiceCommand  == Text Item 4  )

116           {

117               CALL  TurnRight

118               [] SMART: [] Text Display  =  [Position:(3,3)],[Item:4],[Size:100],[Color:White]

119           }
```

Fig. 11.40 Part 1 of function ProcessCommand

```
131   ELSE IF ( VoiceCommand == 0 )
132   {
133      CALL Stop
134      InvalidCommand = TRUE (1)
135      [ SMART: Text Display ] = 0
136      [ SMART: Text to Speech (TTS) ] = Text Item 7
137      [ SMART: Text Display ] = [Position:(3,3)],[Item:7],[Size:100],[Color:Red]
138      WAIT WHILE ( [ SMART: Text to Speech (TTS) ] != 0 )
139      CALL Menu
140   }
```

Fig. 11.41 Part 2 of function ProcessCommand

7. If "InvalidCommand" happened to be set to TRUE from inside Function ProcessCommand, then "InvalidCommand" was reinitialized to FALSE (Statement 38), and the program flow control JUMPED (Statement 39) to LABEL "Whistle" (Statement 15) to restart the whole algorithm all over again (see Fig. 11.37).

Video 11.9 demonstrated how this algorithm executed in real-time, where we could see that the Voice Recognition process required quite some time to finish so the actual action followed about 3–5 s after the verbal command was issued.

11.6 "Moody Dog Bot" SMART Project

This project was created to allow a comparison between a SCRATCH 2 project vs. a SMART project to achieve similar multimedia goals. The corresponding TASK code was "CM50_MoodyDogBot.tskx". It used the same two Audio files: "Small-dog-barking.mp3" and "Angry-dog.mp3" as the SCRATCH 2 project. It used 4 similar background images: "DogBot.png", "MoodyFella.png", "FriendlyDog.png" and "AngryDog.png".

Figure 11.42 showed Part 1 of the algorithm used:

- First, the CM-50 played Melody 10 (Statements 6–8).
- Statements 10–14 were used to allow the user some time to connect the mobile device to the robot via BlueTooth and for the Background Image 1 "DogBot.png" to display. This part was awkward as compared to just clicking on the Green Flag in SCRATCH 2 code (see Fig. 11.4).

Part 2 of the algorithm (see Fig. 11.43) was an Endless Loop to allow the user to use hand claps to select between the "Friendly" (one clap) and "Angry" (two claps) moods of the DogBot:

1	// Moody DogBot with Multimedia activation on Android device via PLAY 700 app
2	// By C. N. Thai 1/1/2017
3	START PROGRAM
4	{
5	// Standard DogBot section
6	🕐 Buzzer Timer = Melody Time
7	🎵 Buzzer Index = Melody No.10 (10)
8	WAIT WHILE (🕐 Buzzer Timer > 0.0sec)
9	// Waiting for user to catch up on phone app
10	🕐 Timer = 2.048sec
11	WAIT WHILE (🕐 Timer > 0.0sec)
12	📱 SMART: 🖼 Background Image = 1
13	🕐 Timer = 2.048sec
14	WAIT WHILE (🕐 Timer > 0.0sec)

Fig. 11.42 Part 1 of "CM50-MoodyDogBot.tskx"

Fig. 11.43 Part 2 of "CM50-MoodyDogBot. tskx"

16	ENDLESS LOOP
17	{
18	// Moody Fella~ section
19	📱 SMART: 🖼 Background Image = 2
20	🔊 Result of Sound Counter = 0
21	WAIT WHILE (🔊 Result of Sound Counter == 0)
22	Claps = 🔊 Result of Sound Counter
23	
24	// Friendly section
25	IF (Claps == 1)
26	{
27	CALL Friendly
28	}
29	
30	// Angry section
31	IF (Claps == 2)
32	{
33	CALL Angry
34	}
35	}

- The "Moody Fella~" Background Image was displayed (Statement 19).
- Statements 20–22 showed how the user hand claps were captured at run time, using three TASK statements, while SCRATCH 2 had to use four blocks (see Fig. 11.4).
- The two IFs constructs (Statements 25–28 and 31–34) were syntactically different between TASK and SCRATCH 2 but functionally they had the same effects. It was interesting to note that "BROADCAST and WAIT" in SCRATCH 2 was equivalent to calling a Function in TASK and waiting for its return back to the main program. In this case, the "parallel" event programming features of SCRATCH 2 were in a way underutilized.

Figure 11.44 described the FUNCTION Friendly to be compared to the SUB SCRIPT Friendly (Fig. 11.6):

- It used a single statement (Statement 41) to check for the Center IR Sensor against 3 blocks used in SCRATCH 2, because local firmware to the CM-50 already existed for the Center IR Sensor, while SCRATCH 2 had to access it via the R+SCRATCH interface with much overhead costs.
- The use of the SMART Play Audio facility (Statements 45–46) allowed the motors to be in motion "only" and "exactly" during the audio playing period, while SCRATCH2 had to use some "guessing" on the waiting periods (2 + 5 s).

Similar observations could be made between FUNCTION Friendly and SUB SCRIPT Friendly. Video 11.10 showed the actual performance of this SMART project.

Thus was there one package "better" than the other one? The author did not think so as each approach had its own strengths and weaknesses, if run-time response times of the robot were most important to the user then the SMART Project approach needed to be used, otherwise the multimedia richness and event programming features of SCRATCH 2 could not be overlooked.

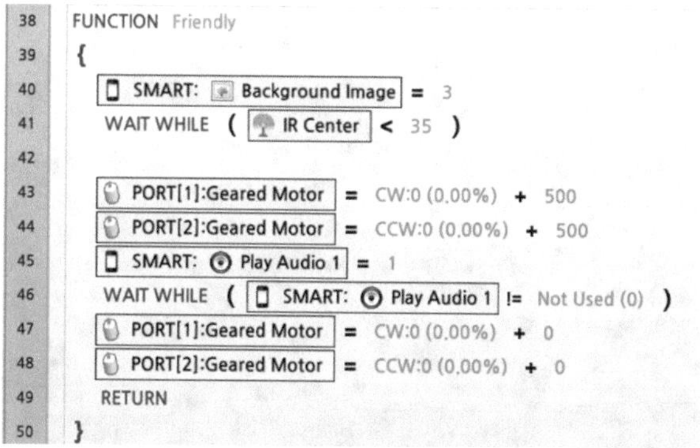

Fig. 11.44 Function "Friendly" of "CM50-MoodyDogBot.tskx"

```
38   FUNCTION Friendly
39   {
40       SMART:  Background Image = 3
41       WAIT WHILE (  IR Center < 35 )
42
43       PORT[1]:Geared Motor = CW:0 (0.00%) + 500
44       PORT[2]:Geared Motor = CCW:0 (0.00%) + 500
45       SMART:  Play Audio 1 = 1
46       WAIT WHILE (  SMART:  Play Audio 1 != Not Used (0) )
47       PORT[1]:Geared Motor = CW:0 (0.00%) + 0
48       PORT[2]:Geared Motor = CCW:0 (0.00%) + 0
49       RETURN
50   }
```

11.7   Other Selected SMART Features

The R+m.PLAY700 App had many other multimedia features that the readers would be encouraged to try out on their own, but the author did have a few favorites that were illustrated in the next three mini SMART projects.

11.7.1   Using Touch Areas

R+m.PLAY700 allowed the monitoring of up to 2 Touch Areas simultaneously. The file "CM50_TouchAreas.tskx" described how to determine which touch area(s) [1 through 25] had been been detected, and how to display either a white circle or a green square at the detected locations. Figure 11.45 showed the main Endless Loop that did the job whereas:

- Statements 10 and 11 saved the actual detected zones into Parameters Touch1 and Touch2.
- Statement 13 cleared the screen of all shapes that were previously displayed.
- Statements 14–17 put a White Circle where the value of Parameter Touch1 pointed to.
- Statements 18–21 put a Green Square where the value of Parameter Touch2 pointed to.

Video 11.11 showed the execution of this mini project.

To close this section let's dwelve into more details of the SMART CONSTANTs that could be used for the RHS of the various SMART commands: Shapes Display, Text Display and Number Display. These SMART CONSTANTs were of 4 bytes in length.

```
8     ENDLESS LOOP
9     {
10        Touch1  =  [ SMART:  Touch Area 1 ]
11        Touch2  =  [ SMART:  Touch Area 2 ]
12        // Clearing screen display
13        [ SMART:  Shapes Display ] =  0
14        IF ( Touch1 > 0 )
15        {
16            [ SMART:  Shapes Display ] = [Position:Touch1],[Item:1],[Size:60],[Color:White]
17        }
18        IF ( Touch2 > 0 )
19        {
20            [ SMART:  Shapes Display ] = [Position:Touch2],[Item:2],[Size:60],[Color:Green]
21        }
22    }
```

Fig. 11.45   Endless loop to monitor and mark up the detected touch areas 1 & 2

The structure of the Shapes Display Constant was defined as follows:

* The first (lowest) byte (bit 0 through 7) was reserved for the SHAPE POSITION [1–25] corresponding to the screen zones defined in Fig. 11.25.
* The second byte (bit 8 through 15) was reserved for the SHAPE NUMBER [0–3]. "1" corresponded to a Circle, "2" corresponded to a Square, "3" corresponded a Triangle and "0" was meant to remove any previous shape used. At present, users cannot add their own shapes.
* The third byte (bit 16 through 23) was reserved for the SHAPE SIZE [0–255].
* The fourth byte (bit 24 through 31) was reserved for the SHAPE COLOR [0–9], where "0" meant "Unknown", "1" was White, "2" was Black, "3" was Red. "4" was Green, "5" was Blue, "6" was Yellow, "7" was Light Gray, "8" was Gray and "9" was Dark Gray.

Similarly, the structure of the Text Display Constant was defined as follows:

* The first (lowest) byte (bit 0 through 7) was reserved for the TEXT POSITION [1–25] corresponding to the screen zones defined in Fig. 11.25.
* The second byte (bit 8 through 15) was reserved for the TEXT ITEM NUMBER [0–200]. "0" meant no text was chosen and [1–199] corresponded to the user-defined TEXT ITEMs in the PROJECT's Display Tool (see Fig. 11.13). "200" was special and pointed to the SMS message received.
* The third byte (bit 16 through 23) was reserved for the TEXT SIZE [0–255].
* The fourth byte (bit 24 through 31) was reserved for the TEXT COLOR [0–9], where "0" meant "Unknown", "1" was White, "2" was Black, "3" was Red. "4" was Green, "5" was Blue, "6" was Yellow, "7" was Light Gray, "8" was Gray and "9" was Dark Gray.

Also similarly, the structure of the Number Display Constant was defined as follows:

* The first (lowest) byte (bit 0 through 7) was reserved for the NUMBER POSITION [1–25] corresponding to the screen zones defined in Fig. 11.25.
* The second byte (bit 8 through 15) was reserved for the NUMBER itself, i.e. its numerical value [0–255].
* The third byte (bit 16 through 23) was reserved for the NUMBER SIZE [0–255].
* The fourth byte (bit 24 through 31) was reserved for the NUMBER COLOR [0–9], where "0" meant "Unknown", "1" was White, "2" was Black, "3" was Red. "4" was Green, "5" was Blue, "6" was Yellow, "7" was Light Gray, "8" was Gray and "9" was Dark Gray.

There were two techniques to manipulate these SMART CONSTANTs:

* The first technique was to use the CUSTOM option (see examples in Figs. 11.19 and 11.45) where any of the 4 bytes could be replaced with a VARIABLE PARAMETER defined elsewhere (but before the considered DISPLAY command of course). In Fig. 11.19, Parameter I was used for the Item Number ("1" or "2" in that particular case). In Fig. 11.45, the POSITION byte was defined by values found in Parameters "Touch1" and "Touch2".

- The second technique was to use integer multiplication and addition, so as to shift these bytes "left" as many bits as needed to form a "combined" SMART CONSTANT where each of the 4 bytes had the "variable" information wanted by the user. This technique would be illustrated in the next project for the SMART CONSTANT that would be normally associated with the MUSICAL INSTRUMENT on the mobile device.

11.7.2   Using Multiple Audio Output Devices

Between the mobile device (i.e. R+m.PLAY700 App) and the CM-50 controller itself, there were four types of audio output devices that could be used simultaneously:

- The built-in buzzer on the CM-50 controller.
- Two independent audio sources on the mobile device.
- One Musical Instrument also on the mobile device.

First, let's dwelve into more details of the structure of the SMART CONSTANT that could be used to play a SMART Musical Instrument. This SMART CONSTANT was 4 bytes in length, but only the lower 3 bytes were used:

- The first (lowest) byte (bit 0 through 7) was reserved for the MUSICAL SCALE or NOTE [1–12]. "1" corresponded to "Do" while "2" corresponded to "Do#", and so forth until "12" corresponding to "Shi".
- The second byte (bit 8 through 15) was reserved for the OCTAVE [1–10].
- The third byte (bit 16 through 23) was reserved for the INSTRUMENT TYPE [1–128]. "1" corresponded to "Acoustic Grand Piano" while "128" to "Gunshot".
- The fourth byte was not used so it was set to zero.

In this project, the plan was to make the controller's buzzer to cycle through its Melody Index from 0 to 24, and for the mobile device to cycle through its Musical Instrument Type from 1 to 128. For the two independent "mobile" audio sources, the plan was to let each of them play its own music selection in a continuous loop.

The "CM50_Cacophony.tskx" file showed how to set up the needed variable parameters and how to activate these 4 audio sources within an Endless Loop. Figure 11.46 detailed the needed initializations):

- After waiting for the user to whistle into the mobile device to get this application started (Statement 6), its display was set to Portrait Mode and its Text Display cleared (Statements 7 and 8).
- Statements 9–19 initialized 11 parameters to be used in the remainder of this program. In particular, MelodyIndex would be cycled from "0" through "24", and "InstrumentType" would be cycled from "1" through "128". The parameters named "Octave", "DoNote", "ReNote" and "MiNote" were just constants. The two temporary Variable Parameters named "InstrumentValueTemp1" and

5	// User needs to make sure that robot is connected to mobile device. User whistles loudy when ready.
6	WAIT WHILE ([ ] SMART: ⌂ Noise (dB) < 50)
7	[ ] SMART: ⟳ Screen Rotation = Portrait Mode (1)
8	[ ] SMART: ▦ Text Display = 0
9	MelodyIndex = 0
10	InstrumentType = 1
11	Octave = 3
12	DoNote = 1
13	ReNote = 3
14	MiNote = 5
15	InstrumentValue1 = 0
16	InstrumentValue2 = 0
17	InstrumentValue3 = 0
18	InstrumentValueTemp1 = 0
19	InstrumentValueTemp2 = 0

Fig. 11.46 Initialization of parameters needed in "Cacophony" application

"InstrumentValueTemp2" would be used to compute intermediary results that would be saved as final values into the Combined Instrument Parameters named "InstrumentValue1", "InstrumentValue2" and "InstrumentValue3". These Combined Instrument Parameters had three types of information combined in them: the Instrument Type [1–128], the Octave Level (fixed at 3 in this application, but could be made variable as needed), and the actual musical notes Do [1], Re [3] and Mi [5] (which could be made variable also).

Figure 11.47 described Part 1 of the main Endless Loop which pertained to the control of the built-in buzzer on the CM-50 controller:

• Statements 24 and 25 displayed on the mobile device the value of the MelodyIndex about to be played by the buzzer. Text Item 1 was set to "Melody".
• Statements 26 and 27 activated the buzzer on the robot controller with the current MelodyIndex value.
• Statements 28–32 incremented MelodyIndex by "1" each time the Endless Loop was executed, however when it reached a value of "25" it would be reset to "0".

Figure 11.48 described Part 2 of the main Endless Loop which pertained to the control of the two Audio Channels on the mobile device:

• Statements 34–37 played Audio Item 3 on the Play Audio channel 1 continuously.
• Statements 38–41 played Audio Item 4 on Play Audio channel 2 continuously.

Figure 11.49 described Part 3 of the main Endless Loop which pertained to the control of the Musical Instrument on the mobile device:

• Statements 43 and 44 displayed the Instrument Type about to be played. Text Item 2 was set to "Instrument".

21	ENDLESS LOOP
22	{
23	// Using Buzzer on Robot Controller - going through 25 melodies [0-24]
24	🗖 SMART: 🖼 Text Display = [Position:(3,1)],[Item:1],[Size:100],[Color:White]
25	🗖 SMART: 🔢 Number Display = [Position:(4,1)],[Item:MelodyIndex],[Size:100],[Color:White]
26	🔊 Buzzer Timer = Melody Time
27	🎵 Buzzer Index = MelodyIndex
28	MelodyIndex = MelodyIndex + 1
29	IF (MelodyIndex == 25)
30	{
31	MelodyIndex = 0
32	}

Fig. 11.47 Part 1 of endless loop – controlling the buzzer

33	// Using Audio Outputs 1 and 2 on Mobile Device playing Audio Items 3 & 4 continuously
34	IF (🗖 SMART: ⏺ Play Audio 1 == Not Used (0))
35	{
36	🗖 SMART: ⏺ Play Audio 1 = Audio Item 3
37	}
38	IF (🗖 SMART: ⏺ Play Audio 2 == Not Used (0))
39	{
40	🗖 SMART: ⏺ Play Audio 2 = Audio Item 4
41	}

Fig. 11.48 Part 2 of endless loop – controlling the two audio channels on the mobile device

- Statement 45 took the current value of "InstrumentType" [1–128] and multiplied it by 65,536 (i.e. shifted it left by 16 bits) and saved this temporary result in "InstrumentValueTemp1".
- Statement 46 took the current value of "Octave" [3] and multiplied it by 256 (i.e. shifted it left by 8 bits) and saved this temporary result in "InstrumentValueTemp2".
- Statement 47 added the current values of "InstrumentValueTemp1" and "InstrumentValueTemp2" and saved this result back into "InstrumentValueTemp1" which now had the combined information on Instrument Type and Octave Level in its value.
- Statement 48 added the current value of "InstrumentValueTemp1" to "DoNote" and saved this result as "InstrumentValue1" which had the combined information on InstrumentType, Octave Level and Do Note. This parameter "InstrumentValue1" would be used later to make the Musical Instrument play a Do note at the Octave level specified and using this particular Instrument Type.
- Statement 49 added the current value of "InstrumentValueTemp1" to "ReNote" and saved this result as "InstrumentValue2" which had the combined information on InstrumentType, Octave Level and Re Note. This parameter "InstrumentValue2"

42	// Using Musical Instrument on Mobil[Edit Code-Block]conds - going through 128 instruments [1-128]
43	☐ SMART: 🖼 Text Display = [Position:(3,3)],[Item:2],[Size:100],[Color:Green]
44	☐ SMART: 🔢 Number Display = [Position:(3,4)],[Item:InstrumentType],[Size:100],[Color:Green]
45	InstrumentValueTemp1 = InstrumentType ✳ 65536
46	InstrumentValueTemp2 = Octave ✳ 256
47	InstrumentValueTemp1 = InstrumentValueTemp1 + InstrumentValueTemp2
48	InstrumentValue1 = InstrumentValueTemp1 + DoNote
49	InstrumentValue2 = InstrumentValueTemp1 + ReNote
50	InstrumentValue3 = InstrumentValueTemp1 + MiNote
51	InstrumentType = InstrumentType + 1
52	IF (InstrumentType == 129)
53	{
54	InstrumentType = 1
55	}
56	☐ SMART: ⏸ Play a musical instrument = InstrumentValue1
57	WAIT WHILE (☐ SMART: ⏸ Play a musical instrument > 0)
58	☐ SMART: ⏸ Play a musical instrument = InstrumentValue2
59	WAIT WHILE (☐ SMART: ⏸ Play a musical instrument > 0)
60	☐ SMART: ⏸ Play a musical instrument = InstrumentValue3
61	WAIT WHILE (☐ SMART: ⏸ Play a musical instrument > 0)
62	}

Fig. 11.49 Part 3 of endless loop – controlling the musical instrument on the mobile device

would be used later to make the Musical Instrument play a Re note at the Octave level specified and using this particular Instrument Type.

- Statement 50 added the current value of "InstrumentValueTemp1" to "MiNote" and saved this result as "InstrumentValue3" which had the combined information on InstrumentType, Octave Level and Mi Note. This parameter "InstrumentValue3" would be used later to make the Musical Instrument play a Mi note at the Octave level specified and using this particular Instrument Type.
- Statement 51 incremented "InstrumentType" by "1" to get ready for the next iteration loop.
- Statements 52 through 55 checked on the current value of "InstrumentType" and if it was equalled to "129", it would be reset to "1".
- Statements 56 through 61 made the SMART Musical Instrument play each Do-Re-Mi note successively, using the current Instrument Type and Octave Level.

Video 11.12 showed how this "Cacophony" project performed.

11.7.3   Intruder Alert Email Service

This next-to-last SMART Project involved a little more setup and would require a valid Google Mail account if the reader wanted to try out the code described in the file "CM50_IntruderAlert.tskx".

First, we must set up the "Send Email" tool for this SMART Project (please refer back to Fig. 11.13 and tap on the "Send Email" area – bottom left in Fig. 11.13). This would open up the Email Setup Screen as shown in Fig. 11.50, where the reader would set up:

(a) The Gmail Security Setting to "ON" for Less Secure Apps (see Fig. 11.51). The author would definitely recommend to reset it back to "OFF" when the reader was done with testing out this application, as this would be a security risk for the reader's Google Mail account.

(b) Next, the reader needed to set up the User Name and Password for an actual G-mail account where the reader would want the images captured via this application to be sent to (assuming it to be the reader's own account of course).

(c) Do the "Sending Email Test" to verify that your provided G-mail account did receive an email from yourself.

Figure 11.52 described Part 1 of the code needed for this project:

- Statement 6 waited on the user to whistle into the mobile device to get the application going.
- Statements 8–10 set up the mobile device's Screen into its Portrait mode, used the Back Camera and set the Camera Sensor into its Motion Detection mode.
- Statements 11–12 made the mobile device wait for 2 s to make sure that all previous settings had enough time to take effects.

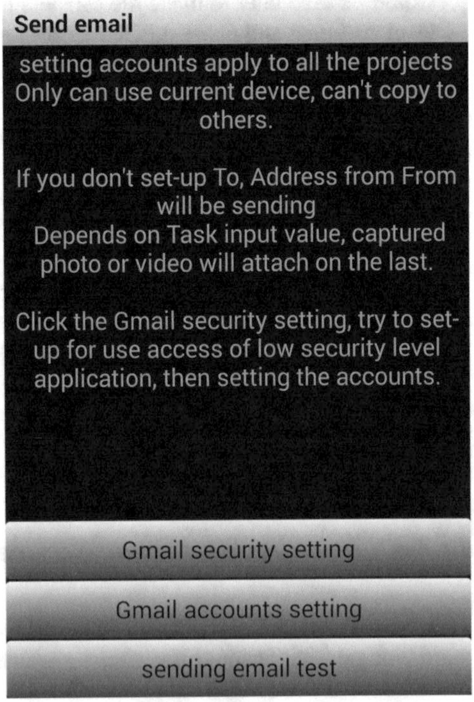

Fig. 11.50 Settings for "Send Email" tool

Send email

setting accounts apply to all the projects
Only can use current device, can't copy to others.

If you don't set-up To, Address from From will be sending
Depends on Task input value, captured photo or video will attach on the last.

Click the Gmail security setting, try to set-up for use access of low security level application, then setting the accounts.

Gmail security setting

Gmail accounts setting

sending email test

Fig. 11.51 Temporarily turning ON access for less secure apps

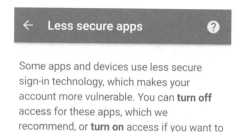

← **Less secure apps**   ?

Some apps and devices use less secure sign-in technology, which makes your account more vulnerable. You can **turn off** access for these apps, which we recommend, or **turn on** access if you want to use them despite the risks. Learn more

Access for less secure apps

○ Turn off

● Turn on

5	// User needs to make sure that robot is connected to mobile device. User whistles loudy when ready.
6	WAIT WHILE ([☐ SMART: 🔊 Noise (dB)] < 50)
7	// Setting Back Camera to Motion Detection Mode & wait for 2 seconds
8	[☐ SMART: 🔄 Screen Rotation] = Portrait Mode (1)
9	[☐ SMART: 📷 Camera Selection] = Back Camera (1)
10	[☐ SMART: 📷 Camera Sensor] = Motion Detection Mode (3)
11	[☐ SMART: ⏱ Smart Timer] = 2.0sec
12	WAIT WHILE ([☐ SMART: ⏱ Smart Timer] > 0.000sec)

Fig. 11.52 Part 1 of "CM50_IntruderAlert.tskx"

Part 2 (Fig. 11.53) contained the Endless Loop that performed the tasks needed:

- Read in the current value for the "DetectedZone" parameter (Statement 16).
- If "DetectedZone" was non-zero, this meant that "Motion" had been detected somewhere in the 25 zones of the Camera Sensor (see Fig. 11.25). Then a Photo was captured from the Camera and a 0.5 s "La" was also produced by the controller's buzzer (Statements 19–23).
- Next, the mobile device waited for 1 s. While the captured photo got saved in the proper location of the mobile device (Statements 25–26).
- Statements 27–28 sent the Captured Photo via email to the G-mail account previously set up and waited until this job was done, and then back to the Endless Loop at Statement 16. Statement 28 used an "OR" combination of 2 logical statements checking on the E-mail Send Status. The "Not Used" status was checked because most likely the E-mail service would not have the time to get started completely yet, when the WAIT WHILE construct was tested for the first time, right after the execution of Statement 27. The "Not Used" condition would not be TRUE during subsequent iterations of the WAIT WHILE loop and this was when the "Sending" condition got into play to make sure that this WAIT WHILE loop terminated when both "Not Used" and "Sending" conditions were no longer TRUE.

14	ENDLESS LOOP
15	{
16	DetectedZone = ☐ SMART: ✖ Motion Detection Area
17	IF (DetectedZone != 0)
18	{
19	☐ SMART: ⓐ Photo Capture = Capture (1)
20	⏰ Buzzer Timer = 0.5sec
21	♪ Buzzer Index = La# (1)
22	WAIT WHILE (⏰ Buzzer Timer > 0.000sec)
23	WAIT WHILE (☐ SMART: ⓐ Photo Capture != 0)
24	
25	☐ SMART: ⏰ Smart Timer = 1.0sec
26	WAIT WHILE (☐ SMART: ⏰ Smart Timer > 0.000sec)
27	☐ SMART: ✉ E-Mail Function = Send Captured Photo (1)
28	WAIT WHILE (☐ SMART: ✉ E-Mail Send Status == Not Used (0) ‖ ☐ SMART: ✉ E-Mail Send Status == Sending (1))
29	}
30	}

Fig. 11.53 Part 2 of "CM50_IntruderAlert.tskx"

The author did check that this mini project had performed properly but no video was prepared for it.

If the reader had experiences with Embedded C in using Arduino and the OpenCM IDE for the controller OpenCM-9.04C, the next section would be a good read as it described how to adapt the same SMART Project of "Taking Pictures" (Sect. 11.3) to an Arduino Sketch using the ROBOTIS OpenCM IDE tool.

11.8   Using OpenCM IDE with R+m.PLAY700

First, please refer to Sects. 9.1 and 9.2 for basic information about the OpenCM IDE tool. Next, we need to review the concept of "Dynamixels". The reader might recall from Chap. 2 (and really throughout all previous chapters) that a ROBOTIS robot could be considered as a network of Dynamixels which could have sensors attached to their various ports and, in a way, the "robot" functioned as programmed by sending appropriate messages to each other within this network. In this world view, the various hardware controllers (CM-XXX and OpenCM-XXX) that had been mentioned in this book had the special Dynamixel ID of "200", which the readers might have already noticed whenever they used the MANAGER tools. The other special Dynamixel ID was "100" which was assigned to the co-controller or "remote device" – if you will. This "remote device" was the Windows PC in the case of R+SCRATCH and the Android or iOS devices in the case of R+m.PLAY700.

Chapter 7 showed applications of one type of these messages which was the "Remocon" packet. The second type of messages were called "Dynamixel" packets which were further sub-divided into Protocol 1.0 (or DXL 1.0) and Protocol 2.0 (DXL 2.0). DXL 1.0 covered the "older" controllers such as CM-530 and earlier, and also the "older" actuators and sensors such as AX-12, MX-28, AX-S1 and IRSA, please see this web link for the detailed descriptions of Protocol 1.0 (http://support.robotis.com/en/product/actuator/dynamixel/dxl_communication.htm).

DXL 2.0 covered the latest generation of controllers (such as CM-50/150/200 and OpenCM-7.00/9.04) and actuators (such as XL-320, Pro and X series). Please see this link for more details about Protocol 2.0 (http://support.robotis.com/en/product/actuator/dynamixel_pro/communication.htm). In Chap. 12, we would deal with the byte-by-byte construction of these packets, but in this section, we would still treat these Dynamixel packets as an abstract object carrying three pieces of "numerical" information:

1. The "ID" of the intended Dynamixel regardless of its true function as a hardware controller or as a physical actuator/sensor, or as a remote device. The ID's size would be a byte (i.e. 8 bits).
2. The "Address" of the function that we wanted the intended Dynamixel to perform whether it would be to go a certain servo position or to capture a picture for us from the phone's back camera. This kind of relationships were usually listed in specific Control Tables for each type of Dynamixel, whether it was an actuator such as the XL-320 (http://support.robotis.com/en/product/actuator/dynamixel_x/xl_series/xl-320.htm), or a hardware controller such as the OpenCM-9.04 (C:\Program Files (x86)\ROBOTIS\RoboPlus 2.0\R+ Scratch\R+Scratch_Data\Config\Controller\Controltable\2.0\OpenCM9.04.en.xml), or a remote device (http://support.robotis.com/ko/software/mobile_app/r+smart/smanrt_manual.htm#Actuator_Address_0B3 – this link is in Korean so the reader needs to use Chrome so that it can translate the contents into English as needed). The Address' size would be a word (i.e. 2 bytes or 16 bits).
3. The "Data" that the intended Dynamixel would need in order to function as we needed it to do, from the "Address" information previously sent to it. The Data's size would vary from a byte to a double word (i.e. 4 bytes or 32 bits) depending on the type of function used (please refer the particular control table as needed).

The readers might also have noticed that in the case of using SCRATCH2-R+SCRATCH, Dynamixel 100 (i.e. the Windows PC) was the supervising Dynamixel in charge of its multimedia services while Dynamixel 200 (i.e. the CM-50) was in charge of its actuators and sensors, but under the initiations from Dynamixel 100. However this situation was reversed with the case of TASK-R+m. PLAY700: Dynamixel 100 (i.e. the Android/iOS device) was still in charge of its multimedia services which were however initiated by Dynamixel 200 this time around. Similarly in this section, the OpenCM IDE would make the OpenCM-9.04 the supervising controller deciding on what and when "remote" services would be needed on the SMART devices via the R+m.PLAY700 tool.

Originally, the author wanted to perform a straight conversion of the existing file CM904_TakePictures.tskx (see Fig. 11.22) into an equivalent Arduino sketch, but this turned out not to be possible as an Arduino method equivalent to a "Custom" command inside TASK could not be found among the method libraries provided by the current version 1.04 of the OpenCM IDE. Thus the original solution was modified slightly to result in two pieces of codes that were generated in parallel to provide the same overall functionalities of a robotic system as shown in Fig. 11.54: "CM904_TakePictures_IDE.tskx" and "CM904_TakePictures.ino".

Fig. 11.54 OpenCM-9.04
CarBot

3	START PROGRAM
4	{
5	ENDLESS LOOP
6	{
7	// User needs to make sure that robot is connected to mobile device. Using IR sensor as CM-904 has no buzzer
8	WAIT WHILE (🔲 PORT[1]:IR Sensor < 50)
9	☐ SMART: 🔁 Screen Rotation ⚌ Portrait Mode (1)
10	☐ SMART: 🖬 Text Display ⚌ 0
11	☐ SMART: 🔊 Text to Speech (TTS) ⚌ Text Item 1
12	☐ SMART: 🖬 Text Display ⚌ [Position:(3,3)],[Item:1],[Size:100],[Color:Green]
13	☐ SMART: 📷 Camera Selection ⚌ Back Camera (1)
14	☐ SMART: ◉ Camera Zoom ⚌ 0
15	WAIT WHILE ([☐ SMART: 🔊 Text to Speech (TTS) != 0)

Fig. 11.55 Beginning of "CM904_TakePictures_IDE.tskx"

Figure 11.55 showed the beginning of the "CM904_TakePictures_IDE.tskx"
TASK program, while Fig. 11.56 showed the corresponding beginning of the
"CM904_TakePictures.ino" Arduino sketch.

The TASK program in Fig. 11.55 used the usual Endless Loop whereas:

1. First give the user some time to setup the R+m.PLAY700 app on the remote
   device with Statement 8, basically by waiting for the user to trigger a signal
   (>=50) on the IR sensor to start the whole process.
2. Next came a series of SMART commands sent to the remote device:

   (a) Rotate the remote device's screen to a vertical layout (Statement 9).
   (b) Clear the screen of all text (if present) with Statement 10.
   (c) Start the Text-To-Speech service using Text Item 1 (i.e. "Ready!") with
       Statement 11.
   (d) Statement 12 displayed Text Item 1 at Position (3.4) with a font Size = 100
       and in Green color.
   (e) Statements 13 and 14 selected the Back Camera and set its zoom to 30.
   (f) Then the program waited for the Text-To-Speech service on the remote
       device to end with Statement 15.

```
17   #include <OLLO.h>
18   OLLO myOLLO;
19
20      /* Serial device defines for dxl bus */
21   #define DXL_BUS_SERIAL1 1  //Dynamixel on Serial1(USART1)  <-OpenCM9.04
22   #define DXL_BUS_SERIAL2 2  //Dynamixel on Serial2(USART2)  <-LN101,BT210
23   #define DXL_BUS_SERIAL3 3  //Dynamixel on Serial3(USART3)  <-OpenCM 485EXP
24
25   Dynamixel Smart(DXL_BUS_SERIAL2);  // phone connects to 904 via BT-210 on SERIAL2
26
27   void setup() {
28      // Dynamixel 2.0 Protocol -> 0: 9600, 1: 57600, 2: 115200, 3: 1Mbps
29      Smart.begin(1);
30      myOLLO.begin(1,IR_SENSOR);
31   }   // end of setup()
32
33   void loop() {
34
35      while(myOLLO.read(1, IR_SENSOR) < 50) {} // Waiting for user to get set up on the phone side
36      Smart.writeByte(100, 10010, 1);   // set phone screen to portrait mode
37      Smart.writeDword(100, 10160, 0);   // clear text display
38      Smart.writeDword(100, 10180, 1);   // TTS item 1
39      Smart.writeDword(100, 10160, 73662733);   // text display item 1 at position (3,3) Size=100, Color=Green
40      Smart.writeByte(100, 10020, 1);   // using rear camera
41      Smart.writeByte(100, 10030, 0);   // set camera zoom to 0
42      while (Smart.readByte(100, 10180) != 0){} // waiting for TTS Item 1 to finish
```

Fig. 11.56 Beginning of "CM904_TakePictures.ino"

On the other hand, the Arduino sketch in Fig. 11.56 showed that a bit more work was first needed with several global definitions (Statements 17 through 25):

1. The "myOLLO" object was needed because an "OLLO" IR Sensor on Port 1 would be used (Statements 17–18).
2. Statement 25 was quite important as it created a Dynamixel object named "Smart" and attached it to Communication Port SERIAL2. This corresponded to the physical remote device using a BT-210 module to connect to the OpenCM-9.04 on its 4-pin communication port (see Fig. 11.54).
3. Next the Setup() method was used to initialize both objects "Smart" and "myOLLO" with statements 29 and 30.
4. Statements 35–42 in Fig. 11.56 would match one-to-one "logically" with Statements 8–15 in Fig. 11.55, but of course the syntax was very different.

No doubt, the readers would have noticed that using SMART commands inside TASK shielded the programmer from much details that had to be specified out when using OpenCM IDE, for example:

1. Setting the phone screen to Portrait mode would require the IDE programmer to find out from the remote device control table (http://support.robotis.com/ko/software/mobile_app/r+smart/smanrt_manual.htm#Actuator_Address_0B3) that the "Address" parameter corresponding to this function was "10010" (first column), and that the byte length of its "Data" parameter was one (seventh column), which specified that a writeByte() method would need to be used. The parameter "ID" of the Dynamixel intended to receive this command/message was of course "100" which corresponded to the remote device in the ROBOTIS Dynamixel world view. And finally the last column showed that "1" was the actual numerical value to use if ones wanted to put the phone screen in its Portrait mode, thus the resulting Statement 36 – "Smart.writeByte(100, 10010, 1);"
2. To clear the phone screen of all text (Statement 37), the same control table showed that "10160" would have to be used for the "Address" parameter, and as the corresponding length of its "Data" parameter was 4 bytes (i.e. a double

word), a writeDword() method had to be used. The actual numerical value to be used for its "Data" was "unclear" from the information provided in this control table, and the IDE programmer would have to visit this web site for additional information   (http://support.robotis.com/ko/software/roboplus/roboplus_task/ programming/parameter/remoteterminal/smart_parameter.htm) (it is in Korean thus once again use Chrome to translate into English). Once there, the user would need to find and click on the link for "Character Display" which brought the user to another web page full of information in Google-translated English! And the user would realize that the actual numerical value of this parameter followed a rather complex formula (= "Position" + "Text Item Number" ×256+ "Font Size" ×65,536+ "Text Color" ×16,777,216). This impressive formula just meant that the "Position" parameter was put into the lowest byte of the 4 bytes used (see Fig. 11.25 for its possible values [1–25] for the user-wanted zones). The "Text Item" number was put into the next higher byte (2nd byte), while the "Font Size" was pushed up higher into the 3rd byte, and the "Text Color" was put the highest 4th byte. To clear the screen of course meant to zero out all four components, thus the "0" value used in Statement 37 "Smart.writeDword(100, 10160, 0);". Please note that the same technique was used in Sect. 11.7.2 in the "Cacophony" project to cycle through the available musical instruments and octave regions, except that the 4th byte was not needed there.

3. Statement 39 was a good application of the formula described in the previous paragraph. Statement 12 in Fig. 11.55 showed that we wanted to display "Text Item 1" (i.e. the string "Ready!") in Position (3,3) (i.e. zone "13") with a "Font Size" of "100" and a "Text Color" of "Green" (i.e. "4"). Thus the formula described in the previous paragraph computed out to be: $13 + 1 \times 256 + 100 \times 65,536 + 4 \times 16777216 = 73662733$ thus resulting in this method call "Smart.writeDword(100, 10160, 73662733);" i.e. Statement 39 in Fig. 11.56. To spare the readers of all this web searching, the author had included in the Springer Extra Materials the file "SmartCommandsControlTable.pdf" which was the original ROBOTIS control table with additional helpful information.

4. Please note that in Fig. 11.56, Statements 36–41 were "Write" commands without checking to see if any status information packet were returned from the remote device during/after their execution. For example, Statement 38 was used to tell the remote device to perform a "Text-To-Speech" (TTS) service to "Text Item 1". As this service would take some time to finish, the programmer "sneaked in" 3 more commands (Statements 39–41) before using Statement 42 which used a readByte() method to inquire from the remote device if the TTS service was finished or not. The completed logic was to use a While loop with the conditional statement "(Smart.readByte(100, 10180)! = 0)", meaning that the remote device would only return a Status Packet containing "0" in its "Data" parameter when it was finished with its previously assigned task/service.

At this point, the author would leave it as an exercise for the readers to go through the rest of the INO program on their own, using the concepts previously described.

Video 11.13 showed how both programs performed on the OpenCM-9.04 CarBot and the readers could observe that there was no difference in the performance of these two approaches, thus why would anyone "labor" to use the OpenCM IDE then? The answer lay in the hardware that ones needed to use with the OpenCM-9.04C for a given robot.

From Chap. 3, we knew that the OpenCM-9.04C could only use the XL-320 actuators on its own board, and if the user needed to use AX, MX or RX actuators, the user needed to use the 485-EXP board also. The current version 2.1.5 of TASK would support switching the Dynamixel Channel on the fly inside a TASK program, however TASK can only support Firmware 2.0 actuators from the X series at this point in time (Spring 2017) on the 485-EXP board. So users wanting to use AX/MX/RX series of actuators on the 485-EXP would be out of luck (also see discussions in Sects. 5.2.2), and their only option would be to use the OpenCM IDE as it was the only ROBOTIS tool, in Spring 2017, to support mixed protocols (DXL 1 and 2) and mixed dynamixel channels by just specifying the communication ports to use: SERIAL1 would be for actuators and sensors on the actual OpenCM-9.04C board, SERIAL2 would be for the remote device and the R+m.PLAY700 app, and SERIAL3 would be for actuators physically attached to the 485-EXP board (see discussions and sample codes in Sects. 9.1 and 9.2), and we even have the USB 2 port remaining available if needed.

Finally, if the reader really had enjoyed using the mobile device with the R+m. PLAY700 App, the next chapter would provide a deeper dive into Android Programming with Android Studio and the OLLOBOT SDK, where the "remote device" would be the "supervising" controller once again.

11.9 Review Questions for Chap. 11

1. How does SCRATCH 2 communicate to the typical robot controller?
2. Where does the SCRATCH 2 code reside at run-time?
3. What are the components of a SCRATCH 2 project?
4. What are the components of a SMART project?
5. To activate a SMART function inside a TASK program, do we need to WRITE to it or READ from it?
6. To check on a status of a SMART function, what programming construct can we use?
7. How does a SMART project divide up the display screen of a typical mobile device for use with its SMART functions of displaying and sensing?
8. When working with a SMART Project, how many hardware TIMERs can a programmer use? How many timers were on the robot controller and how many were on the SMART device?
9. Describe the two techniques that can be used for manipulation of the SMART CONSTANTs.

11.10   Review Exercises for Chap. 11

1. When you run the SCRATCH 2 project "**CM50_MoodyDogBot_1.sb2**", what happened when, for example, the user clapped ONCE, and then TWICE a second or two after?
2. When you run the SCRATCH 2 project "**CM50_MoodyDogBot_2.sb2**", what happened when, for example, the user clapped ONCE, and then TWICE a second or two after?
3. Discuss the results obtained between Exercises 1 and 2.
4. In the program "CM50_TakingPictures.tskx" – Part 1 (Fig. 11.14), does it matter if the current WAIT WHILE statement 19 was moved earlier in the logic flow?
5. In the program "CM50_TakingPictures.tskx" – Part 2 (Fig. 11.18), was the current WAIT WHILE at Statement 26 really necessary? Wouldn't the current WAIT WHILE at Statement 35 be sufficient?
6. For the "Color Line Follower" SMART Project, please run the "Top Screen Scan" code, i.e. "CM50_ColorLineFollower_T.tskx", with the mechanical configuration of the "Bottom Screen Scan" case (i.e. Fig. 11.31). You'll get to use the Assistive LED Light and how was the performance of the robot in this case? Discuss any difference in the performances obtained.

References

Ford JL Jr (2014) Scratch 2.0 programming for teens. CENGAGE Learning PTR, Boston
Warner TL (2015) Teach yourself Scratch 2.0 in 24 hours. SAMS, Indianapolis

Chapter 12
Android® Programming with OLLOBOT SDK

In late Summer of 2016, ROBOTIS released for beta tests the OLLOBOT kit (see Fig. 12.1) and its OLLOBOT SDK (https://github.com/ROBOTIS-GIT/OLLOBOT). And similarly to the R+m.PLAY700 app, the OLLOBOT SDK turned out to be adaptable to all Firmware 2.0 controllers with appropriate modifications to the original code for each situation. It was released as an Eclipse project for the Android OS, and as it would be a while before a wider audience can access this kit, the author's goal for this Chapter was to explore the application of this SDK to various Firmware 2.0 systems, in particular, CM-50, CM-150 and OpenCM-9.04C as they would be available to a wider community at present (Spring 2017).

This chapter's main topics are listed below:

- Memory management of ROBOTIS hardware controllers and its software tools TASK, MOTION and OpenCM IDE.
- Using the OLLOBOT SDK with Android Studio® V.2.2.3.
- Application to CM-50.
- Application to CM-150.
- Application to OpenCM-9.04C.

12.1 Memory Management of Hardware Controllers and DXL 2.0 Packets

First, we would need to have a closer look at how memory was managed in typical ROBOTIS hardware controllers. Figure 12.2 showed how each controller partitioned its available memory space for each software component: Boot Loader, Proprietary Firmware, TASK and MOTION.

Unfortunately, the author so far could not find any definitive documentation for the CM-50's memory map, except that it has 64 KB total, but in practice it was known that the CM-50 could only run TASK programs, thus its memory management

© Springer International Publishing Switzerland 2017   311
C.N. Thai, *Exploring Robotics with ROBOTIS Systems*,
DOI 10.1007/978-3-319-59831-4_12

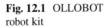

Fig. 12.1 OLLOBOT robot kit

	Controller Type					
	CM-50		**CM-150**		**OpenCM-9.04C**	
Software Component	Start Address	*Size in Bytes*	Start Address	*Size in Bytes*	Start Address	*Size in Bytes*
Boot Loader	No Documentation Found Can only run TASK		0x08000000	*12288*	0x08000000	*12288*
ROBOTIS Firmware			0x08003000	*47104*	0x08003000	*43008*
TASK			0x0800E800	*4096*	0x0800D800	*8192*
MOTION					0x0800F800	*65536*

Fig. 12.2 Memory management in selected ROBOTIS hardware controllers

would be similar to the CM-150's, but with a smaller firmware as it had no expansion port for any other external sensor or servo motor like the CM-150.

When a typical ROBOTIS controller got powered on, the BOOT LOADER would become active first, next the ROBOTIS Proprietary Firmware would get activated also and it would run the TASK program that was previously loaded in its memory space. TASK might also need to use some previously downloaded MOTION pages (which could be considered as simply data for TASK to work on).

The ROBOTIS Proprietary Firmware was designed so that it would normally send Instruction Packets to its actuators and sensors using DXL 1.0 or 2.0 protocol depending on the Firmware level of the actual controller. The OLLOBOT was of the Firmware 2.0 type and please visit this web link for the detailed descriptions of the various fields used in DXL 2.0 (http://support.robotis.com/en/product/actuator/dynamixel_pro/communication/instruction_status_packet.htm). These Instruction

Header			Reserved	Packet ID	Packet Length		Instruction
0xFF	0xFF	0xFD	0x00	ID	LEN_L	LEN_H	Instruction

Fig. 12.3 Byte structure of a DXL 2.0 instruction packet (Part 1)

Packets could be of 12 types, but only READ, WRITE or STATUS types would be considered in this Chapter. The actual numerical value in the INSTRUCTION field of a packet determined its purpose (see Fig. 12.3 for Part 1 of the packet):

- 0x02 for READ – for example, to inquire about the Current Position of a given actuator, the controller firmware would construct a proper READ packet then send it to that actuator, then it would wait for that actuator to return a STATUS packet which it then parsed to find the actual value of the Current Position of the wanted actuator.
- 0x03 for WRITE – for example, to set the speed of an actuator to "CW 512", the controller firmware would again prepare the proper "WRITE" packet and then send it to the actuator. However the controller would not wait on a STATUS packet to return.
- 0x55 for STATUS – for example, the Hardware Controller (ID = 200) returning a value from a Color sensor, connected to its Port 2, to the Remote Device (ID = 100).

Users could now appreciate the work involved in the byte-by-byte construction and handling of these packets that were performed by TASK for the users in the background. At this point, the discerning readers would notice that a typical controller would only **receive STATUS packets** when dealing with actuators and sensors, but what would happen when this controller also communicated with the Remote Device (ID = 100) which could be considered as a co-controller:

- In the case of using R + m.PLAY700 on the Remote Device with TASK running on the Hardware Controller side (Chap. 11), the Hardware Controller was the one issuing WRITE and READ commands to the Remote Device which only returned appropriate STATUS packets to the Hardware Controller. The Remote Device essentially functioned as an "advanced" actuator/sensor to the Hardware Controller.
- In the case of using SCRATCH2/R + SCRATCH on the PC (Chap. 11), the users might have noticed that, as soon as R + SCRATCH connected to the Hardware Controller, whatever TASK code that was running on the Hardware Controller at the time would immediately cease its operation. This TASK code could only then be reactivated after a power reset (see Sect. 11.1 and Video 11.1). This was because ROBOTIS' firmware had been designed such that, once the Hardware Controller received a READ or WRITE command from a Dynamixel in its network, the Hardware Controller immediately disabled its TASK related functions and switched permanently (until power was reset) to a mode whereas it essentially let all INSTRUCTION packets pass through it (if they were not intended for itself (ID = 200)). For those INSTRUCTION packets destined for the Hardware Controller, it would still process them appropriately and would send back STATUS

packets if required. Essentially, the Hardware Controller became just another actuator/sensor to the Remote Device. If we had taken a closer look at the menu for the OLLOBOT app as shown in Fig. 12.1, we would realize that all those displayed commands/buttons were all WRITE or READ commands. Thus the OLLOBOT SDK app would function radically different from the R + m.PLAY700 app, although both of them ran on the same Android device.

The Video 12.1 illustrated an interesting situation:

- The TASK code "CM904_TakePictures_IDE.tskx" was running on the OpenCM-9.04C, and worked fine with the R+m.PLAY700 app.
- Then the R+m.PLAY700 was closed and removed from the working memory of the Android device, and an OLLOBOT app was turned on instead. Ones could see that the TASK code was still operating, as the IR sensor could still be triggered to send WRITE and READ commands to the Remote Device. But this OLLOBOT app was not programmed to handle graphics and camera related services, so all it could do was to display the WRITE/READ packets that it received from the Hardware Controller.
- However, if any of the UDLR buttons on the OLLOBOT app's menu was pushed, then ones could see that the Remote Device was actually controlling the XL-320s directly and moving the CarBot around (although the TASK program did not have such code). However, the IR sensor did not seem to work anymore because all TASK functions were disabled by that time.

In particular, the OpenCM-9.04C would additionally be able to use the OpenCM IDE which would load itself at memory address 0x08003000, effectively removing the TASK and MOTION tools completely. Thus the IDE programmer would now be responsible for all communications channels and packets handling between:

- SERIAL1 Port, for actuators and sensors physically connected to the OpenCM-9.04C board itself (for example, XL-320s and LEDs, IR and Color sensors).
- SERIAL2 Port, for all communications coming from the Remote Device.
- SERIAL3 Port, for Dynamixels physically connected to the 485-EXP board.
- USB SERIAL Port would also be available to the programmer as needed.

Video 12.2 showed how the sketch "CM904_TakePictures.ino" worked with R+m.PLAY700 and then with the previously described OLLOBOT app:

- There was no difference in behavior of this INO with R+m.PLAY700 as compared to the "CM904_TakePictures_IDE.tskx".
- However, when switching to the OLLOBOT app like before, the behavior of the OpenCM-9.04C/Android phone was quite different. The packets kept on flowing from the Hardware Controller to the Android Device, and the UDLR buttons were also non-functional! This was because the ROBOTIS Firmware was no longer there to handle those direct DXL 2.0 packets from the UDLR buttons, and also because in this INO, there was no code written to handle those commands properly.

The remaining sections of this chapter would get into the subtle details of these results/issues regarding the applications of the OLLOBOT SDK to various ROBOTIS controllers.

12.2   Usage of OLLOBOT SDK with Android Studio®

In order to use the OLLOBOT SDK, the user needs to install the Java Development Kit® (http://www.oracle.com/technetwork/java/javase/downloads/index.html) and the Android Studio (https://developer.android.com/studio/index.html). The author used JDK SE 8u121 and AS 2.2.3. The set up tutorial (and other learning materials) for AS can be found at https://developer.android.com/studio/install.html. For readers new to Android Programming, the author is recommending two books, Smyth (2016) or Phillips et al. (2017), and the on-line AS API guide at https://developer.android.com/guide/index.html. Java tutorials are also available at http://docs.oracle.com/javase/tutorial/tutorialLearningPaths.html and the Java Language Basics reference materials can be accessed at https://docs.oracle.com/javase/tutorial/java/nutsandbolts/index.html.

12.2.1   Importing OLLOBOT SDK into Android Studio

After installing the JDK and AS packages, the user should also download the OLLOBOT SDK from GitHub (https://github.com/ROBOTIS-GIT/OLLOBOT), and let's say that the user unzipped this package onto the root of C: drive, i.e. into the folder "C:\OLLOBOT-master" folder, as shown in Fig. 12.4.

As this was an Eclipse project, the user would have to choose the "Import Project" option in the Android Studio tool when it was started. AS would then start importing this project, and would likely "complain" that "missing platforms" and "syncing project" were necessary, the user would just need to agree to all its suggestions to fix these problems. And after a few minutes, the user would be able to see the "import-summary" text file, and a message saying that the Graddle Build was successful (see Fig. 12.5).

Fig. 12.4 OLLOBOT-Master SDK unzipped (Eclipse project)

OS (C:) › OLLOBOT-master

Name

- libs
- res
- src
- .classpath
- .project
- AndroidManifest.xml
- ic_launcher-web.png
- project.properties
- README.md

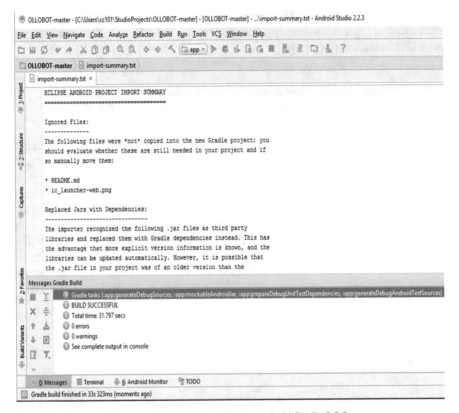

Fig. 12.5 Successful import of OLLOBOT SDK into Android Studio 2.2.3

This new AS Project for the OLLOBOT SDK was actually stored in this folder
C:\OLLOBOT-master1. Next, if the user would click on the top left tab for "Project",
a panel similar to the one shown in Fig. 12.6 would display, showing the many components of this project. The components that would be discussed or modified for the
various adaptations performed in the remainder of this chapter were:

- AndroidManifest.xml.
- Layouts – activity_main.xml, activity_device_list.xml.
- Java – MainActivity.
- Utils – OLLOBOT, Dynamixel, Constants.
- Service – BTConnectionService.

12.2.2   AndroidManifest.xml

The "AndroidManifest.xml" file described the various components of an Android
application, for the OLLOBOT SDK case it showed in Fig. 12.7 that there were:

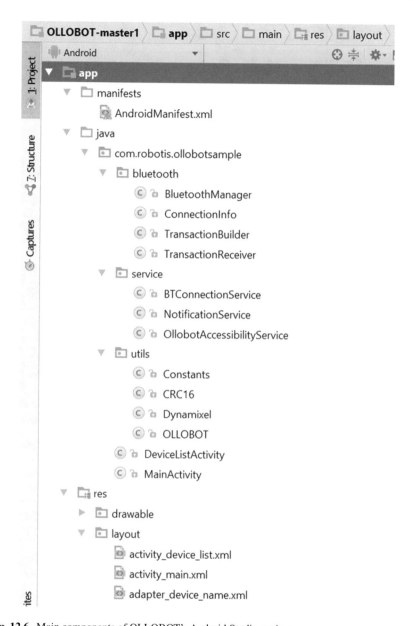

Fig. 12.6 Main components of OLLOBOT's Android Studio project

- Two activities: MainActivity (line 23) and DeviceListActivity (line 33). MainActivity could launch other activities and services (line 29).
- Three services: BTConnectionService (line 41), NotificationService (line 47), and OlloBotAccessibiltyService (line 54).

```
22      <activity
23          android:name="com.robotis.ollobotsample.MainActivity"
24          android:label="OLLOBOT Sample"
25          android:launchMode="singleTask"
26          android:configChanges="keyboardHidden|orientation|screenSize" >
27          <intent-filter>
28              <action android:name="android.intent.action.MAIN" />
29              <category android:name="android.intent.category.LAUNCHER" />
30          </intent-filter>
31      </activity>
32
33      <activity android:name="com.robotis.ollobotsample.DeviceListActivity"
34              android:label="Select a device to connect"
35              android:theme="@android:style/Theme.Dialog"
36              android:configChanges="keyboardHidden|orientation|screenSize" >
37      </activity>
38
39      <!-- Service -->
40      <service
41          android:name= "com.robotis.ollobotsample.service.BTConnectionService"
42          android:icon="@drawable/ic_launcher"
43          android:label= "BT Connection Service"
44          android:configChanges="keyboardHidden|orientation|screenSize" >
45      </service>
46      <service
47          android:name="com.robotis.ollobotsample.service.NotificationService"
48          android:permission="android.permission.BIND_NOTIFICATION_LISTENER_SERVICE" >
49          <intent-filter>
50              <action android:name="android.service.notification.NotificationListenerService" />
51          </intent-filter>
52      </service>
53      <service
54          android:name="com.robotis.ollobotsample.service.OllobotAccessibilityService"
55          android:permission="android.permission.BIND_ACCESSIBILITY_SERVICE" >
56          <intent-filter>
57              <action android:name="android.accessibilityservice.AccessibilityService" />
58          </intent-filter>
59          <meta-data
60              android:name="android.accessibilityservice"
61              android:resource="@xml/accessibilityservice" />
62      </service>
```

Fig. 12.7 Manifest file for OLLOBOT SDK

12.2.3   Layouts: Main Activity

There were three layout files, but we would be interested only in the "activity_main. xml" file which could be displayed as a graphical object (see Fig. 12.8) or an XML text file (see Fig. 12.9) which would be the more useful form as we would want to "reprogram" some of the buttons at a later time (lines 53, 60 and 67).

12.2.4   Utilities: Constants, Dynamixel, OLLOBOT

The utilities files were Constants.java, CRC16.java, Dynamixel.java and OLLOBOT. java. The CRC16.java class created the 16 bit CRC (Cyclic Redundancy Check) used to check for the integrity of the received packets and it would not need to be

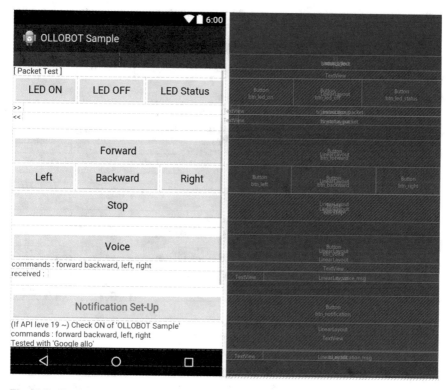

Fig. 12.8 Graphical View of the "activity_main.xml" file

```
52              <Button
53                  android:id="@+id/btn_led_on"
54                  android:layout_width="wrap_content"
55                  android:layout_height="wrap_content"
56                  android:layout_weight="1"
57                  android:text="LED ON" />
58
59              <Button
60                  android:id="@+id/btn_led_off"
61                  android:layout_width="wrap_content"
62                  android:layout_height="wrap_content"
63                  android:layout_weight="1"
64                  android:text="LED OFF" />
65
66              <Button
67                  android:id="@+id/btn_led_status"
68                  android:layout_width="wrap_content"
69                  android:layout_height="wrap_content"
70                  android:layout_weight="1"
71                  android:text="LED Status" />
72          </LinearLayout>
```

Fig. 12.9 Text View of the "activity_main.xml" file

Parameter			16bit CRC	
Parameter1	...	ParameterN	CRC_L	CRC_H

Fig. 12.10 Byte structure of a DXL 2.0 instruction packet (Part 2)

Register	Address	Size [byte]	Name	Description	Access	Default Value	Min Value	Max Value
0	0x0	2	Model Number	Model number	R	440	--	--
7	0x7	1	ID	DYNAMIXEL ID	R	200	--	--
79	0x4F	1	Green LED	Green LED on/off	RW	0	0	1
80	0x50	1	Blue LED	Blue LED on/off	RW	0	0	1
97	0x61	1	Input Voltage	Battery input voltage (unit: 0.1V)	R	--	0	255
112	0x70	1	Controller X-Axis Value	X-axis coordinate of joystick	RW	0	-100	100
113	0x71	1	Controller Y-Axis Value	Y-axis coordinate of joystick	RW	0	-100	100
128	0x80	1	Port 1 Servo Mode	Change left motor mode (wheel or joint)	RW	0	0	1
129	0x81	1	Port 2 Servo Mode	Change right motor mode (wheel or joint)	RW	0	0	1
136	0x88	2	Port 1 Motor Speed	Speed of left wheel	RW	0	-1024	1024
138	0x8A	2	Port 2 Motor Speed	Speed of right wheel	RW	0	-1024	1024
156	0x9C	2	Port 1 Servo Position	Left wheel servo position	RW	--	-1024	1024
158	0x9E	2	Port 2 Servo Position	Right wheel servo position	RW	--	-1024	1024

Fig. 12.11 Corrected OLLOBOT control table

modified in later derivations of this SDK (see Fig. 12.10 for Part 2 of a typical DXL 2.0 packet which contained the 16 bit CRC).

The Constants.java class contained constants representing Message Keys, Preferences, Message Types and Request Codes.

The Dynamixel.java class contained detailed methods for incorporating the appropriate Parameters, i.e. "Register/Address" and "Value" for a particular function that the user wanted to be performed by the receiving OLLOBOT (see Fig. 12.9 and the OLLOBOT Control table at this web link http://www.robotis.us/ollobotsdk/ which had some typos on the Addresses for the functions "Port 2 Motor Speed" and "Port 2 Servo Position"). Figure 12.11 showed the corrected OLLOBOT control table for these functions. This class provided "Read" and "Write" methods for different sizes of the "Value" parameter: byte (8 bits), word (2 bytes), and double word (4 bytes), and also some byte and word manipulation methods.

The OLLOBOT.java class contained Addresses and Lengths of various functions built-in for the OLLOBOT such as LEDs, Servo Position and Speed Settings (which were also listed in Fig. 12.11 – columns 1, 2 and 3).

The last three classes (Constants, Dynamixel and OLLOBOT) would need to be modified to accommodate other controllers, and other actuators and sensors used in robots different from the OLLOBOT.

12.2.5 Main Activity Class

The "MainActivity.java" class was of course the starting point of this Android application. Its definition statement on line 58 in Fig. 12.12 showed that it comes from the superclass Activity of the Android OS, and that it implemented OnClickListener,

```
58 ☒      public class MainActivity extends Activity implements OnClickListener {
59
60             // Debugging
61             private static final String TAG = "ROBOTIS.MainActivity";
62
63             // Context, System
64             private BTConnectionService mService;
65             private ActivityHandler mActivityHandler;
66
67             private ImageView mImageBT = null;
68             private TextView mTextStatus = null;
69
70             private TextView mTvInstructionPacket = null;
71             private TextView mTvStatusPacket = null;
72             private TextView mTvNotificationMsg = null;
73             private TextView mTvStatusMsg = null;
74             private TextView mTvVoiceMsg = null;
75
76             private boolean mIsServiceBound = false;
77
78             private BroadcastReceiver mStatusbarBr;
79             private BroadcastReceiver mNotificationBr;
```

Fig. 12.12 Beginning of "MainActivity.java" file

```
85             @Override
86 ☉↑   +      protected void onCreate(Bundle savedInstanceState) {...}
234
235            @Override
236 ☉↑   +     public synchronized void onStart() { super.onStart(); }
239
240            @Override
241 ☉↑   +     public synchronized void onPause() { super.onPause(); }
244
245            @Override
246 ☉↑   +     public void onStop() { super.onStop(); }
249
250            @Override
251 ☉↑   +     public void onDestroy() {...}
261
262            @Override
263 ☉↑   +     public boolean onCreateOptionsMenu(Menu menu) {...}
267
268            @Override
269 ☉↑   +     public boolean onOptionsItemSelected(MenuItem item) {...}
278
279            @Override
280 ☉↑   +     public void onBackPressed() { super.onBackPressed(); }
283
284            @Override
285 ☉↑   +     public void onConfigurationChanged(Configuration newConfig){...}
```

Fig. 12.13 Override methods in "MainActivity.java" file

```
293        /**
294         * Service connection
295         */
296        private ServiceConnection mServiceConn = new ServiceConnection() {...};
313
314        /**
315         * Start service if it's not running
316         */
317        private void doStartService() {...}
323
324        /**
325         * Stop the service
326         */
327        private void doStopService() {...}
336
337        /**
338         * Initialization / Finalization
339         */
340        private void initialize() {...}
351
352        /**
353         * Launch the DeviceListActivity to see devices and do scan
354         */
355        private void doScan() {...}
359
360        /**
361         * Showing google speech input dialog
362         * */
363        private void promptSpeechInput() {...}
```

Fig. 12.14 Private Methods in "MainActivity.java" file

meaning that it would monitor the "clicking" of those various buttons (actually touch areas on the phone's screen) defined in the layout shown in Fig. 12.8.

Next were a series of Override Methods as shown in Fig. 12.13.

Figure 12.13 showed that the "onCreate" method was quite important as it extended from line 86 to line 233. We'll go into the details of this method in a later section of this chapter.

Then, next were a series of Private Methods as shown in Fig. 12.14.

These methods were involved in finding and setting up BlueTooth devices and services, and also in prompting the user regarding the Speech Recognition service. They would not need to be changed in our subsequent planned modifications of this SDK.

The last section of codes in the MainActivity.java class are for one Public Class "ActivityHandler" and two Public Methods "onActivityResult()" and "onClick()", as shown in Fig. 12.15. They would be modified extensively to accommodate different "button" functional assignments, different actuator and sensor behaviors, and also new verbal commands.

```
381      /*****************************************************
382       *   Public classes
383       *****************************************************/
384
385      /**
386       *  Receives result from external activity
387       */
388 ●↑  ·   public void onActivityResult(int requestCode, int resultCode, Intent data) {...}
450
451
452
453      /*****************************************************
454       *  Handler, Callback, Sub-classes
455       *****************************************************/
456
457   ·    public class ActivityHandler extends Handler {...}   // End of class ActivityHandler
521
522         @Override
523 ●↑  ·   public void onClick(View v) {...}
589
```

Fig. 12.15 Public Class and Methods in "MainActivity.java" file

12.3   Selective Code Tracings Through OLLOBOT SDK

Very early in the writing of this chapter, the author realized that it would not be possible to explain away each and every line of code in the OLLOBOT SDK, thus he aimed for a modest goal of conveying a path of enquiry such that the readers could use to modify the features of this SDK while keeping its overall structure intact.

This enquiry approach was to simply trace through the current OLLOBOT code to show what happened when:

1. The three "LED" buttons were pushed by the user.
2. The Remote Control buttons F-B-L-R-S were pushed by the user.
3. The Speech Recognition service was used.

12.3.1   How "LED" Buttons Were Set Up and Used

First, let's have a closer look again at Fig. 12.9 (i.e. file "activity_main.xml") for lines 53 and 57 which showed that the text "LED ON" was bound to the resource ID'd as "btn_led_on". Similarly, "LED OFF" (line 64) was bound to "btn_led_off" (line 60), and "LED Status" (line 71) was bound to "btn_led_status".

Next, let's find where "btn_led_on", "btn_led_off" and "btn_led_status" would show up again, and they could be found in "MainActivity.java" inside the method "onCreate()" (see Fig. 12.16).

In Fig. 12.16, lines 109 and 121 registered the particular rectangular area with the text "LED ON" (see Fig. 12.9) with "onClickListener" so that during the app's run time, if the user tapped in this area, the Android OS would generate an integer ID that got passed to the method onClick() which could process this ID further (see Fig. 12.17) to result in some concrete actions (i.e. send an appropriate Instruction Packet to the OLLOBOT).

```
108     // Setup button for H/W test
109     Button btnLedOn = (Button) findViewById(R.id.btn_led_on);
110     Button btnLedOff = (Button) findViewById(R.id.btn_led_off);
111     Button btnLedStatusf = (Button) findViewById(R.id.btn_led_status);
112     Button btnForward = (Button) findViewById(R.id.btn_forward);
113     Button btnBackward = (Button) findViewById(R.id.btn_backward);
114     Button btnLeft = (Button) findViewById(R.id.btn_left);
115     Button btnRight = (Button) findViewById(R.id.btn_right);
116     Button btnStop = (Button) findViewById(R.id.btn_stop);
117     Button btnVoice = (Button) findViewById(R.id.btn_voice);
118     Button btnNotification = (Button) findViewById(R.id.btn_notification);
119     Button btnAccessibility = (Button) findViewById(R.id.btn_accessibility);
120
121     btnLedOn.setOnClickListener(this);
122     btnLedOff.setOnClickListener(this);
123     btnLedStatusf.setOnClickListener(this);
124     btnForward.setOnClickListener(this);
125     btnBackward.setOnClickListener(this);
126     btnLeft.setOnClickListener(this);
127     btnRight.setOnClickListener(this);
128     btnStop.setOnClickListener(this);
129     btnVoice.setOnClickListener(this);
130     btnNotification.setOnClickListener(this);
131     btnAccessibility.setOnClickListener(this);
```

Fig. 12.16 Registering various Buttons to the Event Listener "onClickListener"

```
523     public void onClick(View v) {
524         byte[] packet = null;
525         switch (v.getId()) {
526             // See OLLOBOT.java for details.
527             // See OLLOBOT.java for details.
528             // See OLLOBOT.java for details.
529             case R.id.btn_led_on:
530                 packet = Dynamixel.packetWriteByte(OLLOBOT.ID, OLLOBOT.Address.BLUE_LED, 1);
531                 break;
532             case R.id.btn_led_off:
533                 packet = Dynamixel.packetWriteByte(OLLOBOT.ID, OLLOBOT.Address.BLUE_LED, 0);
534                 break;
535             case R.id.btn_led_status:
536                 packet = Dynamixel.packetRead(OLLOBOT.ID, OLLOBOT.Address.BLUE_LED, OLLOBOT.Length.BLUE_LED);
537                 break;
```

Fig. 12.17 How method onClick() processed the user's tapping on the LED buttons at run time

Figure 12.17 showed that this work was done by a Switch structure based on v. getID(). If the "LED ON" button was tapped, the case "R.id.btn_led_on" would be triggered, and line 530 would be executed where a "packet" (i.e. a byte array named "packet", see line 524 for its definition and initialization) would be "appropriately" constructed:

1. First, why use the method "packetWriteByte()" from the class "Dynamixel. java"? The reader had to refer back to the OLLOBOT Control Table shown in Fig. 12.11, whereas the function to turn on the Blue LED had a Size of 1 byte, thus "WriteByte".
2. The method "Dynamixel.packetWriteByte()" required three parameters which were listed in the class OLLOBOT.java (see Fig. 12.18), whereas the first two parameters were OLLOBOT.ID = 200 (line 4), and OLLOBOT.Address.BLUE_LED = 80 (line 14).

```
3      public class OLLOBOT {
4          public static final int ID = 200;
5
6          public class Address {
7              // Bluetooth 2.0 connection pin number is '0000'.
8              // The first packet is ignored after the controller is turned on.
9
10             // About the packet. see the below link.
11             // http://support.robotis.com/en/techsupport_eng.htm#product/actuator/dynamixel_
12
13             public static final int GREEN_LED = 79;              // 0 ~ 1
14             public static final int BLUE_LED = 80;               // 0 ~ 1
15             public static final int INPUT_POWER_VOLTAGE = 97;
```

Fig. 12.18 Selected address' constants definitions in OLLOBOT.java

```
39         public class Length {
40             public static final int GREEN_LED = 1;
41             public static final int BLUE_LED = 1;
42             public static final int INPUT_POWER_VOLTAGE = 1;
43             public static final int CONTROLLER_X_AXIS_VALUE = 1;
44             public static final int CONTROLLER_Y_AXIS_VALUE = 1;
45             public static final int PORT_1_SERVO_MODE = 1;
46             public static final int PORT_2_SERVO_MODE = 1;
47             public static final int PORT_1_MOTOR_SPEED = 2;
48             public static final int PORT_2_MOTOR_SPEED = 2;
49             public static final int PORT_1_SERVO_POSITION = 2;
50             public static final int PORT_2_SERVO_POSITION = 2;
51         }
```

Fig. 12.19 Length's Constants definitions in OLLOBOT.java

3. The third parameter for method "Dynamixel.packetWriteByte()" was "1" because we wanted to turn ON the Blue LED. This was the so-called "Value" associated with the "Address" = 80, and the two right most columns in Fig. 12.11 showed the proper range for this number [0,1].

Similarly, if the user had tapped into the area "LED OFF" instead, line 533 of Fig. 12.17 would then be executed, using a value of "0" for the third parameter to turn off the Blue LED (but still using the method "Dynamixel.packetWriteByte()").

On the other hand, if the user had tapped into the "LED STATUS" button, meaning that the user was inquiring about the status of the Blue LED (i.e. whether it was ON or OFF), line 536 would be executed, as now a READ packet would be needed. Thus, the method "Dynamixel.packetRead()" was used, whereas the two first parameters ID and Address were the same as for the "WriteByte()" method, but the third parameter now had a different meaning (= memory length to read), OLLOBOT. Length.BLUE_LED = 1 byte (see Fig. 12.19 at line 41).

Eventually, a "break" statement was executed, and the program pointer would be sent to line 581 of Fig. 12.20, whereas:

- If "packet" was not an empty byte array, it would be sent to method "mService. sendMessageToRemote()" (line 582) and off it went via BlueTooth Service to the

```
581                 if (packet != null) {
582                     mService.sendMessageToRemote(packet);
583                     mTvInstructionPacket.setText(Dynamixel.packetToString(packet));
584                 } else {
585                     mTvInstructionPacket.setText("");
586                 }
587                 mTvStatusPacket.setText("");
588             }
```

Fig. 12.20 Sending "packet" to remote device (i.e. OLLOBOT)

OLLOBOT (process not shown here). Also the same "packet" got transformed into a string to be displayed on the "Instruction Packet" field (i.e. just below the "LED ON" button and to the right of the characters ">>" in Fig. 12.8). The interested reader could find this resource defined in the layout file "activity_main. xml" as "tv_instruction_packet" on lines 82 and 85.

- Else, i.e. "packet" was equal to "null", the "Instruction Packet" field got cleared (line 585). This resource was named as "tv_status_packet" in the file "activity_ main.xml".
- The final action in this Switch structure (and of the onClick() method) was to clear the "StatusPacket" field, situated below the "InstructionPacket" field and just to the right of the characters "<<" in Fig. 12.8.

Please see Video 12.3 for a demonstration of these OLLOBOT capabilities.

12.3.2   How "F-B-L-R-S" Buttons Were Set Up and Used

Similarly to the procedure spelled out in Sect. 12.2.4, we needed to start out in the layout file "activity_main.xml", but now to look respectively at lines 113, 126, 133, 140 and 153 in Fig. 12.21 for the Resource IDs: "btn_forward", "btn_left", "btn_ backward", "btn_right" and "btn_stop".

The next step was to find the registration of these Resource IDs with "onClick-Listener" in the file "MainActivity.java", i.e. back to Fig. 12.16, and they could be found in lines 112/124, 113/125, 114/126, 115/127 and 116/128.

Then onwards to method onClick() to check on the Switch structure handling these F-B-L-R-S buttons. Figure 12.22 showed the code for the cases "btn_forward" and "btn_forward" where we were provided with three choices to move the OLLOBOT forward:

1. Go and Stop – i.e. to use motors on Ports 1 and 2 in their servo modes (i.e. Position Control mode).
2. Keep going forward when the Forward button was pushed "once".
3. To use the FBLR layout as if they were a virtual joystick (X-Y controller).

As options 1 and 3 were particular to the OLLOBOT only, they won't discussed further.

```
112                          <Button
113                              android:id="@+id/btn_forward"
114                              android:layout_width="wrap_content"
115                              android:layout_height="wrap_content"
116                              android:layout_weight="1"
117                              android:text="Forward" />
118                      </LinearLayout>
119
120                      <LinearLayout
121                          android:layout_width="match_parent"
122                          android:layout_height="wrap_content"
123                          android:orientation="horizontal" >
124
125                          <Button
126                              android:id="@+id/btn_left"
127                              android:layout_width="wrap_content"
128                              android:layout_height="wrap_content"
129                              android:layout_weight="1"
130                              android:text="Left" />
131
132                          <Button
133                              android:id="@+id/btn_backward"
134                              android:layout_width="wrap_content"
135                              android:layout_height="wrap_content"
136                              android:layout_weight="1"
137                              android:text="Backward" />
138
139                          <Button
140                              android:id="@+id/btn_right"
141                              android:layout_width="wrap_content"
142                              android:layout_height="wrap_content"
143                              android:layout_weight="1"
144                              android:text="Right" />
145                      </LinearLayout>
146
147                      <LinearLayout
148                          android:layout_width="match_parent"
149                          android:layout_height="wrap_content"
150                          android:orientation="horizontal" >
151
152                          <Button
153                              android:id="@+id/btn_stop"
```

Fig. 12.21 Layout definitions for the F-B-L-R-S buttons

```
538         case R.id.btn_forward:
539             // go and stop.
540     //      packet = Dynamixel.packetWriteDWord(OLLOBOT.ID, OLLOBOT.Address.PORT_1_SERVO_POSITION, -512 + (512 << 16));
541             // keep going.
542             packet = Dynamixel.packetWriteDWord(OLLOBOT.ID, OLLOBOT.Address.PORT_1_MOTOR_SPEED, -512 + (512 << 16));
543     //      packet = Dynamixel.packetWriteWord(OLLOBOT.ID, OLLOBOT.Address.CONTROLLER_X_AXIS_VALUE, 0 + (30 << 8)); // X:0, Y:30
544             break;
545         case R.id.btn_backward:
546     //      packet = Dynamixel.packetWriteDWord(OLLOBOT.ID, OLLOBOT.Address.PORT_1_SERVO_POSITION, 512 + (-512 << 16));
547             packet = Dynamixel.packetWriteDWord(OLLOBOT.ID, OLLOBOT.Address.PORT_1_MOTOR_SPEED, 512 + (-512 << 16));
548     //      packet = Dynamixel.packetWriteWord(OLLOBOT.ID, OLLOBOT.Address.CONTROLLER_X_AXIS_VALUE, 0 + (-30 << 8)); // X:0, Y:-30
549             break;
```

Fig. 12.22 How onClick() processed the user's tapping on the forward or backward buttons at run time

```
24          // About Motor_speed.
25          // + for clockwize, - for counterclockwise.
26          public static final int PORT_1_MOTOR_SPEED = 136;      // -1024 ~ 1024
27          public static final int PORT_2_MOTOR_SPEED = 138;      // -1024 ~ 1024
```

Fig. 12.23 Motor speed Ports 1 and 2 addresses in OLLOBOT.java

Line 542 showed that the method "Dynamixel.packetWriteDWord()" was used to set the "PORT_1_MOTOR_SPEED" with the value "-512 + (512 < < 16)" – which should have raised some eye brows among the readers!

Let's have a more systematic look at this process. Checking back with the OLLOBOT's Control Table in Fig. 12.11 which showed that the Address for "PORT_1_MOTOR_SPEED" was at "136" and this was consistent with the information listed in "OLLOBOT.java" (line 26 in Fig. 12.23). However "PORT_1_MOTOR_SPEED" had a length of 2 bytes (i.e. a word length) as listed in Fig. 12.19 at line 47, so why would a "Write Packet" with a "double word" length be needed with the use of the method "Dynamixel.packetWriteDWord()"? The answer was in the value "-512 + (512 < < 16)" to be written starting at Address 136. (512 < < 16) meant that the number 512 was shifted left 16 bits, i.e. now resided in the high word area, while (-512) stayed in the low word area, meaning that (-512) would be written to "PORT_1_MOTOR_SPEED" (=136) and (512) would be written to "PORT_2_MOTOR_SPEED" (=138, i.e. 1 word, or 2 bytes, higher than 136).

So now we could understand that it was a technique to set the MOTOR SPEED parameters to PORT 1 and PORT 2 with a single packet (i.e. to initiate motion in both motors at the same time). Here the reader may remember the delay issue encountered in using the SCRATCH2/R + SCRATCH setup in Sect. 5.6.2 because there was a time delay between the actual times when the two motors received their commands to move forward.

In Fig. 12.22, line 547 was just the reverse speed settings to make the OLLOBOT go backwards. The author would leave the job of verifying that the same technique was applied correctly to make the OLLOBOT go left or right. Please see Video 12.4 for a demonstration of these OLLOBOT capabilities.

```
166              <Button
167                  android:id="@+id/btn_voice"
168                  android:layout_width="match_parent"
169                  android:layout_height="wrap_content"
170                  android:layout_weight="1"
171                  android:text="Voice" />
172
173              <TextView
174                  android:layout_width="match_parent"
175                  android:layout_height="wrap_content"
176                  android:text="commands : forward backward, left, right" />
177
178              <LinearLayout
179                  android:layout_width="match_parent"
180                  android:layout_height="wrap_content"
181                  android:orientation="horizontal" >
182
183                  <TextView
184                      android:layout_width="wrap_content"
185                      android:layout_height="wrap_content"
186                      android:text="received :   " />
187
188                  <TextView
189                      android:id="@+id/tv_voice_msg"
190                      android:layout_width="match_parent"
191                      android:layout_height="wrap_content"
192  ■                   android:textColor="#0000FF" />
```

Fig. 12.24 Layout definitions for the Resources "btn_voice" and "tv_voice_msg"

```
102          mTvInstructionPacket = (TextView) findViewById(R.id.tv_instruction_packet);
103          mTvStatusPacket = (TextView) findViewById(R.id.tv_status_packet);
104          mTvNotificationMsg = (TextView) findViewById(R.id.tv_notification_msg);
105          mTvStatusMsg = (TextView) findViewById(R.id.tv_status_msg);
106          mTvVoiceMsg = (TextView) findViewById(R.id.tv_voice_msg);
```

Fig. 12.25 "tv_voice_msg" linked to a TextView object

12.3.3   How Speech Recognition Was Set Up and Used

Same as before, we needed to start out in the layout file "activity_main.xml" (see Fig. 12.24), where the Resources "btn_voice" (line 167) and "tv_voice_msg" (line 189) were defined.

Figure 12.16 showed that the "btn_voice" Button was registered to onClickListener() (lines 117/129), but "tv_voice_msg" was linked to a TextView named "mTvVoiceMsg" (line 106 in Fig. 12.25). Later, "mTvVoiceMsg" would be set to the result found by the Google web-based Speech Recognizer Service (line 444) in the method onActivityResult() which itself was defined at line 388 of "MainActivity.java".

So, let's say that during the apps' runtime the user tapped on the "Voice" Button. This event would be processed by the method onClick(), corresponding to the case "R.id.btn_voice" (line 563 in Fig. 12.26), and the method "promptSpeechInput()" would be called to execute next (Fig. 12.27).

Method promptSpeechInput() would send out an Intent to trigger the Speech Recognizer service which was completely external to the OLLOBOT app (lines 364–373). The method startActivityForResult(intent, Constants.REQ_CODE_

```
563         case R.id.btn_voice:
564             promptSpeechInput();
565             // control in onActivityResult.
566             break;
```

Fig. 12.26 onClick() processing "Voice" button click

```
363     private void promptSpeechInput() {
364         Intent intent = new Intent(RecognizerIntent.ACTION_RECOGNIZE_SPEECH);
365         intent.putExtra(RecognizerIntent.EXTRA_LANGUAGE_MODEL,
366                 RecognizerIntent.LANGUAGE_MODEL_FREE_FORM);
367         intent.putExtra(RecognizerIntent.EXTRA_LANGUAGE, "en-US");
368         intent.putExtra(RecognizerIntent.EXTRA_MAX_RESULTS, 100);
369 //          intent.putExtra(RecognizerIntent.EXTRA_LANGUAGE, Locale.getDefault());
370         intent.putExtra(RecognizerIntent.EXTRA_PROMPT,
371                 "Speak now......");
372         try {
373             startActivityForResult(intent, Constants.REQ_CODE_SPEECH_INPUT);
374         } catch (ActivityNotFoundException a) {
375             Toast.makeText(getApplicationContext(),
376                     "Speech not supported.",
377                     Toast.LENGTH_SHORT).show();
378         }
379     }
```

Fig. 12.27 Method promptSpeechInput()

```
388     public void onActivityResult(int requestCode, int resultCode, Intent data) {
389 //      Log.d(TAG, "onActivityResult " + resultCode);
390
391         switch (requestCode) {
415     case Constants.REQ_CODE_SPEECH_INPUT:
416         if (resultCode == RESULT_OK && null != data) {
417             byte[] packet = null;
418
419             ArrayList<String> result = data
420                     .getStringArrayListExtra(RecognizerIntent.EXTRA_RESULTS);
421
422             for (int i = 0; i < result.size(); i++) {
423                 if ("forward, ford, fort, fought, hot, food, flood".indexOf(result.get(i).toLowerCase()) >= 0) { // for command forward.
424                     packet = Dynamixel.packetWriteDWord(OLLOBOT.ID, OLLOBOT.Address.PORT_1_SERVO_POSITION, -512 + (512 << 16));
425                     break;
426                 } else if ("backward, backwood, banquet, backyard, back, beck, bek".indexOf(result.get(i).toLowerCase()) >= 0) { // for command back.
427                     packet = Dynamixel.packetWriteDWord(OLLOBOT.ID, OLLOBOT.Address.PORT_1_SERVO_POSITION, 512 + (-512 << 16));
428                     break;
429                 } else if ("left, lyft, lift, laugh, lah, loft".indexOf(result.get(i).toLowerCase()) >= 0) { // for command left.
430                     packet = Dynamixel.packetWriteDWord(OLLOBOT.ID, OLLOBOT.Address.PORT_1_SERVO_POSITION, (1024/8) + ((1024/8) << 16));
431                     break;
432                 } else if ("right, white, light, wright, write".indexOf(result.get(i).toLowerCase()) >= 0) { // for command right.
433                     packet = Dynamixel.packetWriteDWord(OLLOBOT.ID, OLLOBOT.Address.PORT_1_SERVO_POSITION, -(1024/8) + (-(1024/8) << 16));
434                     break;
435                 }
436             }
437
438             if (packet != null) {
439                 mService.sendMessageToRemote(packet);
440                 mTvInstructionPacket.setText(Dynamixel.packetToString(packet));
441             }
443 //          Toast.makeText(getApplicationContext(), "[" + result.toString() + "] received.", Toast.LENGTH_SHORT).show();
444             mTvVoiceMsg.setText(result.toString());
445             Log.i("ROBOTIS", "# Voice : [" + result.toString() + "]");
446         }
447         break;
448     } // End of switch(requestCode)
```

Fig. 12.28 Method onActivityResult()

SPEECH_INPUT) was an Android OS level service and please use this web link for more information (https://developer.android.com/reference/android/app/Activity. html#startActivityForResult(android.content.Intent, int)). When this activity exited, it would return this code "Constants.REQ_CODE_SPEECH_INPUT" to the method onActivityResult() which was defined in "MainActivity.java" (see Fig. 12.28), along with a "resultCode" and the best-matching list of strings ("Intent data") that the Recognizer could find.

Method onActivityResult() implemented a Switch structure to handle the case for "Constants.REQ_CODE_SPEECH_INPUT" which saved the "Intent data" into an ArrayList of Strings "result" and started to match them with an IF-ELSE-IF structure (lines 422–436). The arguments used for the IF statements showed that the user could add more words to them to adjust for one's own speech patterns. The implemented actions were to prepare the "appropriate" packets to make the OLLOBOT go forward, backward, left or right. Once again a "WriteDWord()" was used to simultaneously start the two motors. Please also note that the SERVO POSITION mode was used, thus the OLLOBOT would just perform the needed movements for a short time and then stopped.

If "packet" was not empty, it got sent to the OLLOBOT (line 439) and also got displayed as a string at the Instruction Packet field on the main screen of this app (line 440).

The last step for this method was to display the original "result", ArrayList of matching words, on the field for "mTvVoiceMsg" (line 444).

Please see Video 12.5 for a demonstration of the Speech Recognition feature and a reminder that Speech Recognition required web access and would usually take a fair amount of time to do word matching, thus Speech Control would not be good for real-time control or for sharp maneuvers.

The complete OLLOBOT-master1 project folders were zipped into the "OLLOBOT-master1.zip" file and it was included in the Springer Extra Materials file located on http://extras.springer.com/.

Next, to provide further practice for the readers, three projects were created:

1. Adaptation to the CM-50 which had a similar mechanical design with two simple motor ports and three IR sensors (which the OLLOBOT did not have).
2. Adaptation to the CM-150 which had similar capabilities with the CM-50, but with a slightly different control table.
3. Adaptation to the OpenCM-9.04C which had "Dynamixels" (XL-320) for its motors and a variety of ports for its other sensors. Furthermore the OpenCM-9.04C could run on either TASK or OpenCM-IDE programs.

Fig. 12.29 CM-50 CarBot

12.4   Adaptation of OLLOBOT SDK to CM-50

Figure 12.29 showed the CM-50 CarBot that was used in this project with the goal of adapting the OLLOBOT SDK to work with a CM-50. The CM-50 CarBot also had two motors attached to Ports 1 and 2 similarly to the OLLOBOT's, but the CM-50 had three integrated IR sensors: IR Left and IR Right pointing down, while IR Center looked forward (which the OLLOBOT did not have).

12.4.1   Revising Previous LED Buttons and Associated Actions

The plan was to replace the three Buttons "LED ON", "LED OFF" and "LED Status" (and their functions) with "IR LEFT", "IR CENTER" and "IR RIGHT" respectively.

From the work done in Sect. 12.2, we knew that the layout and the registration of these new Buttons to "onClickListener()" (and other related methods and subclasses) would have to be reprogrammed, meaning that we needed to access the Control Table of the CM-50. However, it turned out that the ROBOTIS e-Manual web site did not carry information on the Control Tables of any of their controllers. And it was only by a stroke of luck that the author found an alternative source of these information:

- Users needed to install the R+SCRATCH tool and have it updated to the latest version.
- Go to this folder "C:\Program Files (x86)\ROBOTIS\RoboPlus 2.0\R+ Scratch\ R+Scratch_Data\Config\Controller\Controltable\2.0" and the users can find a series of XML files for all the ROBOTIS controllers currently available, in English, Korean and some in Chinese.
- The use of these control tables was critical for the adaptation of the original OLLOBOT SDK to the chosen controllers CM-50/150/9.04C.

The completed AS project "OLLOBOT-master_CM50.zip" file is available for download as part of the Springer Extra Materials file located on http://extras. springer.com/ using this book ISBN number. The author would suggest users to unzip this file and use Android Studio to open up this project, and follow the subsequent paragraphs describing the changes that were made:

- For the layout file "activity_main.xml", please see Fig. 12.30 (lines 53, 57, 60, 64, 67, and 71) for updating the Resource IDs and the texts associated with the Buttons.
- Next the "OLLOBOT.java" file needed to be revised using information from the CM-50 control table as shown in file "CM-50.en.xml". In Fig. 12.31, the new additions were lines 18–20 for the Addresses of the three IR sensors and lines 57–59 for the Lengths of same IR sensors. Lines 31 and 32 were unchanged as the OLLOBOT and CM-50 used the same addresses for their Ports 1 and 2 (respectively 136 and 138).
- The last steps for adapting the IR sensors were done in the "MainActivity.java" file (see Fig. 12.32). First the "new" Buttons needed to be registered with onClickListener() – lines 109–111, lines 121–123. Next the "onClick()" method needed three of its cases to be revised, as a packet of type READ would be now needed for each of the IR Left/Center/Right Buttons (see Lines 661, 664 and 667).

```
52                      <Button
53                          android:id="@+id/btn_ir_left"
54                          android:layout_width="wrap_content"
55                          android:layout_height="wrap_content"
56                          android:layout_weight="1"
57                          android:text="IR LEFT" />
58
59                      <Button
60                          android:id="@+id/btn_ir_center"
61                          android:layout_width="wrap_content"
62                          android:layout_height="wrap_content"
63                          android:layout_weight="1"
64                          android:text="IR CENTER" />
65
66                      <Button
67                          android:id="@+id/btn_ir_right"
68                          android:layout_width="wrap_content"
69                          android:layout_height="wrap_content"
70                          android:layout_weight="1"
71                          android:text="IR RIGHT" />
```

Fig. 12.30 Changes made to "activity_main.xml" for CM-50 CarBot.

```
17      // IR_LEFT, IR_CENTER, IR_RIGHT sensors added by C. N. Thai for CM-50 and CM-150 2/17/2017
18      public static final int IR_LEFT = 93;
19      public static final int IR_CENTER = 95;
20      public static final int IR_RIGHT = 91;

29      // About Motor_speed.
30      // + for clockwize, - for counterclockwise.
31      public static final int PORT_1_MOTOR_SPEED = 136;        // -1024 ~ 1024 - 136 for CM-50, 152 for CM-150
32      public static final int PORT_2_MOTOR_SPEED = 138;        // -1024 ~ 1024 - 138 for CM-50, 154 for CM-150

56      // IR_LEFT, IR_CENTER, IR_RIGHT sensors added by C. N. Thai for CM-50 2/17/2017
57      public static final int IR_LEFT = 2;
58      public static final int IR_CENTER = 2;
59      public static final int IR_RIGHT = 2;
```

Fig. 12.31 Changes made to "OLLOBOT.java" for CM-50 CarBot

Those were the changes necessary to make the IR Buttons work with the actual IR Sensors located on the CM-50 (see Video 12.6 – Part 1 for a demonstration of these IR Sensors in action).

12.4.2   *Revising Packets for the Motion-Direction Buttons*

In Sect. 12.3.2, it was shown that the OLLOBOT's firmware was capable of setting two PORT_SPEEDs at the same time by using one "double word" packet so as to synchronize the two motors (especially when going forward or backward). This

```
108          // Setup button for H/W test
109          Button btnIRLeft = (Button) findViewById(R.id.btn_ir_left);
110          Button btnIRCenter = (Button) findViewById(R.id.btn_ir_center);
111          Button btnIRRight = (Button) findViewById(R.id.btn_ir_right);

121          btnIRLeft.setOnClickListener(this);
122          btnIRCenter.setOnClickListener(this);
123          btnIRRight.setOnClickListener(this);

656    public void onClick(View v) {
657        byte[] packet = null;
658        switch (v.getId()) {
659            // See OLLOBOT.java for details.
660            case R.id.btn_ir_left:
661                packet = Dynamixel.packetRead(OLLOBOT.ID, OLLOBOT.Address.IR_LEFT, OLLOBOT.Length.IR_LEFT);
662                break;
663            case R.id.btn_ir_center:
664                packet = Dynamixel.packetRead(OLLOBOT.ID, OLLOBOT.Address.IR_CENTER, OLLOBOT.Length.IR_CENTER);
665                break;
666            case R.id.btn_ir_right:
667                packet = Dynamixel.packetRead(OLLOBOT.ID, OLLOBOT.Address.IR_RIGHT, OLLOBOT.Length.IR_RIGHT);
668                break;
```

Fig. 12.32 Changes made to "MainActivity.java" for CM-50 CarBot

```
669            case R.id.btn_forward:
670                // go and stop.
671    //         packet = Dynamixel.packetWriteDWord(OLLOBOT.ID, OLLOBOT.Address.PORT_1_SERVO_POSITION, -512 + (512 << 16));
672                // keep going.
673    //         packet = Dynamixel.packetWriteDWord(OLLOBOT.ID, OLLOBOT.Address.PORT_1_MOTOR_SPEED, -512 + (512 << 16));
674    //         packet = Dynamixel.packetWriteWord(OLLOBOT.ID, OLLOBOT.Address.CONTROLLER_X_AXIS_VALUE, 0 + (30 << 8)); // X:0, Y:30
675                // Using 2 Word packets for CM-50 and CM-150 which cannot handle DWord
676                // Also Ports 1 and 2 are physically flipped on CM-150 as compared to CM-50
677                packet = Dynamixel.packetWriteWord(OLLOBOT.ID, OLLOBOT.Address.PORT_1_MOTOR_SPEED, 1024+512); // CM-50
678    //         packet = Dynamixel.packetWriteWord(OLLOBOT.ID, OLLOBOT.Address.PORT_2_MOTOR_SPEED, 1024+512); // CM-150
679                if (packet != null) {
680                    mService.sendMessageToRemote(packet);
681                    mTvInstructionPacket.setText(Dynamixel.packetToString(packet));
682                } else {
683                    mTvInstructionPacket.setText("");
684                }
685                mTvStatusPacket.setText("");
686                packet = Dynamixel.packetWriteWord(OLLOBOT.ID, OLLOBOT.Address.PORT_2_MOTOR_SPEED, 512); // CM-50
687    //         packet = Dynamixel.packetWriteWord(OLLOBOT.ID, OLLOBOT.Address.PORT_1_MOTOR_SPEED, 512); // CM-150
688                break;
```

Fig. 12.33 Changes made to onClick() (in MainActivity.java) for CM-50 CarBot

"experimental" firmware was also capable of handling positive and negative numbers. Unfortunately, the standard firmware on the CM-50 would not allow writing across functional addresses and the CM-50 could only handle unsigned integers. Thus in order to make the CM-50 move, two packets of with a "word" size had to be sent sequentially, one for Port 1 and another for Port 2. These changes were made in the method "onClick()" (in "MainActivity.java") for the cases "btn_forward", "btn_backwards", "btn_left", and "btn_right". As an example, Fig. 12.33 showed the changes made for the "btn_forward" case:

- Line 677 showed how the packet for PORT_1_MOTOR_SPEED was prepared using a WriteWord() and the Value to be sent was "1024+512". This use needed further explanations. Since Chap. 5, the readers used this type of statement to set the direction and speed of rotation of a typical motor as follows.

- MotorSpeed(ID:1) = CCW + 512 (or CW + 512), where CCW corresponded to "0" and CW corresponded to "1024" in the inner workings of the CM-50 firmware. The decimal number "1024" would be written as "10000000000", in other words the 11th bit from the right was considered to be a sign bit or the direction bit for the speed value.
- A discerning reader might have noticed another discrepancy. Back in Fig. 12.31, Line 30 said that for the OLLOBOT "-512" meant "CCW 512" and this value was set for OLLOBOT's PORT_1 to make it go forward (Line 673 in Fig. 12.31). But from Chap. 2, "1024+512" meant "CW 512" for CM-50, also to make it go forward! So what was going on? The reason was that Ports 1 and 2 were physically changed left to right between the OLLOBOT and the CM-50! So another point that users needed to be aware of when they adapted their codes from one kit to another kit, even though they both came from ROBOTIS.

- Line 680 "mService.sendMessageToRemote(packet)" sent out the first packet to the CM-50 to set Port 1 Motor Speed. The other statements in the group lines 679–685 were used to update the "InstructionPacket" and "StatusPacket" fields on the app layout.
- Line 686 prepared the second packet to be sent to Port 2 Motor Speed, which will be sent by another group of statements similar to Lines 679–685 (not shown here) after the "break" statement on Line 688.
- Those were the changes necessary to make the IR Buttons work with the actual IR Sensors located on the CM-50 (see Video 12.6 – Part 1 for a demonstration of these IR Sensors in action).

Those were the changes necessary to make the Direction Buttons work with the actual Ports 1 and 2 motors attached to the CM-50 (see Video 12.6 – Part 2 for a demonstration of these Direction Buttons in action). The CM-50's performance was not as good as the OLLOBOT's performance in Video 12.4, as there were occasional loss of packets affecting the overall behavior of the CM-50 CarBot.

12.4.3   Revising Voice Recognition Behaviors of CM-50 CarBot

The Voice Recognition process itself was not modified when adapted to the CM-50 CarBot, but as the OLLOBOT was using a "Servo Position" mode to move in response to a voice command, this part was the one that needed to be adapted to the CM-50's motors which were not servo motors.

The codes to be modified were found in the method onActivityResult() (started on Line 388 of "MainActivity.java"). The case to be considered was for "Constants. REQ_CODE_SPEECH_INPUT" started on Line 416. As an example, the code section corresponding to "forward" was captured into Fig. 12.34. The goal was for the CarBot to execute the required maneuver whenever it received a valid voice command, but only for 2 s, and then it would stop its motion.

```
424    if ("forward, ford, fort, fought, hot, food, flood".indexOf(result.get(i).toLowerCase()) >= 0) { // for command forward.
425  //        packet = Dynamixel.packetWriteDWord(OLLOBOT.ID, OLLOBOT.Address.PORT_1_SERVO_POSITION, -512 + (512 << 16)); // OLLOBOT
426          packet = Dynamixel.packetWriteWord(OLLOBOT.ID, OLLOBOT.Address.PORT_1_MOTOR_SPEED, 1024+512); // CM-50
427  //        packet = Dynamixel.packetWriteWord(OLLOBOT.ID, OLLOBOT.Address.PORT_2_MOTOR_SPEED, 1024+512); // CM-150
428          mService.sendMessageToRemote(packet);
429          try {
430              Thread.sleep(15);
431          } catch (InterruptedException e) {
432              // TODO Auto-generated catch block
433              e.printStackTrace();
434          }
435          packet = Dynamixel.packetWriteWord(OLLOBOT.ID, OLLOBOT.Address.PORT_2_MOTOR_SPEED, 512); // CM-50
436  //        packet = Dynamixel.packetWriteWord(OLLOBOT.ID, OLLOBOT.Address.PORT_1_MOTOR_SPEED, 512); // CM-150
437          mService.sendMessageToRemote(packet);
438          try {
439              Thread.sleep(2000);
440          } catch (InterruptedException e) {
441              // TODO Auto-generated catch block
442              e.printStackTrace();
443          }
444          packet = Dynamixel.packetWriteWord(OLLOBOT.ID, OLLOBOT.Address.PORT_1_MOTOR_SPEED, 0);
445          mService.sendMessageToRemote(packet);
446          try {
447              Thread.sleep(15);
448          } catch (InterruptedException e) {
449              // TODO Auto-generated catch block
450              e.printStackTrace();
451          }
452          packet = Dynamixel.packetWriteWord(OLLOBOT.ID, OLLOBOT.Address.PORT_2_MOTOR_SPEED, 0);
453          mService.sendMessageToRemote(packet);
454          break;
```

Fig. 12.34 Changes made to onActivityResult() (in MainActivity.java) for CM-50 CarBot

Essentially, Line 426 prepared the first packet and Line 428 sent it to the CM-50's Port 1. Next, a delay of 15 ms was needed (Line 430) before we could send the second packet due to the limitations of BlueTooth. Unfortunately, the Android OS would judge that a Thread.sleep(15) could create run-time problems for it, so it wrapped the "15 ms delay" inside a try-catch structure.

Line 435 prepared the second packet to be sent to Port 2, then the program waited for 2 s before sending two more packets to set both motor speeds to zero, effectively stopped the robot.

Video 12.6 – Part 3 showed the CM-50's performance with voice commands and there were occasional loss of packets affecting the overall behavior of the CM-50 CarBot as before.

12.5   Adaptation of OLLOBOT SDK to CM-150

The work required to adapt the OLLOBOT SDK to a CM-150 CarBot as shown in Fig. 12.35 was quite similar to the work performed for the CM-50 with the following differences:

- PORT_1_MOTOR_SPEED = 152 (instead of 136 on OLLOBOT and CM-50).
- PORT_1_MOTOR_SPEED = 154 (instead of 138 on OLLOBOT and CM-50).
- The physical layout of the motors Ports 1 and 2 was reversed as compared to the CM-50's layout.

The completed AS project "OLLOBOT-master_CM150.zip" file is available for download as part of the Springer Extra Materials file located on http://extras.springer.com/ using this book ISBN number. The author would suggest users to unzip this file and use Android Studio to open up this project, and adapt the previous

Fig. 12.35 CM-150
CarBot

materials in Sect. 12.4 to verify that the corresponding changes were properly made for the CM-150, as a learner's exercise. Actually Figs. 12.31, 12.32, 12.33, and 12.34 showed the "CM-150" changes already, but they were commented out.

Video 12.7 showed that the CM-150's performance was also poor due to the occasional loss of packets, perhaps even worse than the CM-50's.

12.6 Adaptation of OLLOBOT SDK to OpenCM-9.04C

A CarBot based on the OpenCM-9.04C was constructed using two XL-320s for motors, one IR sensor connected to Port 1, one Color Sensor connected to Port 2, and one LED display connected to Port 3 (see Fig. 12.36).

The overall goal was to use "Left Button" on the OLLOBOT app to get color data from an object using the AUX_COLOR module (Port 2), and to use "Center Button" to turn on the Left LED of the AUX_LED module (Port 3) for 2 s then to turn it off. Similarly, the author wanted "Right Button" to turn on the Right LED of the AUX_LED module for 2 s then also turn it off. The IR Sensor on Port 1 would be used by a TASK program to trigger the sending of various SMART commands from the OpenCM-9.04C to the Android device (i.e. the modified OLLOBOT app).

12.6.1 Revising COLOR/LED Buttons and Associated Actions

This completed AS project "OLLOBOT-master_CM904.zip" file is available for download as part of the Springer Extra Materials file located on http://extras. springer.com/ using this book ISBN number. The author would again suggest users

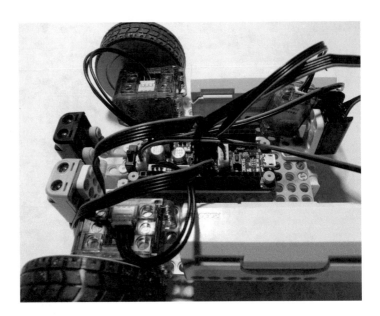

Fig. 12.36 OpenCM-9.04C CarBot

to unzip this file and use Android Studio to open up this project, and follow the subsequent paragraphs describing the changes that were made:

- For the layout file "activity_main.xml", please see Fig. 12.37 (lines 53, 57, 60, 64, 67 and 71) for updating the Resource IDs and the texts associated with the Buttons.
- Next, the "OLLOBOT.java" file needed to be revised using information from the OpenCM-9.04C control table as shown in file "OpenCM9.04.en.xml". In Fig. 12.38, the new additions were lines 22 and 23 for the Addresses of the AUX_COLOR and AUX_LED modules and lines 65 and 66 for the Lengths of same modules. Because the AUX_COLOR and AUX_LED modules could be installed on any of the four Ports available on the OpenCM-9.04C controller board, a new procedure needed to be used to determine the actual address to be used to a particular module connected to a particular port:

  - The general formula to use was as follows. Actual Module Address = (Address of module if it was connected to Port 1) + (Port Number − 1) × (Module Length).
  - The file "OpenCM9.04.en.xml" showed that the address for the AUX_COLOR module would be 408 (if it was hooked to Port 1). But AUX_COLOR was hooked up to Port 2 instead so the address to use was then $408 + (2–1) \times 1 = 409$.
  - The AUX_LED was connected to Port 3, so its actual address was $380 + (3–1) * 1 = 382$.

- The file "OpenCM9.04.en.xml" indicated that the address for an XL-320 was "32" and its length was "2", thus Lines 36 and 67 were created to set its Address and its Length respectively. But the users needed to remember that the Dynamixel ID would still need to be accounted for when we write "packets" to these XL-320 s at a later time.

```
52                              <Button
53                                  android:id="@+id/btn_aux_color"
54                                  android:layout_width="wrap_content"
55                                  android:layout_height="wrap_content"
56                                  android:layout_weight="1"
57                                  android:text="AUX COLOR" />
58
59                              <Button
60                                  android:id="@+id/btn_led_left"
61                                  android:layout_width="wrap_content"
62                                  android:layout_height="wrap_content"
63                                  android:layout_weight="1"
64                                  android:text="LED LEFT" />
65
66                              <Button
67                                  android:id="@+id/btn_led_right"
68                                  android:layout_width="wrap_content"
69                                  android:layout_height="wrap_content"
70                                  android:layout_weight="1"
71                                  android:text="LED RIGHT" />
```

Fig. 12.37 Changes made to "activity_main.xml" for OpenCM-9.04C CarBot

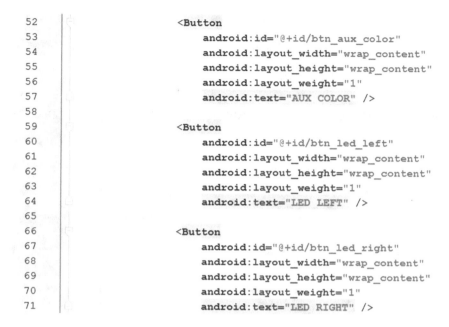

Fig. 12.38 Changes made to "OLLOBOT.java" for OpenCM-9.04C CarBot

- The next steps for adapting the AUX_COLOR and AUX_LED modules were done in the "MainActivity.java" file and with the onCreate() method (see Fig. 12.39). The three "new" Buttons needed to be registered with onClickListener() – lines 112–114, lines 127–129.
- The last steps were done in the "onClick()" method which needed three of its cases to be revised – "btn_aux_color", "btn_led_left" and "btn_led_right" (Fig. 12.40):

   - A packet of type READ was needed for the AUX_COLOR Button (Line 684).
   - A packet of type WriteByte() was needed to turn on the Left LED of the AUX_ LED module (lines 687–688). The thread then waited for 2 s (Lines 689–694), and another "WriteByte" packet was sent to turn off this Left LED (Line 695).
   - Similarly, a packet of type WriteByte() was needed to turn on the Right LED of the AUX_LED module (lines 698–699). The thread then waited for 2 s

```
112          Button btnAuxColor = (Button) findViewById(R.id.btn_aux_color);
113          Button btnLedLeft = (Button) findViewById(R.id.btn_led_left);
114          Button btnLedRight = (Button) findViewById(R.id.btn_led_right);

127          btnAuxColor.setOnClickListener(this);
128          btnLedLeft.setOnClickListener(this);
129          btnLedRight.setOnClickListener(this);
```

Fig. 12.39 Changes made to onCreate() of "MainActivity.java" for OpenCM-9.04C CarBot

```
679  public void onClick(View v) {
680      byte[] packet = null;
681      switch (v.getId()) {
682          // See OLLOBOT.java for details.
683          case R.id.btn_aux_color:
684              packet = Dynamixel.packetRead(OLLOBOT.ID, OLLOBOT.Address.AUX_COLOR, OLLOBOT.Length.AUX_COLOR);
685              break;
686          case R.id.btn_led_left:
687              packet = Dynamixel.packetWriteByte(OLLOBOT.ID, OLLOBOT.Address.AUX_LED, 2);   // turn on left LED
688              mService.sendMessageToRemote(packet);
689              try {
690                  Thread.sleep(2000);
691              } catch (InterruptedException e) {
692                  // TODO Auto-generated catch block
693                  e.printStackTrace();
694              }
695              packet = Dynamixel.packetWriteByte(OLLOBOT.ID, OLLOBOT.Address.AUX_LED, 0);   // turn off left LED
696              break;
697          case R.id.btn_led_right:
698              packet = Dynamixel.packetWriteByte(OLLOBOT.ID, OLLOBOT.Address.AUX_LED, 1);   // turn on right LED
699              mService.sendMessageToRemote(packet);
700              try {
701                  Thread.sleep(2000);
702              } catch (InterruptedException e) {
703                  // TODO Auto-generated catch block
704                  e.printStackTrace();
705              }
706              packet = Dynamixel.packetWriteByte(OLLOBOT.ID, OLLOBOT.Address.AUX_LED, 0);   // turn off right LED
707              break;
```

Fig. 12.40 Changes made to onClick() of "MainActivity.java" for OpenCM-9.04C CarBot

(Lines 700–705), and another "WriteByte" packet was sent to turn off this Left LED (Line 706).

Those were the changes necessary to make the AUX_COLOR and AUX_LED modules operate from the Android device (see Video 12.8 – Part 1 for a demonstration of these modules in action).

12.6.2 Revising Packets for the Motion-Direction Buttons

Similarly as for the CM-50 and CM-150, two packets with a "word" size each had to be sent sequentially, one for the XL-320 with Dynamixel ID = 1 and another for the XL-320 with Dynamixel ID =2. These changes were made in the method "onClick()" (in "MainActivity.java") for all four motion cases, but Fig. 12.41 would only illustrate the case for "btn_forward".

The relevant statements were at Lines 718 (ID = 1) and 729 (ID = 2).

See Video 12.8 – Part 2 for a demonstration of these features. The OpenCM-9.04C seemed to have less packet losses than the other controllers CM-50 and CM-150.

```
708      case R.id.btn_forward:
709          // go and stop.
710 //      packet = Dynamixel.packetWriteDWord(OLLOBOT.ID, OLLOBOT.Address.PORT_1_SERVO_POSITION, -512 + (512 << 16));
711          // keep going.
712 //      packet = Dynamixel.packetWriteDWord(OLLOBOT.ID, OLLOBOT.Address.PORT_1_MOTOR_SPEED, -512 + (512 << 16));
713 //      packet = Dynamixel.packetWriteWord(OLLOBOT.ID, OLLOBOT.Address.CONTROLLER_X_AXIS_VALUE, 0 + (30 << 8)); // X:0, Y:30
714      // Using 2 Word packets for CM-50 and CM-150 which cannot handle DWord
715          // Also Ports 1 and 2 are physically flipped on CM-150 as compared to CM-50
716 //      packet = Dynamixel.packetWriteWord(OLLOBOT.ID, OLLOBOT.Address.PORT_1_MOTOR_SPEED, 1024+512); // CM-50
717 //      packet = Dynamixel.packetWriteWord(OLLOBOT.ID, OLLOBOT.Address.PORT_2_MOTOR_SPEED, 1024+512); // CM-150
718          packet = Dynamixel.packetWriteWord(1, OLLOBOT.Address.XL_320_MOTOR_SPEED, 1024+512); // CM-904
719          if (packet != null) {
720              mService.sendMessageToRemote(packet);
721              mTvInstructionPacket.setText(Dynamixel.packetToString(packet));
722          } else {
723              mTvInstructionPacket.setText("");
724          }
725          mTvStatusPacket.setText("");
726 //      packet = Dynamixel.packetWriteWord(OLLOBOT.ID, OLLOBOT.Address.PORT_2_MOTOR_SPEED, 512); // CM-50
727 //      packet = Dynamixel.packetWriteWord(OLLOBOT.ID, OLLOBOT.Address.PORT_1_MOTOR_SPEED, 512); // CM-150
728          packet = Dynamixel.packetWriteWord(2, OLLOBOT.Address.XL_320_MOTOR_SPEED, 512); // CM-904
729          break;
```

Fig. 12.41 Changes made to onClick() of "MainActivity.java" for "btn_forward" case of OpenCM-9.04C CarBot

```
470          packet = Dynamixel.packetWriteWord(1, OLLOBOT.Address.XL_320_MOTOR_SPEED, 512); // CM-904

480          packet = Dynamixel.packetWriteWord(2, OLLOBOT.Address.XL_320_MOTOR_SPEED, 1024+512); // CM-904
```

Fig. 12.42 Typical changes made to onActivityResult() (in MainActivity.java) for OpenCM-9.04C CarBot

12.6.3   Revising Voice Recognition Behaviors of OpenCM-9.04C CarBot

The algorithm was the same one used in Sect. 12.4.3 for the CM-50, but now all the packets were prepared for each XL-320 (ID = 1) then (ID = 2). Typical changes were illustrated in Fig. 12.42 for going forward.

Video 12.8 – Part 3 showed the OpenCM-9.04C's performance with voice commands and there were occasional loss of packets affecting the overall behavior of the CM-50/150 CarBots as before.

Wrapping up, the OLLOBOT SDK was quite adaptable to all Firmware 2.0 controllers and will be a great tool to advanced users who want to acquire more skills with Android Programming. But there are a few issues to be resolved:

- The packet loss situation could be solved by switching the communication lines – use BlueTooth for developing work between the PC and the Android device which at run time would use wired USB via an LN-101 type of module to connect to ROBOTIS robotics systems (if this route is technically feasible!).
- If the previous option could not be done, the OLLOBOT SDK could be modified to be more like an expanded version of the R + m.PLAY700, which would be a boon for TASK programmers who want to access the multimedia services of a SmartPhone.

But the real technical reason for this option is that basic actuators and sensors should be controlled by a local hardware controller for good performance, and more demanding sensors such as vision can be off-loaded to the remote device. ROBOTIS had this framework already implemented in its OP systems.

12.7   Review Questions for Chap.12

1. What is the memory map for the OpenCM-9.04C controller?
2. What is the memory map for the CM-530 controller?
3. What will happen when the ROBOTIS controller firmware received a STATUS packet?
4. What will happen when the ROBOTIS controller firmware received a READ packet?
5. What will happen when the ROBOTIS controller firmware received a WRITE packet?
6. What will happen when a ROBOTIS actuator or sensor received a STATUS packet?
7. What will happen when a ROBOTIS actuator or sensor received a WRITE packet?
8. What will happen when a ROBOTIS actuator or sensor received a READ packet?
9. What is the Dynamixel ID for a ROBOTIS controller?
10. What is the Dynamixel ID for a Remote Device?

12.8   Review Exercises for Chap. 12

1. The enclosed Android Studio projects implemented a scheme to receive and display READ and WRITE packets from the Hardware Controller to the Android Device (see the beginning of Video 12.8). However the coding changes were not described in this chapter. As a self-learning exercise, the reader is encouraged to re-discover these code fragments in the "OLLOBOT-master_CM904.zip" file (Hint: they are already commented out by the author!).

References

Phillips B et al (2017) Android programming: the Big Nerd Ranch guide. Big Nerd Ranch LLC, Atlanta
Smyth N (2016) Android studio development essentials: android. 7th ed. Neil Smyth/Payload Media. Lexington, KY, U.S.A.

Printed in the United States
By Bookmasters